Electromagnetic and Acoustic Wave Tomography
Direct and Inverse Problems
in Practical Applications

Electromagnetic and Acoustic Wave Tomography

Direct and Inverse Problems
in Practical Applications

Edited by
Nathan Blaunstein and Vladimir Yakubov

CRC Press is an imprint of the
Taylor & Francis Group, an **informa** business

CRC Press
Taylor & Francis Group
6000 Broken Sound Parkway NW, Suite 300
Boca Raton, FL 33487-2742

© 2019 by Taylor & Francis Group, LLC
CRC Press is an imprint of Taylor & Francis Group, an Informa business

No claim to original U.S. Government works

Printed on acid-free paper

International Standard Book Number-13: 978-1-138-49073-4 (hardback)
International Standard Book Number-13: 978-0-4294-8827-6 (eBook)

This book contains information obtained from authentic and highly regarded sources. Reasonable efforts have been made to publish reliable data and information, but the author and publisher cannot assume responsibility for the validity of all materials or the consequences of their use. The authors and publishers have attempted to trace the copyright holders of all material reproduced in this publication and apologize to copyright holders if permission to publish in this form has not been obtained. If any copyright material has not been acknowledged please write and let us know so we may rectify in any future reprint.

Except as permitted under U.S. Copyright Law, no part of this book may be reprinted, reproduced, transmitted, or utilized in any form by any electronic, mechanical, or other means, now known or hereafter invented, including photocopying, microfilming, and recording, or in any information storage or retrieval system, without written permission from the publishers.

For permission to photocopy or use material electronically from this work, please access www.copyright.com (http://www.copyright.com/) or contact the Copyright Clearance Center, Inc. (CCC), 222 Rosewood Drive, Danvers, MA 01923, 978-750-8400. CCC is a not-for-profit organization that provides licenses and registration for a variety of users. For organizations that have been granted a photocopy license by the CCC, a separate system of payment has been arranged.

Trademark Notice: Product or corporate names may be trademarks or registered trademarks, and are used only for identification and explanation without intent to infringe.

Library of Congress Cataloging-in-Publication Data

Names: Blaunstein, Nathan, editor. | Yakubov, Vladimir, editor.
Title: Electromagnetic and acoustic wave tomography : direct and inverse problems in practical applications / editors, Nathan Blaunstein, Vladimir Yakubov.
Description: Boca Raton, FL : CRC Press/Taylor & Francis Group, 2018. | "A CRC title, part of the Taylor & Francis imprint, a member of the Taylor & Francis Group, the academic division of T&F Informa plc." | Includes bibliographical references and index.
Identifiers: LCCN 2018006943| ISBN 9781138490734 (hb : acid-free paper) | ISBN 9780429488276 (ebook)
Subjects: LCSH: Three-dimensional imaging--Mathematics. | Tomography. | Acoustic emission testing. | Electromagnetic waves--Mathematical models. | Remote sensing. | Radar.
Classification: LCC TA1560 .E44 2018 | DDC 681/.2--dc23
LC record available at https://lccn.loc.gov/2018006943

Visit the Taylor & Francis Web site at
http://www.taylorandfrancis.com

and the CRC Press Web site at
http://www.crcpress.com

Contents

Preface vii
Acknowledgments xv
Editors xvii
Contributors..xix

SECTION I THEORETICAL FUNDAMENTALS OF WAVE TOMOGRAPHY

1. Mathematical Fundamentals to Inverse Problems 3
   VLADIMIR YAKUBOV, SERGEY SHIPILOV, AND NATHAN BLAUNSTEIN

2. Theoretical Overview of Wave Tomography..17
   VLADIMIR YAKUBOV, SERGEY SHIPILOV, DMITRY SUKHANOV, AND ANDREY KLOKOV

3. Special Theoretical Approaches in Wave Tomography..........................47
   VLADIMIR YAKUBOV, SERGEY SHIPILOV, DMITRY SUKHANOV, ANDREY KLOKOV, AND NATHAN BLAUNSTEIN

4. Low-Frequency Magnetic and Electrostatic Tomography.....................79
   VLADIMIR YAKUBOV, SERGEY SHIPILOV, DMITRY SUKHANOV, AND ANDREY KLOKOV

5. Eddy Current Tomography 89
   NATHAN BLAUNSTEIN AND ALEXEY VERTIY

SECTION II EXPERIMENTAL VERIFICATION OF WAVE TOMOGRAPHY THEORETICAL FRAMEWORK

6. Radio Tomography of Various Objects Hidden in Clutter Conditions..121
   VLADIMIR YAKUBOV, SERGEY SHIPILOV, DMITRY SUKHANOV, AND ANDREY KLOKOV

7 Proof of Specific Radio Tomography Methods.................................167
VLADIMIR YAKUBOV, SERGEY SHIPILOV, DMITRY SUKHANOV, AND
ANDREY KLOKOV

SECTION III RADIO TOMOGRAPHY PRACTICAL APPLICATIONS

8 Ground-Penetrating and Geo-Radars 203
VLADIMIR YAKUBOV, SERGEY SHIPILOV, ANDREY KLOKOV, AND
NATHAN BLAUNSTEIN

9 Sub-Surface Tomography Applications..225
VLADIMIR YAKUBOV, SERGEY SHIPILOV, ANDREY KLOKOV, AND
NATHAN BLAUNSTEIN

10 UWB Tomography of Forested and Rural Environments...................265
VLADIMIR YAKUBOV, SERGEY SHIPILOV, AND ANDREY KLOKOV

11 Detection of Live People in Clutter Conditions 281
NATHAN BLAUNSTEIN, FELIX YANOVSKY, VLADIMIR YAKUBOV, AND
SERGEY SHIPILOV

SECTION IV NON-CONTACTING ACOUSTIC AND COMBINED RADIO-ACOUSTIC TOMOGRAPHY

12 Applications of Radio-Acoustic Tomography 293
VLADIMIR YAKUBOV, SERGEY SHIPILOV, DMITRY SUKHANOV, AND
ANDREY KLOKOV

SECTION V APPLICATIONS OF LOW-FREQUENCY MAGNETIC AND EDDY CURRENT TOMOGRAPHY

13 Applications of Low-Frequency Magnetic Tomography.....................313
VLADIMIR YAKUBOV AND DMITRY SUKHANOV

14 Eddy Current Tomography Applications 323
ALEXEY VERTIY AND NATHAN BLAUNSTEIN

SECTION VI METHODS OF VISUALIZATION AND RECONSTRUCTION OF OBJECTS

15 Visualization and Reconstruction of Objects......................................335
VLADIMIR YAKUBOV, SERGEY SHIPILOV, DMITRY SUKHANOV, AND
ANDREY KLOKOV

Symbols and Abbreviations 347

Index 353

Preface

This monograph is intended for any researcher, practical engineer, or designer who is concerned with the operation and service of radio and acoustic radar systems of various frequency bands to resolve both direct and inverse problems of radio and acoustic location, with practical applications for the reconstruction and visualization of different elements and objects embedded and hidden in sub-surface clutter environments and structures.

The main goal of this monograph is to introduce the main aspects of novel methods and theoretical frameworks, as well as of advanced technologies and the corresponding radar systems, developed by three groups of researchers, that have separately and jointly been presented during the last three decades. These methods, frameworks, and technologies are based not only on pure research activity, but also on the authors' experience during more than 40 years' teaching of the corresponding courses in these areas for undergraduate and postgraduate students.

The main subject of research presented in this monograph is how to develop radio and acoustic wave tomography methods as a means of remote non-destructive testing, diagnostics of the internal structure of semitransparent media, and reconstruction of the shapes of opaque objects based on multi-angle sounding, presented in such a way that has been covered before in the literature (see references [1–7] and the bibliography therein).

Historically, in the corresponding literature (see bibliography in [1]), tomography means "to write a layer"; that is, to investigate a structure layer by layer. The difference between tomography and other computational diagnostic methods is that information from the same test element is recorded in multiple integral projections, that is, many times from different angles relative to the embedded inhomogeneities.

Since about 50 years ago up until the present day, researchers in this field have learned how to "clean up" these projections and recover the structure of inhomogeneities layer by layer. For the most part, this became possible due to the development of new computational methods and computer technologies. Huge data flows were "smoothly" layered as images of "crosscuts" of the internal structure of objects in a non-destructive fashion. At the present time, computed tomography is rightfully considered as an "absolute" diagnostic technique in medicine. Radio and acoustic wave tomography is similar to X-ray and magnetic resonance tomography, but it deals with electromagnetic radiation in the radio-frequency band and ultrasound waves in the acoustic band. In this case, the wave length is comparable to the size of the inhomogeneities, and diffraction effects and the effects of multiple interactions hold much significance. For that reason, this form of tomography is sometimes called *diffraction tomography*. Without dwelling on all the different methods of and approaches to geometric optics and wave diffraction tomography that are currently available, we focus on active location (detection) wave tomography, which is of vital importance, for example, for security systems applications.

During recent decades, the use of manufactured and homemade improvised explosive devices in acts of terrorism and local armed conflicts has become more frequent. There have been numerous reported cases of the transportation of such devices and other prohibited items in hand luggage and under clothes in airports, and also in stadiums and other crowded places. Put very simply, the problem of designing a highly efficient means for remote detection of prohibited devices and articles is of urgent interest. It is of particular importance in view of different public events being held across the world.

Radio-wave systems are preferable in the development of contactless detection devices for a variety of reasons. In the first place, radio waves are practically harmless to human health. This is their crucial difference from ionizing X-rays. Second, the potential range of application of these systems is quite wide: in crowded public places, in special forces raids for detection and tracking of people hiding behind walls, detection of injured persons after emergency events, and so on. There is also a great demand for contactless and computer-aided systems for quality control in building construction, timber processing, avionics, electronic devices construction, and other industries.

The variety of physical processes involved in radio-wave detection, taking place in natural and simulated complex environments and involving complex objects, underlies the complexity of the mathematical descriptions of such processes and the urgency of solutions of the tomography problem as well.

The main object of this monograph is to describe physical–mathematical models of systems designed to reconstruct images of hidden objects, based on tomographic processing of multi-angle remote measurements of scattered radio and acoustic (ultrasonic) wave radiation. This class of problems is related to so-called *inverse problems*, the mathematical fundamentals of which are briefly described in Chapter 1. One way or another, radio-wave tomography is based on the ray-focusing effect, which enables the inverse transformation of wave projections of test objects and the propagation medium. Multiple effects (scattering and diffraction) of interactions of the wave fields with inhomogeneities of the propagation medium can be considerably reduced by the use of spatiotemporal radiation focusing. The corresponding mathematical and physical theoretical approaches based on the single-scattering (Born) approximation with different ray-focusing and refocusing techniques, and then on multiple-scattering approximations (Rytov and that based on Feynman's path integral), are the subjects of Chapters 2 and 3 of the proposed monograph.

In the course of research on this problem, a number of working laboratory and "Demo" tomography systems have been developed together with the associated software that assist in the assessment of the potential of the system and its key parameters such as resolution, range of action, and response time. The fact that the development of a radio-location tomography is possible is not in doubt—it has been demonstrated by numerous published results and by the co-authors' own research [1–7]. The drawbacks of most current safety systems designed for crowded places and areas with heavy human traffic are as follows:

- Microwave systems are very time consuming as things stand; it takes a lot of time to screen people, as the antennas are moved mechanically.
- The subject being screened must remain motionless for several seconds. Therefore, a doorway or a cabin is required that interrupts the constant human stream.
- It is not difficult to obstruct such a system, for example, by wetting your outer garments with mineral water because the system operates in the millimeter wave length band.

It should be noted that this interference is neutralized in ultra-wide band (UWB) systems where the working frequency spectrum lies within the range from a few MHz to tens of GHz. The use of UWB signals is considered to be the most promising approach from the standpoint of applications. The development of a domestic radio-detection UWB tomography requires the solution of a number of problems.

However, this does not exhaust all of the currently available possibilities. The development of an optimal antenna array configuration is, of necessity, the first step in this process. This development would include the relative placement of the receiving and transmitting antennas and a determination of their minimum number required for tomographic imaging. The solution of this problem is one of the objectives of ongoing research. This includes a determination of the optimal sequence of radar measurements to provide the required data digitization in a short period of time. Moreover, considerable attention must be paid to the development of fast algorithms for real-time 3-D image restoration of scanned objects based on radio sounding data.

To overcome the drawbacks of the existing systems mentioned above, in Chapters 3–5 the theoretical background of 3-D diffraction tomography for high-frequency wave (Chapter 3) and millimeter wave (Chapter 4) tomography were presented, as well as for low-frequency electric and magnetic wave tomography (Chapter 5), generalized by the use of eddy current tomography (Chapter 5). The results obtained from such approaches have then been proven experimentally by use of prototypes performed by the authors of this monograph in Chapters 7, 8, and 14.

Another important subject reflected in Chapters 7 and 8 is the antenna array optimization problem, which we have already alluded to and which also arises from the need to minimize the number of antennas to reduce antenna array costs while holding the number of artifacts to a minimum. From the latter half of the twentieth century onward, several methods of antenna array optimization for radio astronomy and aircraft detection have been developed. The antenna array optimization problem is a natural outgrowth of the problem referred to as *antenna array design with minimum redundancy*. Accordingly, the antennas should be arranged in such a way as to avoid similar measurement data from each antenna for a given search angle of the radar.

Over the past few decades, a great deal of research has been conducted, and field-proven results have been obtained in the far field zone of narrowband antenna arrays. However, a subject of particular interest to the authors is antenna arrays for UWB radiation, which are designed to characterize objects located in the array focusing area. So far, this issue has not been investigated, as there has not been any urgency to research the near field zone of antenna arrays. In this context, the near field zone is considered to be the region of space at a distance of the order of the dimensions of the array. In other words, this refers to the Fresnel diffraction zone.

The next problem is the development of efficient mathematical support to reconstruct tomograms from the wave projections of probed scenes containing hidden objects. In this context, the two following requirements come to the forefront: adequate restoration accuracy and real-time performance of the system. There is some contradiction here, which is formulated as: "Fast is not always best." According to domestic and foreign experts, a resolution that is no worse than 1 cm is an appropriate value for safety systems. Reconstruction should be performed in real-time mode; for instance, within a minute. The task is complicated when it is impossible to obtain complete measurement data. This situation arises when the scanning is performed non-equidistantly, in motion, or when the object to be detected is hidden inside a building or under clothing near the body. This and related kinds of problems stipulate the necessity of a consistent theoretical study, backed up, of course, by experimental work. In other words, the mathematical (theoretical) part cannot be separated from the experimental (measurement) part.

It is critical to stress that wave tomography must utilize phase information from wave projections of the examined objects in an essential way. A phase information record is provided by radio-frequency holography, in which the result of interference of the background wave and the object wave is recorded. The wave projection record is, in fact, a radio-hologram record. From general considerations, it is clear that large aperture synthesis is the preferred method for post-processing these projections, since it provides the highest spatial resolution. There are now a large number of variations of this method under continuous development. All these aspects have been elucidated in Chapters 9–12, where various applications are shown, not only in security, but mostly in civilian areas of human activity.

Electromagnetic radiation interacts with electro-physical inhomogeneities in the propagation medium. In contrast to electromagnetic radiation, acoustic radiation interacts mainly with density inhomogeneities. In this context, using ultrasound for tomography of an inhomogeneous medium provides a wide range of additional possibilities, particularly for determining the type of material that the hidden items are made of. Combining radio-wave and acoustic sounding creates interesting possibilities, for example, for detection of explosive materials and other foreign objects hidden in various clutter environments. Moreover, because ultrasound is strongly attenuated by air, the only ultrasonic sounding system (*son-vision*) that is actually efficient in air is bat echolocation. Some insects also generate ultrasound, but rather for scaring and masking from bats. The practical use of ultrasound in a liquid medium or through immersion liquids (e.g., in metalworking, medicine, etc.) is well known. Preliminary studies hold out hope for the effective use of ultrasound for purposes of near field tomography in air. All these aspects are discussed in Chapter 13.

In Chapter 15, different techniques of visualization and reconstruction of foreign objects embedded in clutter conditions or hidden inside various materials and environments based on the mathematical and theoretical frameworks described in Chapters 1–4 were elucidated.

Finally, it should be emphasized that the general purpose of this monograph is to reflect:

- Novel and advanced physical and mathematical models of an image reconstruction system for an inhomogeneous medium based on tomographic processing of multi-angle projection records of scattered radio-wave and acoustic radiation
- Key elements of the modeling systems: from low-frequency and high-frequency electromagnetic to ultrasound and eddy current
- Subsystems and elements of a multi-angle measuring tool
- Experimental measurement techniques
- Methods and techniques for visualization and reconstruction of objects
- Evaluation of potential and actual performance-based specifications for detection by tomography with different measurement and sounding schemes

The wide practical application of UWB tomography is still limited, on the one hand, by considerable engineering problems in generating and receiving UWB radiation, and by complications, on the other hand, that arise when simultaneously describing and interpreting the manifested physical phenomena of the interaction of the radiation with matter. These phenomena include multiple scattering, diffraction, wave interference, and absorption of ultra-wide bandwidth radiation from arbitrarily placed and randomly oriented inhomogeneities of different size. A multitude of combinations of these effects hinders the solution of the direct problem—a description of the integral effects of the wave disturbance. It should be noted, however, that direct problems, even of high complexity, are solved by nature itself as waves reach the observation points in one way or another. Solution of the inverse problem and reconstruction of the distribution of inhomogeneities in the test volume is problematic under such conditions in any event. Researchers have no choice but to solve such inverse problems. These tasks are generally referred as ill-posed problems, which require the use of regularizing algorithms. The most stable inverse problems are the simplest ones, which account for the dominant mechanisms of the interaction of waves with the propagation media and make it possible to single out (identify) such mechanisms (Chapters 1–6 and 15).

Inverse problems are of crucial importance for applications such as sounding optically opaque media. When radio-frequency radiation penetrates into such a medium, analysis of the transmitted and scattered fields makes it possible to reconstruct its internal structure. This internal structure consists of the spatial distribution of the permittivity. Steep permittivity gradients are typical for interfaces and for immersed objects (Chapters 2, 3, and 9).

A typical example is searching for hidden archaeological graves, underground cables, or antipersonnel landmines (Chapter 9) and ground covers (Chapter 10), as well as detection and identification of prohibited items in stowed luggage and in hand-carried items. Problems of this sort are not simple, but a great number of efficient solutions for them have been developed based on radiation focusing. With regard to a radio-opaque object, radiation hardly penetrates it, so the solution of the inverse problem in this case reduces to reconstructing the shape of the object on the basis of an analysis of scattered (reflected) radiation (Chapters 11–13).

Interesting directions for the future development of radio tomography include techniques for implementing incoherent radiation (Chapters 7 and 8) and low-frequency magnetic fields and eddy current techniques (Chapters 5 and 14). These and other examples of the use of radio-tomographic methods, presented in this monograph, far from exhaust the full spectrum of potential applications. Radio tomography, especially using UWB radiation, is in a state of continuous intensive development. One thing that is new here is the use of UWB radiation for non-linear radio detection (Chapters 2 and 8). The development of visualization techniques for tomographic imaging is of great relevance for applications (Chapter 15).

References

1. Daniels, D. J., *Surface-Penetrating Radar*, London: The Institution of Electrical Engineering, 1996.
2. Yakubov, V. P., S. E. Shipilov, D. Ya., Sukhanov, and A. V. Klokov, *Radiowave Tomography: Achievements and perspectives*, Tomsk, Russia: NTL, 2016 (in Russian).
3. Yakubov, V. P., S. E. Shipilov, *Inverse Problems of Radiophysics*, Tomsk, Russia: NTL, 2016 (in Russian).
4. Turk, A. S., A. Koksal Hocaoglu, and A. A. Vertiy (eds), *Subsurface Sensing*, Hoboken, NJ: Wiley, 2012.

5. Blaunstein, N. Sh., M. B. Sergeev, and A. P. Shepeta, *Applied Aspects of Electrodynamics*, St. Petersburg, Russia: Agaf, 2016 (in Russian).
6. Yakubov, V. P., S. E. Shipilov, D. Ya. Sukhanov, and A. V. Klokov, *Wave Tomography*, Ed. by V. P. Yakubov, Tomsk, Russia: Sci. Tech. Publ. House, 2017.
7. Blaunstein, N., Ch. Christodoulou, and M. Sergeev, *Introduction to Radio Engineering*, Boca Raton, FL: CRC Press, 2017.

Acknowledgments

It is with the kind permission of Vladimir Yakubov, my co-editor and co-author, that I write the acknowledgments for this book.

First, this book would never have seen the light of day had Professor Gennady Ponomarev not introduced and initiated both of us to work in the field of radiophysics and radio-wave propagation in various media at the time when we both joined the radio-physical faculty of Tomsk University (former USSR). He stimulated our interests in solving the direct and inverse problems of radio communication and radiolocation, in which both of us have earned MSc degrees and Vladimir has earned a PhD.

I suspended these investigations during a long period after leaving Tomsk, whereas Vladimir continued to work in these fields and earned the degree of doctor of physical and mathematical sciences. We are grateful for Prof. Ponomarev, who passed away many years ago; all of us, creating a strong scientific school of radiophysics in Tomsk, contributed to basic aspects of radio communication and radar principles, which allowed us to continue separately, and then, from 2016, jointly, with the fundamental problems of radiolocation based on principles of wave tomography.

Separately creating groups of researchers in Russia and Israel, respectively, Vladimir and I started to work in close relation with other scientific groups worldwide.

Creating a research group from former students of Ben Gurion University (Israel), from 1997 I started to work with a group of researchers from Tubitak Scientific Center (Turkey) led by Prof. Alexey Vertiy, with whom we investigated the possibility of using wave tomography and eddy current tomography for the detection of different defects in various materials and wrote Chapters 5 and 14 of this book on that basis. We both remember the past contributions of Dr. Sergey Gavrilov, with his insights on diffraction tomography theory and experimental work to demonstrate applications of eddy current tomography in practice. With the research group from the University of Aviation (Kiev, Ukraine) led by Prof. Felix Yanovsky, Fellow of IEEE, we investigated the possibilities of detecting and finding living people behind walls and under ruins based on their UWB radar and our pre- and post-processing techniques, created by myself and Dr. Gregory

Samelsohn. I thank Prof. Yanovsky for his contribution to this process; he contributed to Chapter 11 of this book.

Professor Vladimir Yakubov, with the contributors to this book, acknowledge that a massive amount of experimental data and the body of their detailed interpretation were obtained within the framework of international collaboration with leading scientists from Magdeburg University (Germany), the Institute for Non-Destructive Testing (Saarbrücken, Germany), and Tohoku University (Sendai, Japan), led by Professors A. Omar, M. Kroning and M. Sato, respectively, with their colleagues, whom they thank for intensive joint work on wave tomography's practical proof and interpretation.

Additionally, both the co-editors and the co-authors extend their appreciation to various companies, such as Geodizond (St. Petersburg, Russia), Transient Technologies LLC (Kiev, Ukraine), Iceberg (Kiev, Ukraine), Siberian Physical-Technical Institute at Tomsk State University (Tomsk, Russia), Radiovision (Tomsk, Russia), and DPES (Israel), for their willingness to use the corresponding technologies and devices in practice in both civilian and applied fields of human activity.

We all appreciate the hard work of the editorial staff of CRC/Taylor & Francis publishing group, and their reviewers who all did their best to present the final text with clarity and precision.

Finally, both co-editors and all contributors are greatly indebted to their families for providing the kind atmosphere in which this book "saw the light."

Editors

Nathan Blaunstein is professor emeritus in the Department of Communication Systems Engineering, Ben Gurion University of the Negev, Beer Sheva, Israel, and chair of the MSc program in the Faculty of Engineering and Computer Science, Jerusalem Technology College, Jerusalem, Israel.

He has published more than 200 articles in international journals and conference proceedings, two of which were awarded best papers, ten books, and two chapters in handbooks. He has five patents in radiophysics and optics and their applications in radar and bio-medicine.

His research interests include inverse problems of radio and optical wave propagation in various media and materials for the purposes of radiolocation (radar), acoustic location (sonar), and optical location (lidar), reconstruction and visualization of any foreign object, defect, and structure hidden or embedded in clutter environments (subsoil, atmospheric, and ionospheric), direct problems in aircraft, and mobile satellite and terrestrial communications, as well as applications in bio-medicine and bio-technology.

Vladimir Yakubov was born in 1948 in the city of Chita, Russia. After graduating from Novosibirsk Secondary Physical and Mathematical School No. 165 in 1965, he entered the radiophysics department of Tomsk University. In 1970, he graduated with honors with a degree in radiophysics and electronics, and he was awarded the medal for the best student scientific work at the all-Russian competition of student work. From 1970, he entered the postgraduate course of the radiophysics department of the radio-physical faculty of Tomsk State University, and in 1977 he defended his thesis for the degree of candidate of physical and mathematical sciences.

From 1973 to 1990, he passed through all the stages of his scientific career, starting as a junior researcher at the Siberian Physical and Technical Institut, finishing the post of associate professor. In 1992, he defended his doctoral dissertation, and in 1993, he was promoted to professor. In 1994, he took up the post of head of the department of radiophysics of the Tomsk State University, radio-physical faculty.

For more than 10 years, Professor Yakubov organized studies of near-Earth and outer space, using automated space stations of the Venera series. Under his leadership, a unique experiment was conducted to study the properties of interplanetary and near-solar plasma in the region of solar wind formation, based on the data of narrowband radio interferometry. In addition, based on the experience of studying the Earth using artificial satellites, he developed a number of advanced methods for remote sensing of inhomogeneous media and objects.

Special mathematical algorithms, created by Professor Yakubov on the base of regularization schemes, allowed a significant expansion of the scope of solved inverse problems, while ensuring the maximization of the amount of extracted information from noisy experimental data. In 2001, his theoretical framework was proved by an automatic radiowave tomograph, which was awarded the gold medal of the Russian Academy of Science. In 2017, a similar award was given to the creation of a metamaterial for radio waves.

Professor Yakubov has many Russian awards, including the title of Honored Worker of Higher Professional Education of the Russian Federation. Under his leadership, 12 candidates and three doctoral dissertations were defended. He has frequently worked by invitation in Germany and Japan to educate and consult with their researchers and postgraduate students.

Contributors

Andrey Klokov
Department of Radiophysics
Tomsk State University
Tomsk, Russia

Sergey Shipilov
Department of Radiophysics
Tomsk State University
Tomsk, Russia

Dmitry Sukhanov
Department of Radiophysics
Tomsk State University
Tomsk, Russia

Alexey Vertiy
Ukrainian Institute of Scientific
and Technology Expertise and
Information
Kiev, Ukraine

Felix Yanovsky
National Aviation University of
Ukraine
Kiev, Ukraine

THEORETICAL FUNDAMENTALS OF WAVE TOMOGRAPHY

I

Chapter 1

Mathematical Fundamentals to Inverse Problems

Vladimir Yakubov, Sergey Shipilov, and Nathan Blaunstein

Contents

1.1 Direct and Inverse Problems: Algebraic Approach 4
1.2 Tikhonov Definition of Well-Posed Problem: Tikhonov Regularization 5
1.3 Least Squares Method 8
1.4 Singular-Value Decomposition: Moore–Penrose Matrix 9
1.5 Singular Regularization 10
1.6 Levenberg–Marquardt Algorithm for Non-Linear Equations Solution 11
1.7 Iteration Algorithms for Solution of Inverse Problems 12
1.8 Solution of Convolution Integral Equation and Wiener Filtering 14
References 15

To understand the problems illustrated by this chapter as well as by the contents of the whole book, let us explain to the reader a principal difference between the direct and inverse problems in investigations of applied electrodynamics, and radio, acoustic, and optical wave physics and their relations with each other.

The *direct problem* deals with finding the wave field for a known distribution of sources of wave radiation, and from knowledge of the character of the response of media where the wave currently propagates, it finds the arriving wave field distribution measured by the corresponding wave detector (radio, optical, or acoustic). The *inverse problem* deals with the determination and localization of the radiated sources' distribution and/or the identification of their shape, form, structure, and type of material, as well as the parameters of media via knowledge of the wave field spatial, temporal, and spectral distribution arriving at the wave receiver and recorded by its detector. The strict analytical solution of such problems is obtained only for a few limited cases of canonical objects and media.

What is important to point out here is that the correct solutions of the direct problem are fully guaranteed by the unified theorem of electrodynamics with unique solutions of Maxwell's equations. Moreover, the natural environment itself always solves the direct problem:

What can be measured experimentally is the exact solution of the direct problem.

As for the *inverse problem*, the researcher does not meet the same situation. The solution of such a problem depends on human effort and intellectual knowledge. In this case, to obtain the solution of the inverse problem, the so-called approximate methods and assumptions are usually used. Correction of any assumption or approximation can be estimated by its coincidence with the corresponding experimental data obtained during the tests [1–11]. Moreover, the strictness of the inverse problem depends on the strictness of the direct problem: the more exact solution of the direct problem that is found, accounting for the dominant physical processes, the stricter the solution of the inverse problem that will be obtained, responding to the first one.

1.1 Direct and Inverse Problems: Algebraic Approach

As a main object of investigations in algebraic theory, a system of linear equations (SLE) is usually considered in the following matrix form [1–7]:

$$\mathbf{A}\mathbf{x} = \mathbf{y} \tag{1.1}$$

where:
  $\mathbf{x} = [x_1, x_2, ..., x_N]^T$ is the column vector of dimension N of the input values
  $\mathbf{y} = [y_1, y_2, ..., y_M]^T$ is the column vector of dimension M of the output values
  $\mathbf{A} = [a_{ij}]$ is the matrix of dimension $N \times M$, where $M \geq N$

All values introduced by Equation 1.1 can also be complex ones. It is obvious that Equation 1.1 describes the so-called *direct problem* where, for the known input data (i.e., according to vector $\mathbf{x}$), it is required to find the output data (described by vector $\mathbf{y}$).

Thus, for the direct problem, we can consider an example of determination of the electromagnetic pulse at the output of any electronic device, namely, at the output of the linear filter, which can be presented in the following manner by use of the convolution integral [7–11]:

$$y(t) = \int_0^t h(\tau) x(t-\tau) d\tau$$

where $h(\tau)$ is the impulse response function of the linear filter. In spectral form, this relation can be presented as:

$$Y(\omega) = H(\omega) X(\omega) \qquad (1.2)$$

Here,

$$H(\omega) = \int_0^t h(\tau) e^{i\omega\tau} d\tau$$

is the transfer function or system function of the filter.

According to Hadamard's definitions, the direct problem is a well-posed problem if the following conditions are valid: (1) the solution exists; (2) the solution is unique; (3) the solution depends continuously on the data (the solution is stable).

Stability means that deviations from the single solution are small during small deviations from the initial data of the problem. This smallness can be stated using Euclid's definition of the *norm* $\|\mathbf{x}\|$ as its length: $\|\mathbf{x}\| = \sqrt{\mathbf{x}^H \cdot \mathbf{x}}$, where H defines the conjugate transpose or Hermitian transpose of a matrix $\mathbf{x}$. If so, the term *small vector* $\mathbf{x}$ determines the vector with small norm $\|\mathbf{x}\|$. In such definitions, we can determine the norm of matrix (or operator) $\mathbf{A}$ as follows:

$$\|\mathbf{A}\| = \sup_{\|\mathbf{x}\|=1} \|\mathbf{A} \cdot \mathbf{x}\| = \sup_{\|\mathbf{x}\|} \frac{\|\mathbf{A} \cdot \mathbf{x}\|}{\|\mathbf{x}\|}. \qquad (1.3)$$

This equation defines the maximum value of all the various unit vectors.

The *inverse problem* is related to the *direct* one, described by Equation 1.1, via computation of vector $\mathbf{x}$ by use of knowledge of measured vector $\mathbf{y}$. It should be noted that the theory of inverse problem solutions gives exact results only when this problem can be deduced from the SLE, that is, when the vector $\mathbf{A}$ is simply a linear operator (matrix) [1–4].

1.2 Tikhonov Definition of Well-Posed Problem: Tikhonov Regularization

As follows from Section 1.1, the inverse problem is a definition of the input process (vector $\mathbf{x}$) via the measured output process (vector $\mathbf{y}$), that is,

$$\mathbf{x} = \mathbf{A}^{-1} \mathbf{y} \qquad (1.4)$$

Definition of the inverse operator (matrix) $\mathbf{A}^{-1}$ is the main aim of the solution of the inverse problem. Unfortunately, in practical applications, the inverse problems are mostly ill-posed (unstable), that is, the Hadamard definition [3] introduced in Section 1.1 is not valid and the full solution becomes incorrect. In other words, small deviations of components of the output vector **y** lead to non-small deviations of those of the input vector **x**, even though the direct problem is initially and always has been stable. As will be shown regarding special applications of the inverse problems, the existence of input noises together with the input signal, even after averaging by a filter of the device, can ultimately be seen in the output signal. At the same time, we can emphasize that the direct problems are usually stable, where a set of numerous noises can be averaged and eliminated in the deviations of the input signals.

This situation can be explained by the following. Schematically, the stability of the direct and inverse problems can be illustrated by the funnel (a) and water from the watering pot (b), as shown in Figure 1.1. The funnel concentrates the throw of the output parameters, whereas the watering pot throws these parameters.

The formal solution of the inverse problem in the frequency domain (Equation 1.2) can be written in the same manner as in the time domain, that is,

$$X(\omega) = \frac{Y(\omega)}{H(\omega)} \qquad (1.5)$$

The following procedure, by use of the inverse Fourier transform (IFT), allows us to obtain the desired solution [5, 6]:

$$x(t) = \frac{1}{2\pi} \int_{-\infty}^{\infty} X(\omega) e^{-i\omega\tau} d\omega$$

Because the function $y(t)$ is usually found by experimental measurement, it definitely includes some errors. In other words, in the frequency domain, its spectral presentation has a lot of high-frequency components, even including situations

Figure 1.1 Schematic presentation of the stability of the direct problem in the form of a funnel (a), and the instability of the inverse problem in the form of a watering pot (b).

where $H(\omega)$ limits to zero. In this case, the solution of the inverse problem will be incorrect.

The objective estimation of the accuracy of the obtained solution $\hat{\mathbf{x}}$ in the situation when the exact solution $\mathbf{x}$ is unknown can be carried out by introducing a special value called the *residual*, defined as:

$$\Phi = \|\mathbf{y} - A\hat{\mathbf{x}}\|^2 \tag{1.6}$$

It was found [2–7] that the smaller the values of Φ, the better the solution of the inverse problem. Or, in other words, the minimum Φ, $\min_{|\hat{\mathbf{x}}|}\Phi(\hat{\mathbf{x}})$, can be achieved for $\mathbf{x} = \hat{\mathbf{x}}$. This statement is seen as elementary in the least squares method and relates to the maximum likelihood estimation in statistics [7–10]. However, simple achievement of the minimum value of the residual (Equation 1.6) does not guarantee the stability of the solution of the inverse problem.

In 1963, Tikhonov [5, 6] introduced a fundamental definition of the regularized algorithm, which became a key statement in the theory of the solution of ill-posed inverse problems [1–4, 7–10]. It is based on the introduction of a special *smoothing functional*, which limits possible variations of the desired solutions; that is, it decreases a class of possible solutions. A simple condition of such a stabilization is

$$\|\hat{\mathbf{x}}\|^2 \leq C < \infty$$

In this case, searching for the minimum of the residual in Equation 1.6, according to the calculus of variations, leads to the minimization of the smoothing functional:

$$\Phi_\alpha = \|\mathbf{y} - A\hat{\mathbf{x}}\|^2 + \alpha \|\hat{\mathbf{x}}\|^2 = \min_{|\hat{\mathbf{x}}|} \tag{1.7}$$

Here, α plays the role of a non-defined Lagrange product and is called in the literature the *parameter of regularization* [5–10]. In this case, the desired solution fully depends on the parameter α: $\hat{\mathbf{x}} = \hat{\mathbf{x}}_\alpha$.

It is clear that, for $\alpha = 0$, the task of solution estimation is definitely non-stabilized. But, for $\alpha \longrightarrow \infty$, the problem under consideration becomes fully stable. The selection of the regularization parameter is an independent problem. In references [5, 6], the algorithm of selection of such a parameter α was proposed, for which the residual of the solution is similar to the variance (standard deviation σ) of the measurements:

$$\Phi_\alpha = \|\mathbf{y} - A\hat{\mathbf{x}}\|^2 = \sigma^2 \tag{1.8}$$

It was proved in [5, 6] that the solution of Equation 1.8 is unique and unified. In practical applications, one difficulty was found—the variance of noises during measurements (σ^2) is not always known from the experiment. Therefore, the parameter of regularization is often determined by an ad hoc method.

At the same time, as found in references [5, 6], a problem, based on the algorithm of regularization, was stated as Tikhonov well-posed, also called *conditionally well-posed*. By introducing some distortions into the ideal model (Equation 1.1), we can obtain small deviations from the exact solution, but the solution itself will remain enough stable.

1.3 Least Squares Method

Let us present the smoothing functional equation (Equation 1.7) in another form during the definition of the norm according to Euclid [8–10]:

$$\Phi_\alpha = \|\mathbf{y} - \mathbf{A}\hat{\mathbf{x}}\|^2 + \alpha\|\hat{\mathbf{x}}\|^2 = (\mathbf{y} - \mathbf{A}\hat{\mathbf{x}})^H (\mathbf{y} - \mathbf{A}\hat{\mathbf{x}}) + \alpha\hat{\mathbf{x}}^H\hat{\mathbf{x}}$$
$$= (\mathbf{y}^H - \hat{\mathbf{x}}^H\mathbf{A}^H)(\mathbf{y} - \mathbf{A}\hat{\mathbf{x}}) + \alpha\hat{\mathbf{x}}^H\hat{\mathbf{x}}$$

The extremum of this functional is achieved by setting the gradient of $\hat{\mathbf{x}}^H$ to zero, that is,

$$-\mathbf{A}^H(\mathbf{y} - \mathbf{A}\hat{\mathbf{x}}) + \alpha\hat{\mathbf{x}} = 0$$

Rewriting this equation, we finally obtain:

$$(\mathbf{A}^H\mathbf{A} + \alpha\mathbf{I})\hat{\mathbf{x}} = \mathbf{A}^H\mathbf{y}$$

Here, **I** is the diagonal unit matrix (*identity matrix*). There is a square non-zero matrix on the left-hand side of this equation and, therefore, its solution can be expressed as:

$$\mathbf{x} = \mathbf{A}^+\mathbf{y} \qquad (1.9)$$

where the matrix **A**⁺ equals:

$$\mathbf{A}^+ = (\mathbf{A}^H\mathbf{A} + \alpha\mathbf{I})^{-1}\mathbf{A}^H \qquad (1.10)$$

This is called the *pseudo-inverse matrix* to matrix **A**. It is clearly seen that $\mathbf{A}^+\mathbf{A}\hat{\mathbf{x}} = \hat{\mathbf{x}}$ always. Equation 1.9 presents a solution of the least squares method with regularization according to the Tikhonov method [5, 6]. In the case of $\alpha = 0$, this method is simply the least squares method that gives, for the square matrix **A**, its inverse matrix $\mathbf{A}^{-1}$.

1.4 Singular-Value Decomposition: Moore–Penrose Matrix

Use of the pseudo-solution in Equation 1.9 is not always effective, even for the correct selection of the regularization parameter defined in Section 1.2, because, during the computations of the pseudo-inverse matrix in the form of Equation 1.10, a lot of computations need to be carried out, during which computational noises have the tendency to quickly create a cumulative effect and, ultimately, destroy the correct desired solution. To obey this effect, a so-called singular-value decomposition of the matrix **A** is usually used, presented as a product of three matrices:

$$\mathbf{A} = \mathbf{U}^H \cdot \mathbf{S} \cdot \mathbf{V} \qquad (1.11)$$

where **U** and **V** are unitary matrices of dimensions of $(M \times M)$ and $(N \times N)$, respectively. This means that $\mathbf{U}^H \cdot \mathbf{U} = \mathbf{I}$ and $\mathbf{V}^H \cdot \mathbf{V} = \mathbf{I}$, where **I** is the identity matrix introduced in Section 1.3. The columns of matrices **U** and **V** are called, in algebraic theory, the *left* and *right singular vectors*, respectively. Matrix $\mathbf{S} = [s_i \delta_{i,j}]$ is a quasi-diagonal matrix, consisting of the singular values s_i, which are strictly positive and can usually be obtained one-by-one in decreasing of each value manner, and $\delta_{i,j}$ are the Kronecker symbols.

We notice that the non-zero eigenvalues of matrices $\mathbf{A} \cdot \mathbf{A}^H$ and $\mathbf{A}^H \cdot \mathbf{A}$ are similar to the eigenvalues of matrix **A**, that is, $s_i = \sqrt{\lambda_i}$, where λ_i are the eigenvalues of matrix **A**. The rank of any matrix **A** equals the rank of diagonal matrix **S**.

Introducing into Equation 1.10 the decomposition Equation 1.11 for $\alpha = 0$, we obtain [9, 10]

$$\mathbf{A}^+ = \mathbf{V}^H \cdot \mathbf{S}^+ \cdot \mathbf{U} \qquad (1.12)$$

where $\mathbf{S}^+ = \left\{ \dfrac{1}{s_i} \delta_{i,j} \right\}$. In such a form, the pseudo-inverse matrix $\mathbf{A}^+$ is called the Moore–Penrose matrix. Use of singular-value decomposition allows us to estimate the norm (Equation 1.3) relatively simply for all matrices in Equation 1.11. Thus, because $\|\mathbf{V}^H \cdot \mathbf{V}\| = \|\mathbf{V}\|^2 = \mathbf{I}$, then

$$\|\mathbf{V}^H \mathbf{V}\| = \sup_{\|\mathbf{x}\|} \frac{\|\mathbf{x}^H \mathbf{V}^H \mathbf{V} \mathbf{x}\|}{\|\mathbf{x}^H \mathbf{x}\|} = \|\mathbf{V}\|^2 = \mathbf{I}$$

and $\|\mathbf{V}\| = 1$ and $\|\mathbf{U}\| = 1$. In the same manner, the norm of the diagonal matrix **S** can be found:

$$\|\mathbf{S}\|^2 = \sup_{\|\mathbf{x}\|} \frac{\sum_j x_j^2 s_j^2}{\sum_j x_j^2} \leq \max_j \left(x_j^2 \right) \qquad (1.13)$$

from which it follows that $\|\mathbf{S}\| = \max_{j|}(s_j) = s_{max}$. From the decomposition equation (Equation 1.11), it follows that:

$$\|\mathbf{A}\| = \|\mathbf{U}^H\| \cdot \|\mathbf{S}\| \cdot \|\mathbf{V}\| \leq \max_{j|}(s_j) \text{ and } \|\mathbf{A}\| = \max_{j|}(s_j) = s_{max}$$

For the analogous Moore–Penrose matrix, we obtain:

$$\|\mathbf{A}^+\| = \max_{j|}(s_j^{-1}) = s_{min}$$

Let us now estimate the accuracy of the solution of the inverse problem using the Moore–Penrose matrix. If we denote deviations of the solutions and the initial data as $\|\Delta\mathbf{x}\|$ and $\|\Delta\mathbf{y}\|$, respectively, then Equation 1.9 yields a constraint:

$$\|\Delta\mathbf{x}\| \leq \|\mathbf{A}^+\| \|\Delta\mathbf{y}\|.$$

Finally, from the initial equation (Equation 1.1), we obtain:

$$\frac{\|\Delta\mathbf{x}\|}{\|\mathbf{x}\|} = \|\mathbf{A}\| \|\mathbf{A}^+\| \frac{\|\Delta\mathbf{y}\|}{\|\mathbf{y}\|} = cond(\mathbf{A}) \frac{\|\Delta\mathbf{y}\|}{\|\mathbf{y}\|}, \qquad (1.14)$$

where

$$cond(\mathbf{A}) = \|\mathbf{A}\| \|\mathbf{A}^+\| = \frac{\max\ s}{\min\ s} = \frac{s_{max}}{s_{min}}$$

is called the *condition number* of matrix **A**, which demonstrates the accuracy of the inverse problem with respect to measurement. Only for the orthogonal transform, when *cond*(**A**) = 1, is the error of the solution not increased. We should notice that the solution of the least squares problem is most effective with use of the Moore–Penrose matrix.

1.5 Singular Regularization

As follows from the conditional expression Equation 1.14, to increase accuracy of the pseudo-solutions one needs to decrease, first of all, the condition number of the Moore–Penrose matrix **A**. To do that, one should simply decrease a set of singular values; for example, by introducing some small limit value τ to account for only those singular values for which $s_i \geq \tau$. In this case, the Moore–Penrose matrix $\mathbf{S}^+$ should be changed to [9, 10]:

$$\mathbf{S}^+ = \left\{ \frac{\chi(s_i - \tau)}{s_i} \delta_{i,j} \right\} \qquad (1.15)$$

where:

$\chi(s_i - \tau)$ is the Heaviside step function: $\chi(s_i - \tau) = 1$ if $s_i \geq \tau$
$\chi(s_i - \tau) = 0$ if $s_i < \tau$

The limit value τ according to Equation 1.14 should be not less than $\tau_0 = (\|\Delta \mathbf{y}\|/\|\mathbf{y}\|) s_{\max}$. In this case, we have guaranteed that the error of the solutions of the inverse problem will be small enough, if $(\|\Delta \mathbf{x}\|/\|\mathbf{x}\|) \leq 1$.

More smoothing regularization can be proposed by taking regularization according to Equation 1.10, as proposed by Tikhonov [5, 6]. To account for the singular-value decomposition equation (Equation 1.11), the pseudo-inverse matrix $\mathbf{S}^+$ defined by Equation 1.15 should be changed to:

$$\Sigma^+ = (\mathbf{S} + \alpha \mathbf{S}^+)^+ = \left\{ \left(s_i + \frac{\alpha}{s_i} \right)^{-1} \delta_{i,j} \right\}. \tag{1.16}$$

Such a decomposition for the Moore–Penrose matrix (Equation 1.12) takes into account Tikhonov's regularization. The norm of matrix Σ^+ can be described as follows:

$$\|\Sigma^+\| = \min_{i|} \left(\left(s_i + \frac{\alpha}{s_i} \right)^{-1} \right) = \frac{1}{2\sqrt{\alpha}}$$

Selection of the regularization parameter will be effective if, according to the conditions of Equation 1.14, the following constraint is achieved:

$$\frac{\|\Delta \mathbf{y}\|}{\|\mathbf{y}\|} \frac{s_{\max}}{2\sqrt{\alpha}} \leq 1 \tag{1.17}$$

In this case, the following condition should be valid: $s_{\min} \leq \sqrt{\alpha} \leq s_{\max}$.

1.6 Levenberg–Marquardt Algorithm for Non-Linear Equations Solution

The Levenberg–Marquardt method is valid for the solution of systems of non-linear equations, and can be written in the following form:

$$\mathbf{y} = \mathbf{F}(\mathbf{x}) \tag{1.18}$$

where $\mathbf{F}(\mathbf{x})$ is any vector function of vector $\mathbf{x}$. The aim of this method is to transfer the general Equation 1.18 to an SLE. For this purpose, let us suppose that the initial approximate vector $\mathbf{x}_0$ for a vector $\mathbf{x}$ is known. Then, the solution will be presented in the following manner:

$$\mathbf{x} = \mathbf{x}_0 + \Delta\mathbf{x}$$

Assuming a small error $\Delta\mathbf{x}$, we obtain, according to [9],

$$\Delta\mathbf{y} = \mathbf{y} - \mathbf{F}(\mathbf{x}_0) = \mathbf{F}(\mathbf{x}) - \mathbf{F}(\mathbf{x}_0) = \mathbf{A}\Delta\mathbf{x} \qquad (1.19)$$

Here, $\mathbf{A} = \dfrac{\partial \mathbf{F}(\mathbf{x}_0)}{\partial \mathbf{x}_0} = \left[\dfrac{\partial F_i(\mathbf{x}_0)}{\partial \mathbf{x}_{0j}}\right]$ is the Jacobian matrix, and $i = 1, \ldots, M$; $j = 1, \ldots, N$ are the numbers of the corresponding vectors' elements. The system of equations is the desired SLE for definition of the error $\Delta\mathbf{x}$.

The solution of the vector equation (Equation 1.18) can be found by use of the Moore–Penrose pseudo-inverse matrix, that is,

$$\Delta\hat{\mathbf{x}} = \mathbf{A}^+ \Delta\mathbf{y} \qquad (1.20)$$

When found in such a manner, the solution of Equation 1.14 allows correction of an initial approximate vector $\mathbf{x}_0$, that is, $\mathbf{x}_0 = \hat{\mathbf{x}} - \Delta\hat{\mathbf{x}}$. Now, exchanging $\mathbf{x}_0$ with $\hat{\mathbf{x}}$, one can obtain a new approximation of the Jacobian matrix. Then, using the Moore–Penrose matrix, we repeat an iteration procedure (Equation 1.20). This iteration procedure is called in the literature the *Levenberg–Marquardt algorithm*. It may be noticed that the Moore–Penrose matrices allow the use of the regularization procedure at each step of the iteration. The rules for stopping the iteration procedure are as follows:

- When solution $\hat{\mathbf{x}}$ becomes non-changing
- When the residual $\Phi = \|\mathbf{y} - \mathbf{F}(\hat{\mathbf{x}})\|^2$ for the vector equation (Equation 1.18) becomes stable or achieves a given level (e.g., a value of variance of the accuracy of measurements of the vector $\mathbf{y}$)
- When the current solution becomes similar to the desired solutions obtained in Section 1.2

1.7 Iteration Algorithms for Solution of Inverse Problems

Often, it is very complicated to achieve the differentiation operation of the Jacobian matrices according to the Levenberg–Marquardt algorithm described in Section 1.6. In this case, the iteration algorithms are usually used.

Let us introduce a definition of the conjugate operator (matrix) $\mathbf{A}^H$. The conjugate matrix is that for which the following equation is valid:

$$\mathbf{A}^H \mathbf{y} \cdot \mathbf{x} = \mathbf{y} \cdot \mathbf{A}\mathbf{x} \tag{1.21}$$

for arbitrary vectors $\mathbf{x}$ and $\mathbf{y}$. Using the conjugate operator for Equation 1.1, we finally obtain:

$$\mathbf{A}^H \mathbf{y} = \mathbf{A}^H \mathbf{A}\mathbf{x}$$

or $\mathbf{A}^H \mathbf{y} - \mathbf{A}^H \mathbf{A}\mathbf{x} = 0$, which yields

$$\mathbf{x} = \mathbf{x} + \lambda \left(\mathbf{A}^H \mathbf{y} - \mathbf{A}^H \mathbf{A}\mathbf{x} \right) \tag{1.22}$$

for arbitrary parameter λ. This parameter defines the degree of *convergence* of the current solution obtained during the iteration procedure.

Let us rewrite the latter equation in the following form by introducing the unit operator (matrix) $\mathbf{I}$, that is,

$$\mathbf{x} = \lambda \mathbf{A}^H \mathbf{y} + \left(\mathbf{I} - \lambda \mathbf{A}^H \mathbf{A}\mathbf{x} \right) \mathbf{x}$$

Here, we should add an additional operator called the *limited operator*, which can be defined as [9]:

$$\mathbf{L}\mathbf{x} = \begin{cases} \mathbf{x}, x_i \in (a,b) \\ 0, x_i \notin (a,b) \end{cases}, a \leq b \tag{1.23}$$

Accounting for Equation 1.18, we finally obtain the following equation:

$$\mathbf{x} = \lambda \mathbf{A}^H \mathbf{y} + \left(\mathbf{I} - \lambda \mathbf{A}^H \mathbf{A} \right) \mathbf{L}\mathbf{x} \tag{1.24}$$

The iteration solution of Equation 1.19 can be presented as follows:

$$\mathbf{x}^{(k+1)} = \lambda \mathbf{A}^H \mathbf{y} + \left(\mathbf{I} - \lambda \mathbf{A}^H \mathbf{A} \right) \mathbf{L}\mathbf{x}^{(k)} \tag{1.25}$$

To obtain a successful solution for Equation 1.20, it is necessary to take precisely the initial approximation $\mathbf{x}^{(0)}$, and to take correctly the convergence parameter λ. Initially, the convergence can be obtained only if $\mathbf{A}^H$ is the conjugate operator of matrix $\mathbf{A}$.

1.8 Solution of Convolution Integral Equation and Wiener Filtering

The convolution integral equation was introduced in Section 1.1. We present it again for convenience, to understand the matter:

$$y(t) = \int_0^t h(\tau) x(t-\tau) d\tau \qquad (1.26)$$

Its formal solution (Equation 1.5) in the presence of errors during measurement leads to significant errors in the solutions of an inverse problem. Therefore, the Tikhonov regularization technique is usually used. According to Tikhonov's methodology [5, 6], a general residual should be minimized (see Equation 1.7):

$$\|\mathbf{y} - \mathbf{A}\mathbf{x}\|^2 + \alpha \|\mathbf{x}\|^2 = \min_{|\mathbf{x}}. \qquad (1.27)$$

In the case of integrated functions $f(t)$, their norm can be defined through the scalar product:

$$\|f\|^2 = \int_{-\infty}^{\infty} f^*(t) f(t) dt = \frac{1}{2\pi} \int_{-\infty}^{\infty} \hat{f}^*(\omega) \hat{f}(\omega) d\omega = \frac{1}{2\pi} \|\hat{f}\|^2 \qquad (1.28)$$

Here, $\hat{f}(\omega)$ is the Fourier transform of the function $f(t)$. Equation 1.28 is known as Parseval's theorem. Using Equation 1.28, we can rewrite a general residual equation (Equation 1.27) in clear form:

$$\frac{1}{2\pi} \int_{-\infty}^{\infty} \left[\hat{y}(\omega) - H(\omega) \hat{x}(\omega) \right]^* \left[y(\omega) - H(\omega) x(\omega) \right] d\omega$$

$$+ \alpha \int_{-\infty}^{\infty} \hat{x}^*(\omega) \hat{x}(\omega) d\omega = \min_{\mathbf{x}}$$

Taking the derivative of $\hat{x}^*(\omega)$ and limiting it to zero, we finally obtain the following equation:

$$-H^*(\omega) \left[\hat{y}(\omega) - H(\omega) \hat{x}(\omega) \right] + \alpha \hat{x}(\omega) = 0 \qquad (1.29)$$

The spectrum of the defined solution of Equation 1.29 can be found as follows:

$$\hat{x}(\omega) = \frac{\hat{y}(\omega) H^*(\omega)}{H(\omega) H^*(\omega) + \alpha} = \frac{\hat{y}(\omega)}{H(\omega)} R_\alpha(\omega) \qquad (1.30)$$

Here, $R_\alpha(\omega) = \dfrac{H(\omega)H^*(\omega)}{H(\omega)H^*(\omega)+\alpha}$ is the regularization product according to Tikhonov. It may be noticed that for $\alpha = 0$, we get $R_\alpha(\omega) = 1$ and then the formal solution (Equation 1.5) of the inverse problem. A final solution can be written by use of the IFT:

$$x(t) = \frac{1}{2\pi} \int_{-\infty}^{\infty} \frac{\hat{y}(\omega)}{H(\omega)} R_\alpha(\omega) e^{-i\omega t} d\omega. \qquad (1.31)$$

Such a presentation is known as Wiener filtering with regularization. We should outline that selection of the regularization parameter α essentially influences the accuracy of solution of the inverse problem, such as the problem of reconstruction of images in optics by rejection of so-called *blur*, which in radiophysics is usually called the *clutter noise* [8, 10].

References

1. Ivashov, V. K., V. V. Vasin, and V. P. Tanana, *Theory of Linear Incorrect Problems and its Applications*, Moscow: Science, 1978 (in Russian).
2. Lavrent'ev, M. M., V. G. Romanov, and S. P. Shishatzkii, *Incorrect Problems of Mathematical Physics and Analysis*, Moscow: Science, 1980 (in Russian).
3. Bakushinskii, A. B. and A. V. Goncharskii, *Interactive Methods of Incorrect Problems Solution*, Moscow: Science, 1989 (in Russian).
4. Bakushinskii, A. B. and A. V. Goncharskii, *Incorrect Problems Numerical Methods and Applications*, Moscow: MGU, 1989 (in Russian).
5. Tikhonov, A. N., A. V. Gonchrskii, V. V. Stepanov, and A. G. Yagola, *Numerical Methods of Solution of Incorrect Problems*, Moscow: Science, 1990 (in Russian).
6. Tikhonov, A. N., A. S. Leonov, and A. G. Yagola, *Nonlinear Incorrect Problems*, Moscow: Science, 1995 (in Russian).
7. Sizikov, V. S., *Inverse Problems and MatLab*, Saint Petersburg, Russia: Lan', 2011 (in Russian).
8. Yakubov, V. P., *Doppler Ultra-Large-Ranged Interferometry*, Tomsk: Vodole', 1997 (in Russian).
9. Yakubov, V. P., S. E. Shipilov, S. D. Ya., and A. V. Klokov, *Radiowave Tomography: Achievements and Perspectives*, Tomsk: NTL, 2016 (in Russian).
10. Yakubov, V. P., S. E. Shipilov, *Inverse Problems of Radiophysics*, Tomsk: NTL, 2016 (in Russian).
11. Fisher, S. C., R. S. Steward, and M. J. Harry, Processing ground penetrating radar data, *Proceedings of 5th International Conference on Ground Penetrating Radar*, Kitchener, Canada, 1994, 9 pages.

Chapter 2

Theoretical Overview of Wave Tomography

Vladimir Yakubov, Sergey Shipilov,
Dmitry Sukhanov, and Andrey Klokov

Contents

2.1 Method of Inverse Projections 18
2.2 Method of Fourier Synthesis Based on Projections of Shadowing 19
2.3 Method of Double Focusing 23
2.4 Radio Wave Tomography Synthesis: Stolt's Method 28
2.5 Transmission Tomography of the Semitransparent Media Based on the Kirchhoff Approximation 33
2.6 Transmission Tomography of Opaque Objects 36
2.7 Linear-Frequency-Modulated Radiation Tomography Technique 39
2.8 Incoherent Tomography for Reconstruction of Objects Hidden in Clutter 42
References 45

The word *tomography* consists of two separate Greek words: τομoσ (slice, section) and γροφοσ (to write). So, tomography is similar to a "write sections" procedure, that is, a division of the structure of the object under investigation into slices. Wave tomography records information on each element of the object under investigation by use of numerous integral projections under various angles with respect to the direction of the investigated object.

Radio, optical, and acoustic wave tomography are separate novel directions of physical tomography, and have recently been used for sounding, visualization, and reconstruction of various objects, features, and processes presented and occurring in the environment, as well as anomalous occurrences on and in the human body (tumors, cancers, skin nevus, etc.). In all types of wave tomography, the wave nature of any radiation is usually used during investigations carried out to understand deeply the inner and outer nature of the objects under investigation. This chapter generalizes results obtained in references [1–23].

2.1 Method of Inverse Projections

Generally speaking, tomography is based on a definition such as *shadowing projection*. It can be defined by the value of attenuation of the radiation intensity of the wave passing from the transmitter (denoted in Figure 2.1 by point 1) to the receiver (denoted by point 2) [13, 14]:

$$f_{12} = \int_{L_1}^{L_2} g(L) \, dL \qquad (2.1)$$

Here, $g(L)$ determines the distribution of obstructions (or elements of the object under investigation) along the direction of illumination of the desired object.

In Equation 2.1, it is supposed that the attenuation of the wave intensity is directly proportional to the distributions of the elements of the object. Putting this trajectory onto an arbitrary coordinate system, we see that the trajectory will pass through the cells (or will touch some of them, as shown in Figure 2.1 by shadow [dark] cells).

In the method of inverse projections, it is assumed that the value of filling of cells, denoted by the dark color, is identical and can be found via the average value, $\bar{L}$, that is,

$$g(\bar{L}) = f_{12} / (L_2 - L_1) \qquad (2.2)$$

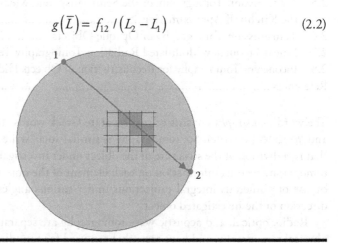

Figure 2.1 Explanation of method of inverse projections.

With numerous projections taken from different angles of view, the value of filling of each cell can be computed as a simple average value over all angles of view. As a result, a reconstruction of filling the cells, consisting of the elements of the object under investigation, can be carried out. Such a technique is called the *method of inverse projections*, and it is comparatively simple (with respect to those methods of recovering the objects under investigation discussed in this chapter). Moreover, its advantage is that the recovery of the elements of the object begins and remains parallel with measurements. The quality of images (tomograms) depends strongly on the division of cells and on the number of angles of view. It can be improved by increasing the number of angles of view.

2.2 Method of Fourier Synthesis Based on Projections of Shadowing

In wave tomography (radio, optical, or acoustic), the basic technique to recover tomograms (pictures, photos, images, etc.) is the method of Fourier synthesis (FS), which is based on slicing the reverse of multiple specular projections shadowing the object by the use of the direct and the inverse Fourier transforms (DFT and IFT, respectively).

Let us briefly describe this method, as follows [13, 14]. During a photo snapshot of the shadow projection (SP), the measured characteristics, that is, the intensity of radiation absorption, $f(y, \varphi)$, can be obtained as the result of accumulation of the sliced intensities along the desired straight line, that is,

$$f(y,\varphi) = \int_{-\infty}^{\infty} g(x\cos\varphi - y\sin\varphi, x\sin\varphi + y\cos\varphi)dx \qquad (2.3)$$

This straight line (the direction of the wave incidence on the object) has an angle φ to the initial coordinate system (xoy), as shown in Figure 2.2.

We can define the shadow function $f(y, \varphi)$ as the integrated projection of the spatial distributed energy density of the object (body), $g(X, Y)$, illuminated by the wave. If we now compute the DFT of the SP,

$$F(\kappa,\phi) = \int_{-\infty}^{\infty} f(y,\phi)e^{i\kappa y}dy \qquad (2.4)$$

then the final function obtained will be similar to the spatial spectrum (SS) of the body under illumination by the wave, that is, to:

$$F(\kappa,\varphi) = G(u = -\kappa\sin\varphi, v = \kappa\sin\varphi)$$

20 ■ Electromagnetic and Acoustic Wave Tomography

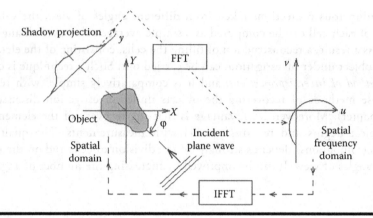

Figure 2.2 Problem of Fourier synthesis.

The latter can be determined as:

$$G(u,v) = \int_{-\infty}^{\infty} \int_{-\infty}^{\infty} g(X,Y) e^{-i(uX+vY)} dX dY \qquad (2.5)$$

A difficulty is that one of the spectral functions, $F(\kappa, \varphi)$, is written in the polar coordinate system, whereas the second one, $G(u, v)$, is written in the Cartesian coordinate system. A simple transform of the data from one coordinate system to another can overcome this problem. Thus, if the values of $F(\kappa, \varphi)$ are known at the knots of the polar coordinate system $\{F_1, F_2, F_3, F_4\}$, and it is necessary to recover a value of the point in the other coordinate space (let say, by use of the Cartesian system) that is denoted by a star in Figure 2.3, the latter value can be calculated by the following relationship:

$$G = \frac{F_1/d_1 + F_2/d_2 + F_3/d_3 + F_4/d_4}{1/d_1 + 1/d_2 + 1/d_3 + 1/d_4}$$

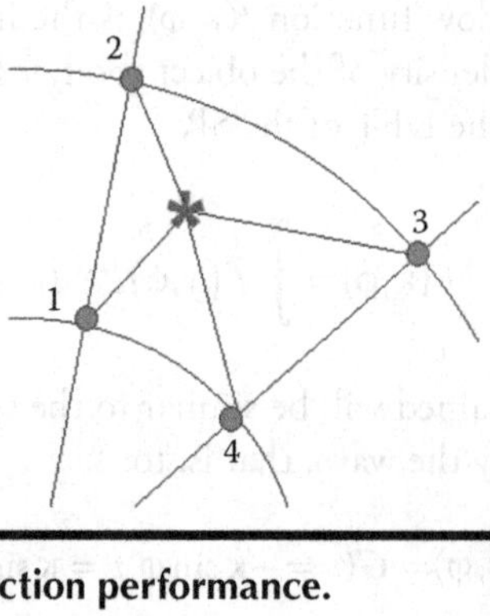

Figure 2.3 Shadow projection performance.

Here, $\{d_1, d_2, d_3, d_4\}$ are the distances between the desired Cartesian point, denoted by a star, and the knots of the polar coordinate system close to it.

Finally, with the help of the 2-D IFT, we may obtain the desired characteristic function (CF) of the object under investigation:

$$g(X,Y) = \frac{1}{(2\pi)^2} \int_{-\infty}^{\infty} \int_{-\infty}^{\infty} G(u,v) e^{-i(uX+vY)} du dv \qquad (2.6)$$

Such a technique is usually used in magnetic and X-ray tomography, where the information about the phase of the wave response is not used and the collimated systems of measurements of SPs are usually used. In Figures 2.4 through 2.7, we illustrate the steps of the proposed technique, described by Equations 2.3 through 2.6, for example, for the object having a tube form.

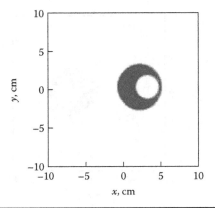

Figure 2.4 Original CF of object $g(x, y)$.

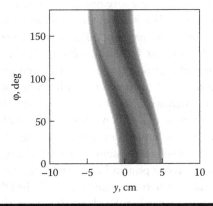

Figure 2.5 Shadow of object.

22 ■ *Electromagnetic and Acoustic Wave Tomography*

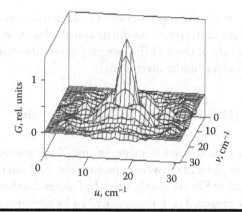

Figure 2.6 Reconstructed spatial frequency spectrum for object $G(u, v)$.

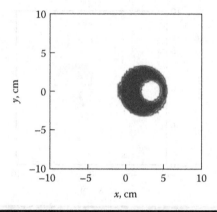

Figure 2.7 Reconstructed CF of object $g(x, y)$.

It should be noticed that the necessary operations were carried out with usage of fast algorithms and in real time. This is one of the main advantages of the FS method. It can be useful to analyze the semitransparent object, as was done for the objects shown in Figures 2.4 through 2.7, and also for opaque objects. But, in all these examples, the scenario where the source of radiation and the receiver were located on both sides of the test object was analyzed, and the process of diffraction from the object's contour and the corresponding interference of the wave were not taken into account. These are the main points that limit usage of the FS method in radio and acoustic wave tomography. At the same time, after some additional modifications, this method can be used for impedance tomography and transmission wave tomography, which will be shown in Sections 4.2 and 6.4. It has one principal advantage—the recovery procedure of the desired object under testing can start simultaneously with the filling of the corresponding matrix by the initial data regarding the object. However, nowadays the FS procedure is begun after finishing the scanning process of the object under investigation.

2.3 Method of Double Focusing

In wave tomography, where the wave length of radiation is of the same order of the dimensions of the obstructions and the objects surrounding the transmitter and the receiver, the effects of diffraction and interference cannot be ignored, because they become so efficient. In this case, the projections and images of any objects become blurred. In the general case, the sounding of the object is multi-specular; that is, the object under investigation is seen from different positions within the space. Therefore, the problem of sounding becomes very complicated, both for the *direct problem* and for the *inverse problem*, mentioned in Chapter 1.

The correct solutions of the direct problem are fully guaranteed by the uniqueness of Maxwell's equations. Moreover, the natural environment itself always solves the problem: what is measured experimentally is the exact solution of the direct problem. With the inverse problem, the same situation does not exist. The solution of such a problem depends on human effort and knowledge. In this case, to obtain the solution of the inverse problems, the so-called *approximate methods and assumptions* are usually used. Correction of any assumption or approximation is made to coincide with the corresponding experimental data obtained during the tests. Moreover, the strictness of the inverse problem depends on that of the direct problem—if a more exact solution of the direct problem is found, accounting for the dominant physical processes, a much stricter solution of the inverse problem will be obtained in response to the former.

Let us explain the problem and its initial conditions. If $j(\mathbf{r}_1)$ is the spatial distribution of the outer sources (namely, currents) of the wave located inside the volume V_1, the field created by these currents in the point $\mathbf{r}$ inside this volume, consisting of homogeneous media, can be defined as [13, 14]:

$$E(\mathbf{r}) = \iiint_{V_1} j(\mathbf{r}_1) G_0(\mathbf{r}_1 - \mathbf{r}) d\mathbf{r}_1 \qquad (2.7)$$

Here, $G_0(\mathbf{r}) = \exp(ik|\mathbf{r}|)/(4\pi|\mathbf{r}|)$ is the Green function in free space, that is, the solution of the Helmholtz equation for free space [15, 16]:

$$\Delta G_0(\mathbf{r}) + k^2 G_0(\mathbf{r}) = -\delta(\mathbf{r}) \qquad (2.8)$$

where:

$k = 2\pi\sqrt{\varepsilon}/\lambda = 2\pi f \sqrt{\varepsilon}/c$ is the wave number in any environment under consideration (for free space $k \equiv k_0 = 2\pi/\lambda = 2\pi f/c$)

$\varepsilon = \varepsilon_0(\varepsilon'_r + \varepsilon''_r)$ is the complex permittivity of any environment under consideration

$\varepsilon_0 = (1/36\pi) \cdot 10^{-9}$ F/m is the dielectric constant of free space

Here, we consider the non-magnetic media, when the magnetic permeability $\mu = \mu_0 \mu_r \equiv 4\pi \cdot 10^{-7}$ H/m is simply the magnetic constant of free space.

The main goal of the simple direct problem is to find the wave field $E(\mathbf{r})$. The aim of the inverse problem is to find a solution of the integral equation (Equation 2.7) and then to reproduce the distribution $j(\mathbf{r}_1)$ according to measured values of $E(\mathbf{r})$. If this wave field can be measured at all points of the 3-D space, the solution will be presented by the operation inverse to convolution in Equation 2.7 (this procedure is called *deconvolution*). Usually, the field $E(\mathbf{r})$ is measured at the selected surface S surrounding the volume V_1; that is, the 3-D problem is converted to the 2-D problem. The distribution $j(\mathbf{r}_1)$ can be found by a special procedure called the *inverse focusing of radiation* (*IFR*). According to this procedure, the recorded radiation is focused (collected) at any point of the space $\mathbf{r}_F$ as a result of the following integration [13, 14]:

$$U(\mathbf{r}_F) = \iint_S W(\mathbf{r}_F, \mathbf{r}_1) E(\mathbf{r}_1) dS = \iiint_{V_1} j(\mathbf{r}_1) Q(\mathbf{r}_F, \mathbf{r}_1) d^3 \mathbf{r}_1 \qquad (2.9)$$

where:
$W(\mathbf{r}_F, \mathbf{r}_1)$ is the arbitrary focusing function
$Q(\mathbf{r}_F, \mathbf{r}_1)$ is the kernel of a new integral equation for $j(\mathbf{r}_1)$

$$Q(\mathbf{r}_F, \mathbf{r}_1) = \iint_S W(\mathbf{r}_F, \mathbf{r}_1) G_0(\mathbf{r}_s - \mathbf{r}_1) dS \qquad (2.10)$$

Function $Q(\mathbf{r}_F, \mathbf{r}_1)$ plays the role of the system function.

Let us now obtain the system function based on Equation 2.8 for the Green function. As is known [13–17], this equation has two solutions:

- For the outgoing wave (propagating to infinity): $G_0(\mathbf{r}) = e^{ik|\mathbf{r}|} / (4\pi |\mathbf{r}|)$
- For the incoming wave (propagating from infinity): $G_0^*(\mathbf{r}) = e^{-ik|\mathbf{r}|} / (4\pi |\mathbf{r}|)$

Next, we assume that the sources for these two types of waves are different. Then, Equation 2.8 yields:

$$G_0(\mathbf{r} - \mathbf{r}_1) \nabla^2 G_0^*(\mathbf{r} - \mathbf{r}_2) - G_0^*(\mathbf{r} - \mathbf{r}_2) \nabla^2 G_0(\mathbf{r} - \mathbf{r}_1)$$
$$= G_0^*(\mathbf{r} - \mathbf{r}_2) \delta(\mathbf{r} - \mathbf{r}_1) - G_0(\mathbf{r} - \mathbf{r}_1) \delta(\mathbf{r} - \mathbf{r}_2)$$

We suppose that both sources are located inside the finite volume V and will integrate over this volume, accounting for Green's theorem [13–16]. Finally we obtain:

$$G_0(\mathbf{r}_1 - \mathbf{r}_2) - G_0(\mathbf{r}_2 - \mathbf{r}_1)$$
$$= \oiint_S \left\{ G_0(\mathbf{r}_s - \mathbf{r}_1) \frac{dG_0^*(\mathbf{r}_s - \mathbf{r}_2)}{d\mathbf{n}} - G_0^*(\mathbf{r}_s - \mathbf{r}_2) \frac{dG_0^*(\mathbf{r}_s - \mathbf{r}_1)}{d\mathbf{n}} \right\} dS \qquad (2.11)$$

Here, the integration is carried out over surface S covering volume V; **n** is the normal unit vector to the surface S and outside the volume V. The left-hand side of the integral equation (Equation 2.11) is a simple *sinc-function* (also called the *point spread function*). That is:

$$G_0(\mathbf{r}_1 - \mathbf{r}_2) - G_0(\mathbf{r}_2 - \mathbf{r}_1) = \frac{1}{2i}\delta_k(|\mathbf{r}_1 - \mathbf{r}_2|) = \frac{1}{2i}\frac{\sin(k|\mathbf{r}_1 - \mathbf{r}_2|)}{\pi|\mathbf{r}_1 - \mathbf{r}_2|}$$

For $k \to \infty$, this function limits to the regular Dirac δ-function, $\delta(|\mathbf{r}_1 - \mathbf{r}_2|)$, and the integrand in Equation 2.11 can be simplified:

$$\delta_k(|\mathbf{r}_1 - \mathbf{r}_2|) = \frac{4}{i}\oiint_S \left\{ G_0(\mathbf{r}_s - \mathbf{r}_1)\frac{dG_0^*(\mathbf{r}_s - \mathbf{r}_2)}{dn} \right\} dS \qquad (2.12)$$

This expression can be interpreted as an orthogonality of the outgoing and incoming waves. In such an interpretation, the function $G_0(\mathbf{r}_s - \mathbf{r}_1)$ can be considered as an outgoing wave that is observed at the surface S from the source located at the point $\mathbf{r}_1$. If so, the second product in the integrand of Equation 2.12 can be considered as a following focusing function:

$$W(\mathbf{r}_F, \mathbf{r}_s) = \frac{4}{i}\frac{dG_0^*(\mathbf{r}_s - \mathbf{r}_2)}{dn} = W(\mathbf{r}_F - \mathbf{r}_s) \qquad (2.13)$$

which, after timing on $G_0(\mathbf{r}_s - \mathbf{r}_1)$ and integration over the surface S, as a lens in optics, collects the field of the outgoing wave back to the point $\mathbf{r}_2$:

$$\oiint_S \left\{ G_0(\mathbf{r}_s - \mathbf{r}_1)W(\mathbf{r}_s - \mathbf{r}_2) \right\} dS = \delta_k(\mathbf{r}_2 - \mathbf{r}_1)$$

All operations stated by Equations 2.11 through 2.13 are illustrated geometrically in Figure 2.8a, and a form of the sinc-function is shown in Figure 2.8b.

If the surface S is not closed, as follows from the integrals in Equations 2.11 and 2.12 and shown in Figure 2.8, the localization principle will only be particular and the localization function will have an elongated form at the direction normal to the surface of the aperture (see Figure 2.9). In this case, the system function will be [13]:

$$Q(|\mathbf{r}_F - \mathbf{r}_1|) = \frac{4}{i}\iint_S \left\{ G_0(\mathbf{r}_s - \mathbf{r}_1)\frac{dG_0^*(\mathbf{r}_F - \mathbf{r}_s)}{dn} \right\} dS \qquad (2.14)$$

The selection of the focusing function $W(\mathbf{r}_F, \mathbf{r}_s)$ described was physically proved and corresponds to the inverse focusing procedure, when the radiation of the partial waves arrives from a given point in space $\mathbf{r}_F$ (called the *point of focusing*) and is summated in the same phase (i.e., *coherently*).

26 ■ Electromagnetic and Acoustic Wave Tomography

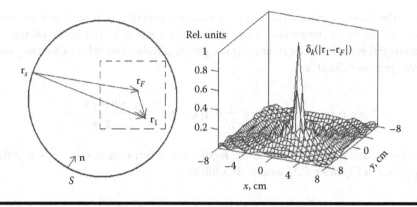

Figure 2.8 System function in the case of complete focusing by a closed aperture: point spread function.

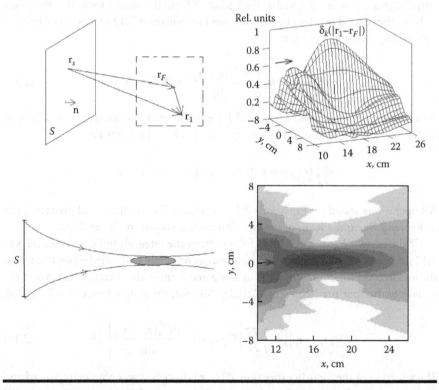

Figure 2.9 System function in the case of partial focusing by an aperture of finite size.

To equalize the partial waves, the converged (to infinity) Green function $G_0^*(\mathbf{r}_F - \mathbf{r}_s)$ was used. We note that the form of the system function is the result of interference recorded in proximity to the point of focusing $\mathbf{r}_F$. Using monochromatic radiation for sounding of objects under test and accounting for the limitations in dimension of the receiving antenna's aperture, we find that the system function has an elongated form (see Figure 2.9). This means that better resolution can be achieved in the cross-direction to the surface of the aperture (with the scale $l_\perp \sim \lambda = c/f$), but its pure resolution is in the longitudinal direction to the aperture. In the general case, the form of the system function from Equation 2.14 depends strongly on the radiation frequency, which determines the oscillating character of this function and its spatial deviations.

By use of pulse (or wideband signal [16]) radiation, the localization of the system function, as well as its spatial resolution, can be increased, mostly along the wave path, using a principle of reciprocity. It may be noticed that the above-mentioned focusing procedure can be useful not only for the radiated aperture, but also for the received aperture. At the same time, the obtained crossing focusing procedure (both at the transmitter and at the receiver) gives good localization of the source of radiation and its interaction with the test object or materials, as shown in Figure 2.10.

Use of the crossing focusing procedure essentially decreases the effect of interaction of radiation on the environment or the material of the object under test. Using the double focusing procedure, the resulting signal, after arriving at the receiver aperture, will be proportional to the intensity of the obstruction (inhomogeneity) located in the illuminated region of overlapping of the two focusing functions, both for the radiated field and for the received field. The total region of localization obtained in such a manner can be seen as a result of the product of the corresponding focusing functions discussed. It should be pointed out that realization of the described double focusing procedure is possible both at the hardware level by usage of lenses, mirrors,

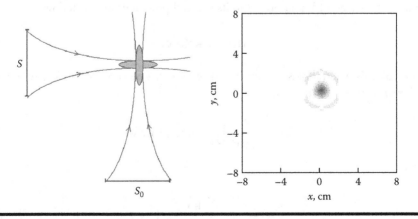

Figure 2.10 **Cross-focusing of radiation.**

or prisms, and at the computer level in conditions of measurement of the wave field accounting for its phase structure. In other words, for the realization of the focusing effect, it is important, first of all, to have a knowledge of the wave phase. During interference, partial waves are summarized in-phase (i.e., coherently), which increases the intensity of the total wave by many times. Out of focus, the partial waves are summarized incoherently and amplification of the wave intensity does not occur or occurs partially. Such an explanation is in good agreement with the concept of the *Fresnel zone*, which analyzes special zones significant for electromagnetic wave propagation.

If the transmitting and the receiving apertures of the sounding or located system are arranged together (the monostatic case of location: see reference [15]), multimode interactions of the partial waves become weaker only in the cross-directions with respect to the direction of radiation, as shown in Figure 2.9. Longitudinal multimodal (e.g., multiray) interactions can be attenuated either by increase of dimensions of the antenna aperture (as is done with synthesized aperture radar [SAR]) or by use of the impulse radar system.

It should be mentioned that direct realization of the method of double focusing, as described, is a complicated operation requiring significant time either for mechanical scanning or for the numerous multidimensional summations during the computing procedure. Moreover, for a given focusing function $W(\mathbf{r}_F, \mathbf{r}_s)$, it is important to correctly define the phase, which could support a coherency in partial wave summation from the focusing point. As for the amplitude, it does not play such a significant role in the achievement of the focusing effect.

2.4 Radio Wave Tomography Synthesis: Stolt's Method

The method of radio wave tomographic synthesis (RWTS) is the generalization of the FS in the case of existence of wave fields. This means that the proposed method of transformation of Equation 2.9 to its spectral presentation is as follows:

$$\hat{U}(\mathbf{\kappa}) = \hat{j}(\mathbf{\kappa})\hat{Q}(\mathbf{\kappa}) \qquad (2.15)$$

Based on this spectral presentation and using the inverse convolution (called *deconvolution*) with regularization,

$$j(\mathbf{r}) = \frac{1}{(2\pi)^3} \iiint \hat{j}(\mathbf{\kappa}) \exp(i\mathbf{\kappa}\mathbf{r}) d\mathbf{\kappa}$$

where

$$\hat{j}(\mathbf{\kappa}) = \frac{\hat{Q}^*(\mathbf{\kappa})}{\hat{Q}(\mathbf{\kappa})\hat{Q}^*(\mathbf{\kappa}) + \alpha} \hat{U}(\mathbf{\kappa}) \qquad (2.16)$$

we can present the spectral functions, as [13]:

$$\hat{j}(\kappa) = \iiint j(\mathbf{r})\exp(-i\kappa\mathbf{r})d\mathbf{r} \quad (2.17a)$$

$$\hat{Q}(\kappa) = \iiint Q(\mathbf{r})\exp(-i\kappa\mathbf{r})d\mathbf{r} \quad (2.17b)$$

$$\hat{U}(\kappa) = \iiint U(\mathbf{r})\exp(-i\kappa\mathbf{r})d\mathbf{r} \quad (2.17c)$$

Here, α is the parameter of regularization. The solution proposed here is very complicated, because it needs a lot of computational time and has a weakly localized focusing effect in the longitudinal direction to the object under investigation. To overcome this problem, a similar but simpler solution based on direct spectral analysis of the distribution of received wave radiation written in the form of Equations 2.17 can be used.

To solve this problem, we will consider a simple case of a homogeneous background medium, and will present the Green function in the homogeneous background medium in the form of a 2-D spectrum of plane waves according to Weyl's expansion of plane waves (called *Weyl's formula*, which was presented by Ya. Alpert from the former USSR in 1967: see reference [16]):

$$G_0(\mathbf{r}) = \frac{\exp(ik|\mathbf{r}|)}{4\pi|\mathbf{r}|} = \frac{1}{(2\pi)^2}\int_{-\infty}^{\infty}\int_{-\infty}^{\infty}\frac{\exp\{i(\kappa_\perp \mathbf{r}_\perp + \kappa_z z)\}}{2\kappa_z}d\kappa_\perp \quad (2.18)$$

where:

$|\kappa_\perp| = \sqrt{k^2 - k_z^2} \equiv k_x^2 + k_y^2$ is the transverse projection of the wave vector
k_z is its longitudinal projection
$|\mathbf{r}_\perp| = \sqrt{r^2 - z^2} \equiv \sqrt{x^2 + y^2}$ is the transverse projection of the vector **r** in the Cartesian coordinate system
z is its longitudinal projection

If this is true, according to Weyl's formula, a wave field observed at the plane (*xoy*) for $z=0$ (as shown in Figure 2.11) and registered at the receiver can be presented by the following formula:

$$E(\mathbf{r}) = \frac{1}{(2\pi)^2}\int_{-\infty}^{\infty}\int_{-\infty}^{\infty}\frac{\exp\{i(\kappa_\perp \mathbf{r}_0 + \kappa_z z)\}}{2\kappa_z}\hat{j}(\kappa)d\kappa_\perp$$

30 ■ Electromagnetic and Acoustic Wave Tomography

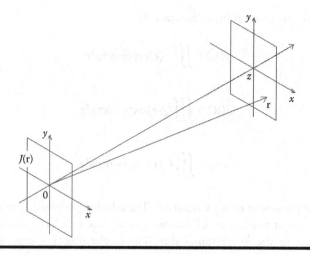

Figure 2.11 Planes of radiation sources and receivers.

Finally, the 2-D Fourier transform of this field simply gives the spectrum of the wave field distribution over spatial frequencies (or wave lengths):

$$E(\kappa_\perp, z) = \frac{1}{(2\pi)^2} \int_{-\infty}^{\infty} \int_{-\infty}^{\infty} \exp\{-i\kappa_\perp r_\perp\} E(r_\perp, z) dr_\perp = \frac{i}{2\kappa_z} \exp\{i\kappa_z z\} \hat{j}(\kappa) \quad (2.19)$$

where $\hat{j}(\kappa)$ is defined by Equation 2.17a. Equation 2.19 allows us to obtain the 3-D spatial spectrum of source (current) distribution versus the 2-D spectrum of the observed field of radiation, that is,

$$\hat{j}(\kappa) = \iiint_{V_1} j(r_1) \exp(-i\kappa r_1) dr_1 = \frac{2\kappa_z}{i} \exp\{-i\kappa_z z\} E(\kappa_\perp, z) \quad (2.20)$$

Relations between components of the vector κ of spatial frequencies (wave lengths) were presented above, and the vector κ lies at the sphere of radius $k = 2\pi f/c$ of wave numbers, which should be selected to fully fill a spectrum of spatial frequencies. According to the proposed procedure, which gives the solution of 3-D tomography, it is enough to take the 3-D IFT and, according to a simple extrapolation from a coordinate system (κ_x, κ_y, k) to a coordinate system (κ_x, κ_y, κ_z), transfer a spectrum of spatial frequencies of sources $\hat{j}(\kappa)$, according to Equation 2.20, to reproduce a spatial distribution of the real sources (currents):

$$j(r) = \iiint_{V_1} \hat{j}(\kappa) \exp(i\kappa r) d\kappa \quad (2.21)$$

For this purpose, the fast algorithms can be used.

Such an approach generalizes and unites the approaches considered in Sections 2.1 and 2.2, and its solution lies in the basis of the Stolt migration [13, 14], the idea of which was to directly relate the spatial spectrum of the received signal (or wave) recorded by the arbitrary aperture to the spatial spectrum of sources of radiation generated by the corresponding signal (or wave). This can be understood by introducing briefly the *Huygens–Fresnel principle*.

Let us suppose that at the arbitrary plane $z = 0$ is a given distribution of the wave field $E_0(\mathbf{r})$ (see Figure 2.12). According to the Huygens–Fresnel principle, the field at the plane $z = 0$ can be obtained according to the following formula [13–17]:

$$E(r) = 2 \iint_S E_0(\mathbf{r}_s) \frac{dG_0(\mathbf{r} - \mathbf{r}_s)}{d\mathbf{n}} ds \tag{2.22}$$

where:
- $ds = dxdy$
- $\mathbf{r}_s = \{x, y, z = 0\}$
- $\mathbf{r} = \{x, y, z\}$

According to Weyl's formula (Equation 2.18), we obtain, according to Equation 2.19, a 2-D spectrum of the wave field for the selected coordinate system:

$$E(\boldsymbol{\kappa}_\perp, z) = \int_{-\infty}^{\infty} \int_{-\infty}^{\infty} \exp\{-i\boldsymbol{\kappa}_\perp \mathbf{r}_\perp\} E(\mathbf{r}_\perp, z) d\mathbf{r}_\perp = \exp\{i\kappa_z z\} \hat{E}_0(\boldsymbol{\kappa}_\perp, z = 0) \tag{2.23}$$

In this expression, the important product is the transfer function $H(\boldsymbol{\kappa})$ in free space (i.e., in background media), $H(\boldsymbol{\kappa}) = \exp\{i\kappa_z z\}$. Therefore, the reproduction of the spatial distribution of the spectra of sources leads to elimination of this product $H(\boldsymbol{\kappa})$. Physically, this procedure means an IFR, as discussed in Section 2.2. Timing Equation 2.23 with the phase product $H^*(\boldsymbol{\kappa}) = \exp\{-i\kappa_z z\}$ means the rejection of spatial spreading of the field and corresponds to the IFR arriving at the plane $z = 0$, discussed in Section 2.2 regarding the inverse focusing procedure. This operation is achieved by the convolution of the field with the focusing function:

$$W(\mathbf{r}_F, \mathbf{r}_s) = \frac{4}{i} \frac{dG_0^*(\mathbf{r}_F - \mathbf{r}_s)}{d\mathbf{n}} \tag{2.24}$$

This is the main task of an approach proposed by Stolt [13, 14]. In Figure 2.13, the result of imitation modeling of the main procedure for recovering of sources by use of focusing suddenly over all depths is shown.

In the top panel, the result of numerical modeling of distribution of the real part of the field from a point source is shown. This direct problem was computed in the following coordinate system: $x = 20$ cm, $y = 0$, $z = 150$ cm. The field distribution was computed along the x-axis in limits of ± 50 cm in the plane $y = 0$ at the frequency

32 ■ *Electromagnetic and Acoustic Wave Tomography*

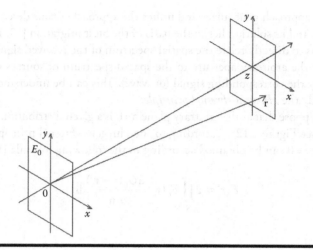

Figure 2.12 Field transfer from one plane to another.

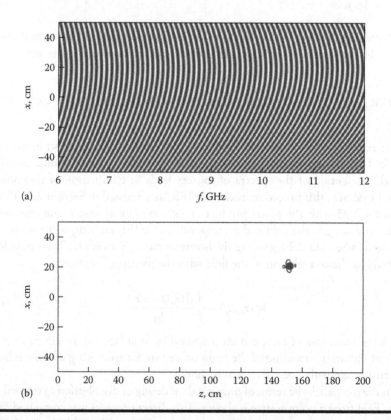

Figure 2.13 Location sounding problem: (a) simulation modeling; (b) result of reconstruction of point-like inhomogeneity using group focusing method.

range from 6 to 12 GHz. The bottom panel shows the results of the solution of the inverse problem, that is, recovery of the tomogram in the plane $y = 0$. It is seen that the reconstruction of the desired test object was carried out clearly enough.

2.5 Transmission Tomography of the Semitransparent Media Based on the Kirchhoff Approximation

Now, we consider the following situation, where the method of shadowing projections discussed in Sections 2.1 and 2.2 (see also several examples discussed in references [18–23]) can be used. This scenario is called *transmission tomography*, when the semitransparent media are illuminated inside it, as seen in Figure 2.9. In other words, transmission radio tomography is a tomography related to cross-media illumination by the probing electromagnetic waves (see Figure 2.14).

The source of radiation is placed at point 1; the receiver is located at point 2. Using radio lenses S_1 and S_2, the radiation wave is focused at point r_F. As a result, in proximity to the focusing point r_F located at the plane S_0, a *working zone* is formatted (denoted in Figure 2.14 by the dark area bounded by curved lines), inside which the wave radiation is concentrated. This zone is more sensitive to interaction with inhomogeneities of the medium under sounding. In Figure 2.15, distributions of the wave field amplitude (in the form of contour lines) and phase (gradations of gray color) are presented around the working zone. The contour lines of wave amplitude distribution are in the form of ellipses. For example, at an amplitude level of 0.5, the ellipse has a length of 30 cm and diameter of 3 cm. Computations were carried out for the operational frequency of 10 GHz. It is seen from Figure 2.15 that the phase front in proximity to the working zone at each cross-section area is quasi-plane.

If the test objects, consisting of the circular and rectangular cylinders, as shown in Figure 2.16a, are now placed inside the working zone, they are illuminated by radiation under multiple angles of view, called the *full-angle image*; the objects are

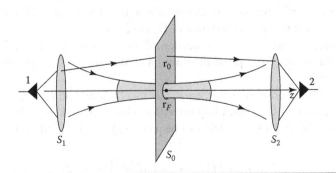

Figure 2.14 Scheme of transmission tomography.

34 ■ *Electromagnetic and Acoustic Wave Tomography*

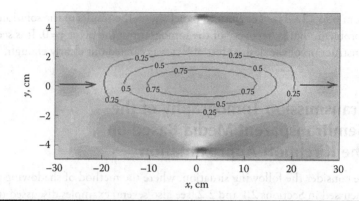

Figure 2.15 Distribution of field amplitude and phase in region of focusing.

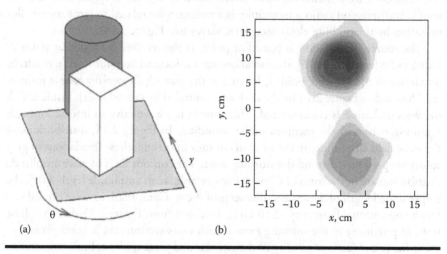

Figure 2.16 (a) Shape of test scene objects and (b) reconstructed tomogram.

simultaneously moved through the working zone and rotated. The corresponding observed picture of wave projections has an image, as presented in Figure 2.17, and the reconstructed tomogram is shown in Figure 2.16b.

In Figure 2.17, there are two projections shown: one for the amplitude distribution, and the second for a phase distribution of the radiation passing through the object under test.

Let us consider briefly the mathematical model of formation of the wave projections. If the source of the sounding wave is a point source, then at the plane S_2 of observation (Figure 2.14), the field can be presented as [13–15, 18–23]:

$$E(\mathbf{r}_2, \mathbf{r}_1) = 2 \iint_{S_0} G_0(\mathbf{r}_{s0} - \mathbf{r}_1) \frac{dG_0(\mathbf{r}_2 - \mathbf{r}_{s0})}{d\mathbf{n}} ds_0 \qquad (2.25)$$

Theoretical Overview of Wave Tomography ■ 35

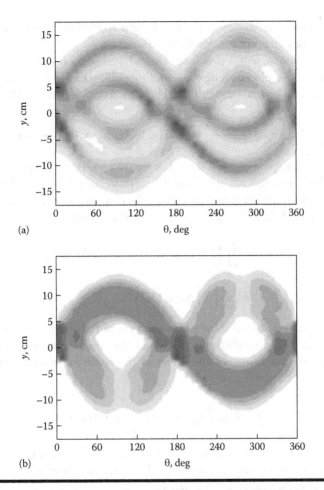

Figure 2.17 Full-angle images of (a) distribution of amplitude attenuation and (b) phase disturbance of wave field when test scene is scanned.

This integral equation can be analyzed by use of the Kirchhoff phase approximation, which states that the main impact in perturbation of the radiation passing the test object introduces the phase changes of the wave in the media. If so, in the frame of geometrical optics, we can present the wave projection of inhomogeneities located in the test region in the following form [13–15, 18–23]:

$$E(\mathbf{r}_2, \mathbf{r}_1) = 2 \iint_{S_0} G_0(\mathbf{r}_{s0} - \mathbf{r}_1) \frac{dG_0(\mathbf{r}_2 - \mathbf{r}_{s0})}{d\mathbf{n}} \exp\left\{ ik \int \Delta n(x') dx' \right\} ds_0 \quad (2.26)$$

Here, in the exponential product, the integration is carried out along the straight line $\mathbf{r}_2 - \mathbf{r}_{s0}$, and Δn is the perturbation of the refractive index, comparable to that for the background homogeneous medium. This approximation is very useful and mostly gives the correct results for short radio waves.

The use of double lenses that focus the radiated wave at the transmitting and receiving terminals, placed along the sounding radio trace, leads it to pass through two convolutional procedures with the following focusing functions [14, 20]:

$$W_1(\mathbf{r}_F, \mathbf{r}_1) = \frac{4}{i}\frac{dG^*(\mathbf{r}_F - \mathbf{r}_1)}{d\mathbf{n}} \text{ and } W_2(\mathbf{r}_F, \mathbf{r}_2) = \frac{2}{i}G^*(\mathbf{r}_F - \mathbf{r}_s)$$

where the focusing point $\mathbf{r}_F$, as is shown in Figure 2.14, lies at the plane S_0. During this procedure, arising from spreading δ-functions (*sinc-functions*), the restriction of the limits of integral in Equation 2.26 in the direction across wave radiation (**x**) can be maintained. In such a procedure, we finally obtain:

$$U(y, \theta) = \iint_{S_1}\iint_{S_2} E(\mathbf{r}_2, \mathbf{r}_1) W_1(\mathbf{r}_F, \mathbf{r}_1) W_2(\mathbf{r}_F, \mathbf{r}_2) ds_1 ds_2$$
$$\approx \exp\left\{ik\int \Delta n(x')dx'\right\} \quad (2.27)$$

This result describes the meaning of the solution of transmission tomography. As mentioned, the *phase distribution* is much more informative for the understanding of the proposed procedure. In the frame of the Kirchhoff phase approximation described, it can be written as:

$$f(y, \theta) = \int \Delta n(x')dx' = \frac{1}{k}\arg\{U(y, \theta)\} \quad (2.28)$$

This expression is the well-known equation of the channel in the frame of the geometrical optics approximation, which, in the Cartesian coordinate system where the test object is located, can be written as:

$$f(y, \theta) = \varphi(y, \theta)/k = \int \Delta n(x\cos\theta - y\sin\theta, x\sin\theta + y\cos\theta)dx \quad (2.29)$$

The form of Equation 2.29 is similar to Equation 2.3 describing the shadowing projections. The non-known function $\Delta n(x, y)$ then can be reconstructed by the method of FS discussed in Section 2.3. We will repeat that the result of reconstruction of two objects is shown in Figure 2.16b. It is seen that the form of the cross-section area of the test objects, consisting of the dielectric cylinder and dielectric block, was reconstructed adequately enough.

2.6 Transmission Tomography of Opaque Objects

When radio waves illuminate the opaque objects, this process is described approximately by elements of the diffraction theory (see also references [13–16]), according to which the diffractive field can be found as (see the similar formula in Equation 2.3b):

$$E(\mathbf{r}) = 2 \iint_{S_{\text{out}}} E_0(\mathbf{r}_s) \frac{dG_0(\mathbf{r}-\mathbf{r}_s)}{d\mathbf{n}} ds \qquad (2.30)$$

Here, the integral is taken along the illuminated part of the surface S_{out}, which is an outer surface surrounding the test object, perpendicular to the direction of the incident wave $E_0(\mathbf{r}_s)$ (see Figure 2.18); $\mathbf{n}$ is the normal unit vector to the surface S_{out}.

The diffracted field defined by Equation 2.30 can also be presented, according to Babinet's principle, as $E(\mathbf{r}) = E_0(\mathbf{r}) + E_1(\mathbf{r})$, where the integral

$$\begin{aligned} E_1(\mathbf{r}) &= 2 \iint_{S_{\text{out}}} E_0(\mathbf{r}_s) \frac{dG_0(\mathbf{r}-\mathbf{r}_s)}{d|\mathbf{n}|} ds \\ &= -2 \int_{-\infty}^{\infty} \int_{-\infty}^{\infty} F(\mathbf{r}_s) E_0(\mathbf{r}_s) \frac{dG_0(\mathbf{r}-\mathbf{r}_s)}{d|\mathbf{n}|} ds \end{aligned} \qquad (2.31)$$

is the secondary field created by the opaque object. The CF $F_0(\mathbf{r}_s)$ limits the area of integration that lies inside the area of the terminator, created by the incident wave $E_0(\mathbf{r}_s)$. In other words: $F(\mathbf{r}_s) = 1$ inside the terminator, and $F(\mathbf{r}_s) = 0$ outside it. An array of such terminators creates a set of shadowing projections, which can be inverted by the use of the FS described in Section 2.3. Recovery of the CF can also be carried out by the use of the method of wave tomographic synthesis, described in Section 2.4. The equivalent method that can be used for reconstruction of the object under test is the method of inverse focusing (see Sections 2.3 and 2.4).

At the next step of transmission tomography is the solution of the inverse problem, that is, the tomography problem, obtained by use of the method of inverse focusing, described in Sections 2.3 and 2.4. Let us consider a complex amplitude

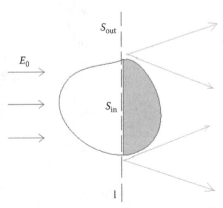

Figure 2.18 Wave scattering on opaque object. *1*: Terminator plane or twilight zone.

$E(x, \varphi)$ of the diffracted wave field that was registered behind the object at the different points of measurements (x) located at the segment of measurements L under different angles of object rotation (φ) with respect to the measured system. Then, the inverse focusing, as was shown in Sections 2.2 through 2.4, leads to computation of the integral according to [13]:

$$U(x_F, y_F) = \int_{-\pi}^{\pi} d\varphi \int_L dx\, M(x_F, y_F; x, \varphi) E(x, \varphi) \qquad (2.32)$$

Here, the focusing function can be defined as:

$$\begin{aligned} M(x_F, y_F; x, \varphi) \\ = \exp\left\{-ik\left(\sqrt{(x_F - X)^2 + (y_F - Y)^2} + \sqrt{(x_F - x_0)^2 + (y_F - y_0)^2}\right)\right\} \end{aligned} \qquad (2.33)$$

where:
$X = R\sin\varphi + x\cos\varphi$
$Y = -R\cos\varphi + x\sin\varphi$
$x_0 = -R_0\sin\varphi$
$y_0 = -R_0\cos\varphi$
R and R_0 are the distances from the line of scanning L and from the source of radiation to the center of rotation
(x_0, y_0) are the coordinates of the source of radiation

The coordinates of focusing (x_F, y_F) were taken relative to the stationary object under investigation (see Figure 2.19a). Imaging of the object computed by the use of Equations 2.32 and 2.33 is shown by Figure 2.19b.

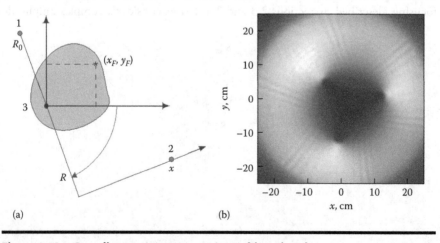

Figure 2.19 Coordinate system (a); computed imaging (b).

The results of numerical simulation of the test object in the form of a triangular cylinder are shown in Figure 2.20a, and its calculated diffractive field is presented in Figure 2.20b.

It is seen that the angles of the object are recovered precisely enough, but the edges of the object are slightly blurred. The latter is related to the fact that the edges are located in shadow zones or half-shadow zones for most of the transmission projections. Moreover, the proposed method does not allow for visualization of objects that possess geometry, where all projections are *a priori* located in shadow zones.

2.7 Linear-Frequency-Modulated Radiation Tomography Technique

Linear-frequency-modulated (LFM) radio waves operating in the millimeter wave length band have recently been used for better reconstruction and recovery of foreign objects hidden in clutter environments or embedded in non-transparent media. All well-known worldwide radiometers work in this wave length band, because by use of such a small wave length range from several to tens of millimeters, it is possible to identify small objects or details of foreign objects of a few millimeters in dimension embedded in clutter conditions. Compared with other methods of wave tomography described in previous sections of Chapter 2, tomography with LFM signals has some peculiarities that will be discussed next.

For construction of the data processing algorithm using SAR technology, let us consider a model of wave interaction with an inhomogeneous medium in an approximation of single-scattering (the Born approximation) [13–15], which we present in another form:

$$E(\mathbf{r}_{\perp 0}, f) = k^2 \iiint_V \Delta\varepsilon_r(\mathbf{r}_{\perp 1}, z_1) G_0^2(R, z_1) d\mathbf{r}_{\perp 1} dz_1 \qquad (2.34)$$

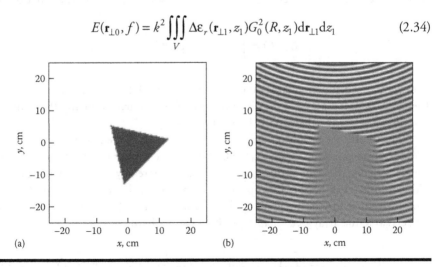

Figure 2.20 (a) Cross-section of opaque triangular cylinder and (b) its (calculated) diffraction field.

where, again, $G_0(R, z_1) = \exp\{ikR\}/(4\pi R)$ is a Green function in free space for radiation on frequency f and of wave number $k = 2\pi f/c$, $R = \sqrt{(\mathbf{r}_{\perp 1} - \mathbf{r}_{\perp 0})^2 + z_1^2}$; $\Delta \varepsilon_r$ characterizes the distribution of the sounded inhomogeneities, for example, the normalized permittivity perturbations in the Born approximation.

In this case, we use the following form of narrowband LFM signal [13, 14]:

$$s_0(t) = A(t) \exp\{2\pi i (f_0 + \alpha t/2)t\} \quad (2.35)$$

The signal received at the mixer output of the receiver for each point of location $\mathbf{r}_{\perp 0}$ can be presented as [13, 14]:

$$s(\mathbf{r}_{\perp 0}, t) \approx \left[k_0 A(t)\right]^2 \iiint_V \Delta\varepsilon_r(\mathbf{r}_{\perp 1}, z_1) G_0^2(R, f_0 + \alpha t) d\mathbf{r}_{\perp 1} dz_1 \quad (2.36)$$

where $k_0 = 2\pi f_0/c$.

Now, using the Fourier transform, one can transfer the integral in Equation 2.36 into the spectral region (e.g., into the joint space–frequency domain), which gives:

$$s(\mathbf{r}_{\perp 0}, f) \equiv \int_{-\infty}^{\infty} s(\mathbf{r}_{\perp 0}, t) \exp\{i2\pi ft\} dt$$
$$\approx k_0^2 \iiint_V \Delta\varepsilon_r(\mathbf{r}_{\perp 1}, z_1) G_0^2(R, f_0) \delta_A(\nu - 2\alpha R/c) d\mathbf{r}_{\perp 1} dz_1 \quad (2.37)$$

Here, in the case of a large enough baseline of the sounding signal, the function

$$\delta_A(f) = \int_{-\infty}^{\infty} A^2(t) \exp\{i2\pi ft\} dt$$

is the *spreading* δ-function (also called the *sinc-function* [16]) over frequency band ν. Therefore, in Equation 2.37, the distance between the source point and the point of sounding is related to the specific frequency ν.

Now, we use the operation of focusing the image over the crossing coordinates of observation points $\mathbf{r}_{\perp 0} = (x_0, y_0)$, which is simply a summation of the received signals with equalization of the phase for waves, scattered by a given focusing point with coordinates $(\mathbf{r}_{\perp F}, z_F)$. This operation is carried out by use of the double convolutional procedure in the plane of the antenna aperture (namely, in the plane of a special antenna array: see Chapter 6). Finally, we obtain:

$$s(\mathbf{r}_{\perp F}, z_F, f) \equiv \iint_{S_0} s(\mathbf{r}_{\perp 0}, f) \exp\left\{-i2\pi k_0 2\sqrt{(\mathbf{r}_{\perp 0} - \mathbf{r}_{\perp F})^2 + z_F^2}\right\} d\mathbf{r}_{\perp 0} \quad (2.38)$$

After changing the integration procedure, the integral in Equation 2.37 can be rewritten as:

$$s(\mathbf{r}_{\perp F}, z_F, f) \equiv \iiint_V \Delta\varepsilon_r(\mathbf{r}_{\perp 1}, z_1) Q(\mathbf{r}_{\perp 1}, \mathbf{r}_{\perp F}, z_1, z_F, f) d\mathbf{r}_{\perp 1} dz_1 \qquad (2.39)$$

where

$$Q(\mathbf{r}_{\perp 1}, \mathbf{r}_{\perp F}, z_1, z_F, f)$$
$$= k_0^2 \iint_{S_0} G^2(R, f) \delta_A(f - 2\alpha R/c) \exp\left\{-i2\pi k_0 2\sqrt{(\mathbf{r}_{\perp 0} - \mathbf{r}_{\perp F})^2 + z_F^2}\right\} d\mathbf{r}_{\perp 0} \qquad (2.40)$$

Numerical computations show that for a large enough size of the plane of observation and for the given frequency $f = f_F \equiv 2\alpha z_F/c$, Equation 2.40 becomes homogeneous versus its arguments, and $Q(\mathbf{r}_{\perp 1}, \mathbf{r}_{\perp F}, z_1, z_F, f) \approx Q_\perp(\mathbf{r}_{\perp 1} - \mathbf{r}_{\perp F}, z_F) \delta_A(z_1 - z_F)$, where function $Q_\perp$, by introducing a new distance to the focusing point, $R_F = \sqrt{(\mathbf{r}_{\perp 0} - \mathbf{r}_{\perp 1})^2 + z_F^2}$, can be written as:

$$Q_\perp(\mathbf{r}_{\perp 1} - \mathbf{r}_{\perp F}, z_F) = k_0^2 \iint_{S_0} G^2(R_F, f_0) \exp\left\{-i2\pi k_0 2\sqrt{(\mathbf{r}_{\perp 0} - \mathbf{r}_{\perp F})^2 + z_F^2}\right\} d\mathbf{r}_{\perp 0} \qquad (2.41)$$

Finally, taking account of Equation 2.41, it can be written that the signal at the focus plane will equal

$$s(\mathbf{r}_{\perp F}, z_F, f_F) \approx \iint \Delta\varepsilon_r(\mathbf{r}_{\perp 1}, z_F) Q_\perp(\mathbf{r}_{\perp 1} - \mathbf{r}_{\perp F}, z_F) d\mathbf{r}_{\perp 1} \qquad (2.42)$$

The kernel of the integral transform (Equation 2.42) is localized enough to present perturbations of the permittivity of inhomogeneities as:

$$\Delta\varepsilon_r(\mathbf{r}_{\perp 1}, z_1) = \frac{s(\mathbf{r}_{\perp F}, z_F, f_F)}{\iint Q_\perp(\mathbf{r}_{\perp 1}, z_F) d\mathbf{r}_{\perp 1}} \qquad (2.43)$$

Equation 2.43 gives the solution to the inverse problem and fully corresponds to the usage of SAR technology.

For a demonstration of the usefulness of this method, a numerical modeling of the scene consisting of three point scatterers was performed. In Figure 2.37a, their positions at the scene are shown for the case of $f_0 = 94$ GHz, $\alpha = 1.2 \cdot 10^{13} c^{-2}$ for the test time of ~5.1 ms. In Figure 2.21b, the result is shown of the numerical modeling of the real part of function $s(\mathbf{r}_0, t)$, that is, the raw radio imaging of these three scatterers.

Figure 2.22a presents the imaging of the real part (i.e., the *cosine quadrature*) of function $s(\mathbf{r}_0, t)$ obtained after the Fourier transform versus the time. The final

42 ■ *Electromagnetic and Acoustic Wave Tomography*

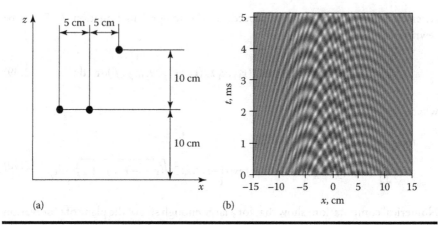

Figure 2.21 (a) Location of three point scatterers and (b) their raw radio image.

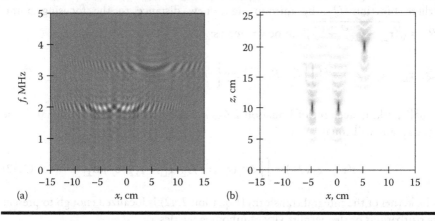

Figure 2.22 (a) Source image spectrum of test objects and (b) its radiotomogram.

radio tomogram is shown in Figure 2.22b. It is seen that the spatial resolution is weaker with respect to the resolution across the irradiated wave pattern. A similar tendency will be shown after data processing of real experiments with three metallic objects (see Chapter 7).

2.8 Incoherent Tomography for Reconstruction of Objects Hidden in Clutter

Incoherent tomography relates to the cases when the transmitter and receiver of radio waves are not synchronized. This fact does not allow us to measure the relative phase shift between the radiated and the detecting signals. For reconstruction or recovery of radio images based on incoherent radiation, there are several methods that have been investigated recently.

Such a situation occurs either when the wave radiation is incoherent due to its nature (solar or galaxy radiation, halogen lamp radiation, etc.), or when it is impossible to control an initial phase of radiation. Physically, the usage of incoherent radiation means that if the radiation can be replaced in the spectrum, for each of the frequencies in the frequency band the initial phase ψ can be considered as a random or a non-known value.

We will draw the reader's attention to the problem occurring during distance sounding of objects placed behind obstructions (walls, subsoil media, fences, etc.), when the radiation registered by the detector is not coherent. In Figure 2.23, a scheme of incoherent sounding of any object placed behind the wall is shown. Generally speaking, the main idea of the approach proposed here is to obtain, during experimental measurements, a relative phase for each spectral component of the radiation spectrum, reflected from the hidden (or embedded) object. The obstruction itself plays the role of the producer of the basic wave. The intensity of the observed interference picture gives a hologram of the reflected wave field.

Averaging of the interference field over all frequencies and subtraction from the general average intensity produces the recovery of the quadrature for each frequency of the spectrum. The IFT gives the reconstruction of the sounding objects' distribution according to the time delay of each incoming signal and, finally, gives the

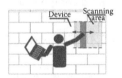

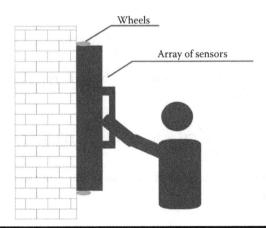

Figure 2.23 Scheme of experimental setup for reconstruction of any object located behind a wall.

reconstruction over the ranges between objects. Let us explain the above-mentioned technique using the corresponding mathematical apparatus.

The complex amplitude of spectral components can be written as [13, 14]:

$$u(k) = A\exp\{i(kz+\psi)\} \qquad (2.44)$$

where $k = 2\pi f/c$ is the corresponding spectral wave number. If Γ is the coefficient of reflection from the obstruction (wall), and $\tilde{\Gamma}$ is the coefficient of reflection from the test object, then the complex amplitude of the interference field beyond the obstruction can be written as:

$$U(k) = u(k)\left[\Gamma + \tilde{\Gamma}\exp(i2kZ)\right] \qquad (2.45)$$

Here, Z is the distance from the obstruction to the sounded object (see Figure 2.39). Supposing that $|\Gamma/\tilde{\Gamma}| \gg 1$, we can write for this field that:

$$W(k) = |U(k)|^2 = A^2\left[\Gamma^2 + 2\Gamma\cdot\tilde{\Gamma}\cos(2kZ)\right] \qquad (2.46)$$

Introducing the average field intensity $W_0 = \langle|U(k)|^2\rangle = A^2\Gamma^2$, and subtracting it from Equation 2.46, we finally obtain its varied component:

$$\delta W(k) = W(k) - W_0 \cong 2A^2\cdot\Gamma\cdot\tilde{\Gamma}\cos(2kZ) \qquad (2.47)$$

Equation 2.47 means that the varied component of wave intensity directly proportional to the quadrature component of the wave field reflected from the object is:

$$C(k) - \frac{\delta W(k)}{2\sqrt{W_0}} = A\cdot\Gamma\cos(2kZ) \qquad (2.48)$$

Now, using the IFT over the frequency, we obtain, in the region of positive time delays ($t > 0$), the following:

$$C(t) = \frac{1}{2\pi}\iint_k C(k=\omega/c)d\omega = \frac{1}{2}A\cdot\Gamma\delta(t-2Z/c) \qquad (2.49)$$

Now, if in Equation 2.49 we change the time delays t on the corresponding distances, $z = tc/2$, then

$$P(t) = \frac{4}{c}C(t) = A\cdot\Gamma\delta(z-Z) \qquad (2.50)$$

In the case where many objects are placed behind an obstructing wall, Equation 2.50 can be generalized as follows:

$$P(t) = A \sum_i \Gamma_i \delta(z - Z_i) \quad (2.51)$$

The corresponding experimental proofs will be presented in Chapter 7.

References

1. Yakubov, V. P. and M. L. Masharuev, Method of double focusing for microwave tomography, *Microwave Optical Technologies Letters*, vol. 13, no. 4, 1996, pp. 187–189.
2. Pimchuk, R., V. P. Yakubov, D. Ya. Sukhanov, et al., Spatial-temporal synthesis in microwave tomography, *Izv. Vuzov, Physics*, vol. 33, no. 9/2, 2010, pp. 108–109 (in Russian).
3. Sukhanov, D. Ya. and V. P. Yakubov, Usage of linear-frequency-modulated signals in the three-dimensional radio tomography, *Journal of Technical Physics (JTP)*, vol. 80. no. 4, 2010, pp. 115–119 (in Russian).
4. Sukhanov, D. Ya. and A. A. Murav'eva, Monochromatic transmission diffraction radiotomography, *Izv. Vuzov, Physics*, vol. 56, no. 8/2, 2013, pp. 193–195 (in Russian).
5. Yakubov, V. P., E. D. Tel'pukhovskii, and G. M. Tsepelev, Radio locational tomography, *J. Optics of Atmosphere and Ocean*, vol. 49, No. 12, 2006, pp. 1081–1086 (in Russian).
6. Yakubov, V. P., K. G. Skolyarchk, R. V. Pinchuk, et al., Radiowave tomography of hidden objects for security systems, *Izv. Vuzov, Physics*, vol. 51, no. 10, 2008, pp. 63–79 (in Russian).
7. Yakubova, O. V., E. D. Tel'pukhovskii, and V. P. Yakubov, Focusing of ultra-wideband radiation by lens from artificial dielectric, *Izv. Vuzov, Physics*, vol. 48, no. 6, 2005, pp. 153–154 (in Russian).
8. Yakubov, V. P., E. D. Tel'pukhovskii, and G. M. Tsepelev, Radio locational tomography of inhomogeneous media, *Izv. Vuzov, Physics*, vol. 49, no. 9, 2006, pp. 20–24.
9. Klokov, A., V. Yakubov, and M. Sato, Ultra-wideband radiolocation of a forest, *Proceedings of the 2010 IEICE General Conference*, March 16–19, 2010, Sendai, Japan, Institute of Electronics, Information and Communication Engineers, USA, pp. 298–299.
10. Cloude, S. R., P. D. Smith, A. Milne, D. Parkes, and K. Trafford, Analysis of time domain UWB radar signals, *SPIE, Pulse Engineering*, vol. 1631, 1992, pp. 111–122.
11. Colton, D. and R. Kress, *Inverse Acoustic and Electromagnetic Scattering Theory*, New York: Springer, 1998.
12. Hamran, S., D. T. Gjessing, J. Hjelmstad, and E. Aarholt, Ground penetrating synthetic pulse radar: dynamic range and modes of operation, *J. Appl. Geophys.*, vol. 33, 1995, pp. 7–14.
13. Yakubov, V. P., S. E. Shipilov, D. Ya. Sukhanov, and A. V. Klokov, *Radiowave Tomography: Achievements and Perspectives*, Tomsk, Russia: NTL, 2016 (in Russian).
14. Yakubov, V. P. and S. E. Shipilov, *Inverse Problems of Radiophysics*, Tomsk, Russia: NTL, 2016 (in Russian).

15. Blaunstein, N. Sh., M. B. Sergeev, and A. P. Shepeta, *Applied Aspects of Electrodynamics*, St. Petersburg: Agaf, 2016 (in Russian).
16. Blaunstein, N., Ch. Christodoulou, and M. Sergeev, *Introduction in Radio Engineering*, Boca Raton, FL: CRC Press, Taylor & Francis, 2017.
17. Shipilov, S. E., Method of aperture synthesis in 3D-radiotomography, *Izv. Vuzov, Physics*, vol. 57, no. 9, 2013, pp. 70–71 (in Russian).
18. Razinkevich, A. K., S. E. Shipilov, and V. P. Yakubov, Radiolocation tomography of farthest objects, *Izv. Vuzov, Physics*, vol. 55, no. 8/2, 2012, pp. 20–23 (in Russian).
19. Shipilov, S. E., A. A. Efremov, and V. P. Yakubov, Recovering of the form of the curved mirror combined antenna, *Izv. Vuzov, Physics*, vol. 51, no. 9/2, 2008, pp. 103–104 (in Russian).
20. Yakubov, V. P., S. E. Shipilov, and S. E. Sukhanov, Microwave tomography of radiopaque objects, *Russian Journal of Nondestructive Testing*, vol. 47, no. 11, 2011, pp. 765–770.
21. Shipilov, S. E., V. P. Yakubov, and S. V. Ponomarev, Radiowave mapping of deformation profile of the parabolic reflector, *Izv. Vuzov, Physics*, vol. 55, no. 9/2, 2012, pp. 274–275 (in Russian).
22. Koshelev, V. I., S. E. Shipilov, and V. P. Yakubov, Usage of the method of genetic functions for recovering of the form of objects in the small-angle ultra-wideband radiolocation, *Radiotekhnika i Radioelectronika*, vol. 45, no. 12, 2000, pp. 1470–1476 (in Russian).
23. Yakubov, V. P., S. E. Shipilov, S. E. Sukhanov, and A. K. Razinkevich, Ultra-broadband tomography of remote objects, *Russian Journal of Nondestructive Testing*, vol. 48, no. 3, 2012, pp. 191–196 (in Russian).

Chapter 3

Special Theoretical Approaches in Wave Tomography

Vladimir Yakubov, Sergey Shipilov, Dmitry Sukhanov, Andrey Klokov, and Nathan Blaunstein

Contents

3.1 Wave Location Tomography..48
    3.1.1 General Solution of Radio-Location Tomography 48
    3.1.2 Matched Filtering Method for Definition of Radiation Sources Distribution 50
    3.1.3 Method of Migration in Spatiotemporal Region by Use of Pulse Signals 51
    3.1.4 Method of Synthesized Focusing 54
    3.1.5 Single Focusing at the Boundary of Two Media 56
    3.1.6 Two-Step Focusing 60
    3.1.7 Group Focusing Technique...63
3.2 Method of Refocusing..65
3.3 Single-Side Monostatic Tomography of Non-Transparent Objects 69
3.4 Diffraction Tomography Techniques..70
    3.4.1 Diffraction Tomography Method Based on Rytov Approximation 72
    3.4.2 Method of Diffraction Tomography Based on Feynman's Path Integral 73
References 75

In this chapter, we discuss the special theoretical techniques performed recently to overcome and obey multiscattering and multidiffraction effects caused by the environment of a signal reflected or passing through the object under investigation buried in the sub-surface media or embedded and hidden in different material structures. The corresponding theoretical frameworks follow the results presented and discussed in references [1–29].

3.1 Wave Location Tomography

3.1.1 General Solution of Radio-Location Tomography

In the case of active sounding of desired environmental space (or material media), when the receiver works also as the transmitter, such a system is called the *transceiver* or *monostatic radar* [15]). In this case, a monostatic radio-location scheme of sounding is realized, where the sources of information radiation are the secondary currents induced in the media by the radiation wave field. The simplest model of such a scenario is the model of single scattering, namely, the Born approximation [15, 16], when a point source of radiation, located at the point r_0, generates secondary currents in the inhomogeneous environment of the form [13, 14–17]:

$$j(r_1) = k^2 \Delta\varepsilon_r(r_1) G_0(r_1 - r_0) \qquad (3.1)$$

Here, $\Delta\varepsilon_r(r_1)$ is the spatial distribution of deviations of the relative permittivity of the environment under investigation (see the definition in Section 2.2). In such definitions, we obtain

$$E(r) = k^2 \iiint_{V_1} \Delta\varepsilon_r(r_1) G_0(r_1 - r_0) G_0(r_1 - r) dr_1 \qquad (3.2a)$$

However, this formula is valid for bi-static locations, where the receiver and the transmitter are separated from each other. To convert this formula to monostatic locations, it should possess both the points of receiver and transmitter location, that is, $r_0 = r$. Then, the received radiation will be described by the following expression:

$$E(r) = k^2 \iiint_{V_1} \Delta\varepsilon_r(r_1) G_0^2(r_1 - r) dr_1 \qquad (3.2b)$$

Usage of the simple differentiation procedure, $E_1(r) = -2\pi i d[E(r)/k^2] dk$ yields:

Special Theoretical Approaches in Wave Tomography ■ 49

$$E_1(\mathbf{r}) = k^2 \iiint_{V_1} \Delta\varepsilon_r(\mathbf{r}_1) G_2(\mathbf{r}_1 - \mathbf{r}) d\mathbf{r}_1 \quad (3.3)$$

Here, $G_2(|\mathbf{r}|) = \exp(2ik|\mathbf{r}|)/(4\pi|\mathbf{r}|)$ is a Green function for free space, but taken with double frequency. For problems of locational tomography, it is interesting to find the spatial distribution of the relative permittivity values deviation $\Delta\varepsilon_r(\mathbf{r}_1)$.

For both bi-static and monostatic radar systems, the case of radiation of the pulse signals of the form of:

$$S_0(t) = \int \hat{S}_0(2\pi f) \exp(-i2\pi f t) df$$

gives the response of the medium to the radiation in the following form:

$$S(\mathbf{r}_0, \mathbf{r}, t) = \int E(\mathbf{r}) \hat{S}_0(2\pi f) \exp(-i2\pi f t) df$$

$$= k_0^2 \iiint_{V_1} \Delta\varepsilon(\mathbf{r}_1) \frac{S_0\left(t - \frac{|\mathbf{r}_1 - \mathbf{r}_0| + |\mathbf{r}_1 - \mathbf{r}|}{c}\right)}{(4\pi)^2 |\mathbf{r}_1 - \mathbf{r}_0| \cdot |\mathbf{r}_1 - \mathbf{r}|} \quad (3.4)$$

Here, $k_0 = 2\pi f/c$ is the wave number for the arbitrary average frequency in the signal spectrum. In the case of a point inhomogeneity under test,

$$\Delta\varepsilon_r(\mathbf{r}_1) = \Delta\varepsilon_{r0} \delta(\mathbf{r}_1 - \mathbf{r}_0) \quad (3.5)$$

we obtain the point reaction of the inhomogeneity under test as

$$S(\mathbf{r}, t) = k_0^2 \iiint_{V_1} \Delta\varepsilon_r(\mathbf{r}_1) \frac{S_0\left(t - \frac{2|\mathbf{r}_1 - \mathbf{r}|}{c}\right)}{(4\pi)^2 |\mathbf{r}_1 - \mathbf{r}|^2} d\mathbf{r}_1 \quad (3.6)$$

The characteristic picture of such a point reaction is shown in Figure 3.1.

Along the vertical axis, the time of the numerical process is denoted, and along the horizontal axis, the coordinate of the observation point $\mathbf{r}$. By gradations of gray color, the form of the pulse is shown for each vertical slice. Such a picture is called the *diffraction hyperbola*.

It should be noted that the simplified Equations 3.4 and 3.6 are similar to Equations 2.7 and 2.9 in Chapter 2. In this case, Stolt's method, discussed in Section 2.4, can be used for the inverse focusing procedure, as well as the method of *matched filtering* that will be considered next.

50 ■ *Electromagnetic and Acoustic Wave Tomography*

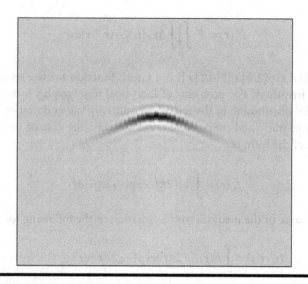

Figure 3.1 Diffraction hyperbola.

3.1.2 Matched Filtering Method for Definition of Radiation Sources Distribution

The inverse problem of reconstruction (recovering) the distribution of sources of radiation is an inverse problem to Equation 2.7. Let us suppose that the sources are distributed at the arbitrary plane z_0 in such a manner that the total current distribution equals $j(\mathbf{r}) = j_s(\mathbf{r}_\perp)\delta(z-z_0)$, where $|\mathbf{r}_\perp| = \sqrt{x^2+y^2}$, $|\mathbf{r}| = \sqrt{x^2+y^2+z^2}$. It is supposed also that the location of the plane is known exactly. If the wave field is registered at another plane $z = const$, then its spatial spectral distribution, according to Weyl's formula (Equation 2.19), can be written as:

$$E(\mathbf{\kappa}_\perp, z) = \frac{i}{2\kappa_z} \hat{j}_s(\mathbf{\kappa}_\perp)\exp\{i\kappa_z(z-z_0)\} \qquad (3.7)$$

Here, the spectral (i.e., in the wave length domain) distribution of the surface currents, as in Section 2.2, can be written:

$$\hat{j}(\mathbf{\kappa}_\perp) = \iint_S j_s(\mathbf{r}_\perp)\exp(-i\mathbf{\kappa}_\perp \mathbf{r}_\perp)d\mathbf{r}_\perp. \qquad (3.8)$$

If we consider the point source $j_s(\mathbf{r}_\perp) = \delta(\mathbf{r}_\perp - \mathbf{r}_{\perp 0})$, then, for the spectrum of the radiated wave, we obtain from Equation 3.7:

$$E(\mathbf{\kappa}_\perp, z) = E_0(\mathbf{\kappa}_\perp, z)\frac{i}{2\kappa_z}\exp\{i\kappa_z(z_0-z)\} \qquad (3.9)$$

This function can be considered as a transfer function (TF), or as the system function mentioned in Section 1.1. The matched filtering method, in assumption of the existence of white Gaussian noise, gives the following solution of the inverse problem:

$$j_s(\mathbf{\kappa}_\perp) = E(\mathbf{\kappa}_\perp,z) \cdot E_0^*(\mathbf{\kappa}_\perp,z)\exp\{i\kappa_z z_F\} = \frac{1}{4\kappa_z \kappa_z^*}\exp\{i\kappa_z z_F\} E(\mathbf{\kappa}_\perp,z) \quad (3.10)$$

Here, the coordinate z_F defines the plane of focusing at which the signal-to-noise (S/N) ratio is maximal (best). Taking now the inverse Fourier transform of Equation 3.10, we will obtain the desired solution of the inverse problem. We should notice the following:

1. If the distance to plane z_0 is not known or known approximately, then the phase distribution of the recovered image $\varphi(\mathbf{\kappa}_\perp)$ will give the reconstruction of the surface profile, $z_s(\mathbf{\kappa}_\perp) = \varphi(\mathbf{\kappa}_\perp)/k$, at which the sources of radiation are distributed. This can be done at the arbitrary frequency $f = k \cdot c/2\pi$ of the spectrum (i.e., of the frequency band) of wave radiation, both for pulse radiation and for monochromatic radiation.
2. The expression Equation 3.6 gives the possibility of obtaining its other solution in the form:

$$j_s(\mathbf{\kappa}_\perp) = \frac{E(\mathbf{\kappa}_\perp,z)}{E(\mathbf{\kappa}_\perp,z)} = -i2\kappa_z \exp\{-i\kappa_z(z_0 - z)\}E(\mathbf{\kappa}_\perp,z) \quad (3.11)$$

This solution differs from the solution of the matched filtering method only by amplitude product. Both solutions can be enjoyed by use of the following expression:

$$j_s(\mathbf{\kappa}_\perp) = \frac{E(\mathbf{\kappa}_\perp,z) \cdot E_0^*(\mathbf{\kappa}_\perp,z)}{E(\mathbf{\kappa}_\perp,z) \cdot E_0^*(\mathbf{\kappa}_\perp,z) + \alpha}\exp\{i\kappa_z z_F\} \quad (3.12)$$

Here, α is the parameter of regularization.

3.1.3 Method of Migration in Spatiotemporal Region by Use of Pulse Signals

Most methods and techniques discussed up until now relate to a case of monochromatic (*narrowband* [15]) signals, dealing only with a single frequency of the carrier signal. If the radiated signal has form of a pulse (or *wideband signal* [15]), that is, it consists of a wide band of carrier frequencies, it is possible either to use its spectral presentation over the given frequency band, or to use the *method of migration*. We will briefly describe the method of migration, giving the reader the meaning of the matter.

During pulse signal radiation, the form of which is described by function $S_0(t)$, the spatiotemporal replica of this signal, registered at the receiver, can be written as [7–10]

$$E(\mathbf{r},t) = \iiint_{V_1} j(\mathbf{r}_1) \frac{S_0\left(t - \frac{2|\mathbf{r}_1 - \mathbf{r}|}{c}\right)}{(4\pi)^2 |\mathbf{r}_1 - \mathbf{r}|^2} d\mathbf{r}_1 \qquad (3.13)$$

This formula describes the dependence of the signal field strength in the joint space and time domains, and the case of the point source, $j(\mathbf{r}_1) = j_0 \delta(\mathbf{r}_1 - \mathbf{r}_0)$, is presented on the left-hand side of Figure 3.2. In this case, the signal strength recorded at the detector will be reduced to the simple form:

$$E_0(\mathbf{r}.t) = j_0 \frac{S_0\left(t - \frac{2|\mathbf{r}_1 - \mathbf{r}|}{c}\right)}{(4\pi)^2 |\mathbf{r}_1 - \mathbf{r}|^2} \qquad (3.14)$$

Such a dependence can be transferred asymptotically to a hyperbolic one and, therefore, it is called the *diffraction hyperbola*. In the general case of spatial distributed sources in the volume V_1, the observed source field equation (Equation 3.13) obtains a form of superposition of multiple shifted hyperbolas, as seen from the right-hand side of Figure 3.2.

It should be pointed out that Equation 3.13 is the solution of the *direct problem*. An inverse problem is the definition of the source's $j(\mathbf{r}_1)$ distribution via knowledge of the registered $E(\mathbf{r}, t)$ distribution, which can be solved by the *method of migration*. The method of migration states that each observed hyperbola is a projection of one point of the source, and all values of the signals at the corresponding hyperbola should be simply integrated (or summated), accounting for their time delay, which should be measured and selected precisely. If the selected point of the

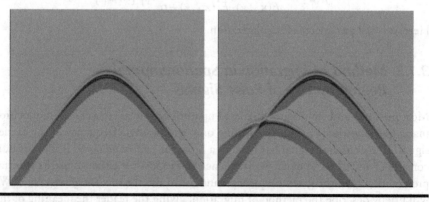

Figure 3.2 Diffraction hyperbolas. Vertical axis: time; horizontal axis: coordinate.

focusing procedure corresponds to the real position of the source, the result will give a greater value of the signal; otherwise, the contribution will not be sufficient.

In the case under consideration, if we select and take a focusing point $\mathbf{r}_F$, the solution by use of the method of migration is simply the following summation:

$$U(\mathbf{r}_F) = \sum_m E(\mathbf{r}_m, t_m = |\mathbf{r}_m - \mathbf{r}_F|/c)$$

Here, $\mathbf{r}_m$ is the points of observation of the signal fields $E(\mathbf{r}_m, t_m)$. To obtain a full image of the sources' distribution, it is enough to scan a full space of interest by a point of focus $\mathbf{r}_F$.

As a result, we obtain:

$$U(\mathbf{r}_F) = \iiint_{V_1} j(\mathbf{r}_1) \delta_S(\mathbf{r}_1, \mathbf{r}_F) d\mathbf{r}_1 \qquad (3.15)$$

where function

$$\delta_S(\mathbf{r}_1, \mathbf{r}_F) = \sum_m \frac{S_0(t_m = |\mathbf{r}_m - \mathbf{r}_F|/c - |\mathbf{r}_1 - \mathbf{r}_m|/c)}{4\pi |\mathbf{r}_1 - \mathbf{r}_m|}$$

is the system function (SF), which is simply a reaction of the point source. For a large enough aperture and a high density of filling of points inside the registered aperture, this SF becomes a function only of difference between the arguments, that is, $\delta_S(\mathbf{r}_1 - \mathbf{r}_F)$.

In this case, Equation 3.15 is simply an convolution integral:

$$U(\mathbf{r}_F) = \iiint_{V_1} j(\mathbf{r}_1) \delta_S(\mathbf{r}_1 - \mathbf{r}_F) d\mathbf{r}_1 \qquad (3.16)$$

The solution of such an integral equation was described in Section 2.4.

It is important to note that, when using a known form of the signal $S_0(t)$, a form of the SF can be computed *a priori* and kept in the computer memory. Then, for the method of migration, the fast algorithms of computation become appropriate. Usually, in realization of the method of migration, the SF is similar to the *smoothen* δ-function (the *point spread function* or *sinc-function* [15]), and it can be rewritten as:

$$j(\mathbf{r}_1) \approx U(\mathbf{r}_F) = \sum_m E(\mathbf{r}_m, t_m = |\mathbf{r}_m - \mathbf{r}_F|/c) \qquad (3.17)$$

We note that, despite the simplicity of the procedure to obtain the expression $U(\mathbf{r}_F)$, it requires essential computing and temporal resources. At the same time, its advantage is in the possibility to create and to correct images parallel to the process of receiving measurement data. Figure 3.3 shows an example of imaging of

54 ■ *Electromagnetic and Acoustic Wave Tomography*

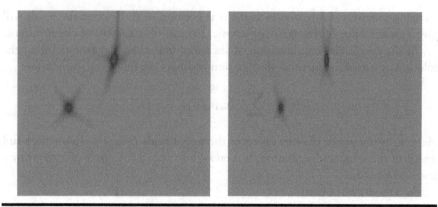

Figure 3.3 Example of reconstruction of two sources of distribution by migration method. Horizontal axis: horizontal coordinate; vertical axis: depth of location.

distribution of two sources by use of the method of migration in the case of particular (~50%, left-hand side) and total (~100%, right-hand side) filling of the matrix of the input data obtained from measurements.

3.1.4 Method of Synthesized Focusing

According to our point of view, from a wide spectrum of the physical approaches proposed in Sections 3.1.1 through 3.1.3, the method of *synthesized focusing*, which is a combination of the migration and scanning methods described in Section 3.1.3, is much more suitable. In this method, parallel or "one-to-another" localization of radiation with a medium (or object) is carried out. This approach is similar to the mathematical approach of inverse projections described in Chapter 2, but differs from it by the use of coherent radiation only. Finally, this procedure gives significant benefits due to much better spatial resolution and much faster computations. Synthesis of the focusing effect is based on usage of focusing phenomena, as the result of multiray interference of coherent waves at the desired point $\mathbf{r}_F$ caused by an in-phase summation of these waves. In the case of monochromatic radiation, this operation is carried out for each point of detection $\mathbf{r}_m$ and for each point of radiation $\mathbf{r}_{0n}$ by timing the complex amplitude of the received wave (signal) field on the focusing function, that is,

$$M(\mathbf{r}_F, \mathbf{r}_m, \mathbf{r}_{0n}) = \exp\{-ik(|\mathbf{r}_F - \mathbf{r}_m| + |\mathbf{r}_F - \mathbf{r}_{0n}|)\}$$

As a result, we get the following function:

$$U(\mathbf{r}_F) = \sum_m \sum_n M(\mathbf{r}_F, \mathbf{r}_m, \mathbf{r}_{0n}) E(\mathbf{r}_m, r_{0n}) \qquad (3.18)$$

The main idea of this operation is to obtain the sum of all partial waves at the given point of focus $\mathbf{r}_F$. Scanning the desired area of sounding at this point, it is possible to obtain the spatial distribution of test objects (e.g., inhomogeneities or anomalies). If sounding is now carried out by the use of a set of frequencies f_l, that is, for different values of wave numbers, $k_l = 2\pi f_l/c$, then the summation of a set of interferential pictures $U(\mathbf{r}_F, k_l)$ will decrease the level of the side lobes and will increase the main maximal lobe of the reradiated pattern from the object at the focusing plane. Mathematically, this procedure can be written as:

$$U(\mathbf{r}_F) = \sum_l U(\mathbf{r}_F, k_l) \tag{3.19a}$$

In the case of use of pulse radiation, the last expression states that the focusing procedure should be spatiotemporal, that is,

$$U(\mathbf{r}_F) = \sum_m \sum_n S(\mathbf{r}_m, \mathbf{r}_{0n}, t_{mn}) \tag{3.19b}$$

where $t_{mn} = \dfrac{|\mathbf{r}_F - \mathbf{r}_m| + |\mathbf{r}_F - \mathbf{r}_{0n}|}{c}$, and c is the wave velocity in free space.

In the private case, shown in Figure 3.2, the procedure of focusing is equivalent to integration (summation) along the hyperbola. Therefore, in the literature, this method of focusing is usually called the method of *summation by diffraction hyperbolas* [13, 14]. As mentioned, this approach is close to that of the use of spatiotemporal matched filtering described in Section 3.1.2, and the focusing function plays the role of TF of the matched filter; so, it is close to the method of inverse focusing (MIF) discussed in Section 3.1.1.

We note that the waiting impact of summands in Equations 3.18 and 3.19 is important for the increase of quality of focusing images of objects or inhomogeneities located in a medium. But, the technique plays a more principal role: how precisely the phase of each wave recorded by the detector was selected. The amplitude weighting of the summands in Equations 3.18 and 3.19 influence the level of the side lobes of the reradiated pattern and the possibility to obtain destroyed images of the objects or inhomogeneties hidden in the media (called *artifacts*).

The wave field formed as the result of the focusing procedure at the plane of focusing has the form of quasi-parallel (collimated) beams of electromagnetic radiation. As shown by the corresponding computations, at the focusing plane the phase and amplitude fronts of the wave are close to those of the plane wave. Therefore, interaction of the wave radiation, monochromatic, or ultra-wide band (UWB) pulse, with media or objects under investigation, can be considered in *collimated beam* approximation. Next, we will discuss the peculiarities of location sounding with single-side illumination of the desired object.

3.1.5 Single Focusing at the Boundary of Two Media

Let us assume that, as the result of measurements at the frequency $f = ck/2\pi$, distribution of the scattered wave $E(r_\perp, f)$ along an arbitrary plane above the surface of the medium under sounding was obtained (see Figure 3.4).

Usage of the method of synthesis aperture radar (SAR) gives the possibility of recovering the 3-D distribution of inhomogeneities located in the testing area by inversion of the integral (Equation 3.2a). For such an inversion, let us focalize the field (Equation 3.2a) at some arbitrary point $\mathbf{r}_{\perp F}$ located at the boundary of two media.

The operation of focusing is based on the in-phase summation of complex amplitudes of the scattered wave field and on the selection of the point of focus. To achieve such a summation, it is necessary to compensate the phase shift at the path from the transmitting antenna to the focusing point and from the focusing point to the receiving antenna. The result of focusing the scattered field at some arbitrary surface point $\mathbf{r}_{\perp F}$ can be written with help of the convolutional integral [13, 14]:

$$F(\mathbf{r}_{\perp F}, f) = \iint_S E(\mathbf{r}_{\perp 0}, f) M(\mathbf{r}_{\perp F} - \mathbf{r}_{\perp 0}, f) d\mathbf{r}_{\perp 0} \tag{3.20}$$

Here, $M(\mathbf{r}_\perp, f) = \exp\left\{-ik_0\left(2\sqrt{\mathbf{r}_\perp^2 + h^2}\right)\right\}$ is the form of the weight function (focusing function), and integration in Equation 3.20 is carried out along the whole plane of observation S, which presents a plane of synthesized aperture. The hidden object presents the desired inhomogeneity, located at the lower half-space and consisting of the refractive index n. Accounting for Equations 3.2 and 3.3, the focusing wave field can be written as:

$$F(\mathbf{r}_{\perp F}, f) = \iiint_{V_1} \Delta\varepsilon_r(\mathbf{r}_{\perp 1}, z_1) Q(\mathbf{r}_{\perp 1}, \mathbf{r}_{\perp F}, z_1, f) d\mathbf{r}_{\perp 1} \tag{3.21}$$

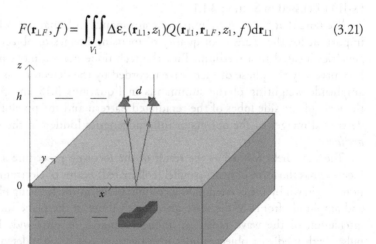

Figure 3.4 Detection of hidden inhomogeneities.

where

$$Q(\mathbf{r}_{\perp 1}, \mathbf{r}_{\perp F}, z_1, f) = \iint_S G_0^2(\mathbf{r}_{\perp 1} - \mathbf{r}_{\perp 0}, z_1) M(\mathbf{r}_{\perp F} - \mathbf{r}_{\perp 0}, f) d\mathbf{r}_{\perp 0} \quad (3.22)$$

Equation 3.22 describes the response of the system at the point inhomogeneity, located at the point $\mathbf{r}_{\perp 1}$; that is, this is the TF of the system operated at frequency f for the focusing procedure occurring at point $\mathbf{r}_{\perp F}$ and lying at the surface of the lower medium (see Figure 3.4).

We should notice that in the case of large aperture S, the field inside the region of focusing becomes homogeneous, and the integral (Equation 3.22) reduces to the convolutional integral. Then we can write that, approximately,

$$Q(\mathbf{r}_{\perp 1}, \mathbf{r}_{\perp F}, z_1, f) = Q(\mathbf{r}_{\perp F} - \mathbf{r}_{\perp 1}, z_1, f)$$

This allows us to apply the TF for one point of focus to many other points. Here, only a knowledge of the distance between the focus and observation points, $r_\perp = |\mathbf{r}_{\perp F} - \mathbf{r}_{\perp 1}|$, is important.

The result of computations of the TF $Q(\mathbf{r}_\perp, z, f)$ for free space (refractive index $n = 1$) at a frequency of $f = 10$ GHz for the synthesis aperture with a length of 50 cm and a height of 30 cm is presented in Figure 3.5a. Here, the dark-colored pixels correspond to the maximal values of obtained amplitude. The same picture is presented in Figure 3.5, but for the lower medium (see Figure 3.4) with a refractive index of $n = 3$. It is seen that with an increase of the refractive index n in the lower medium, the region of maximum response of the system is at the points of scattering spreads downward along the vertical direction. In other words, increase of the phase shift of the incident wave along its geometrical path leads to essential enlargement of the area of focus in the lower medium, where the inhomogeneity under testing is located (see Figure 3.4).

It should be noted that in the procedure presented, the focusing effect is achieved only in the horizontal plane, obeying this procedure in the vertical direction. Generalization of this procedure to the 3-D medium requires complicated mathematical calculations, occupying a lot of computational time [1].

We should outline that the illustrations presented in Figure 3.5a and b are correct only for a 2-D surface for high values of n and for depths of location of the inhomogeneity not far from the boundary surface between the air and the medium under investigation. This effect becomes much stronger with an increase in the refractive index n.

Accounting for a large elongation of the TF along the vertical direction, let us assume that the refracted plane waves propagate in the lower medium perpendicular to the boundary between the air and the medium (see Figure 3.4). The accuracy of such an assumption depends on the parameter n: the higher refractivity of the tested medium, the more precisely such an assumption can be used. In this case, the expression for the wave vector $\mathbf{\kappa}_{1z}$ becomes simpler, that is,

58 ■ *Electromagnetic and Acoustic Wave Tomography*

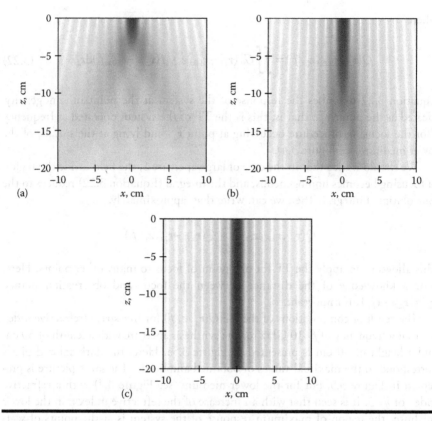

Figure 3.5 TF of the system in the frequency domain: (a) for free space; (b) for a medium with $n=3$; (c) transverse approximation for a medium with $n=3$.

$\kappa_{1z} = \sqrt{k^2 n^2 - \kappa_\perp^2} = kn = k_1$, and a new product, $\sim\exp\{ik_1z\}$, occurring additionally in Equation 3.22, can be removed from the integral; that is,

$$G(\mathbf{r}_\perp, z) = \exp\{ik_1 z\} \int_{-\infty}^{\infty}\int_{-\infty}^{\infty} \frac{iT(\mathbf{k})\exp\{i(\mathbf{k}_\perp \mathbf{r}_\perp) + \kappa_z h)\}}{2(2\pi)^2 \kappa_z} d\mathbf{k}_\perp \quad (3.23)$$

$$\equiv \exp(ik_1 z) G_\perp(\mathbf{r}_\perp)$$

Here, h is the depth of position of the inhomogeneity under test in the lower medium (as shown in Figure 3.4), and $T(\mathbf{k})$ is the transmission coefficient of the plane wave passing the boundary between air and the lower medium.

Changing the steps of integration in Equations 3.21 and 3.22 and using the approximation in Equation 3.23, we get the following expression for the focusing field:

$$F(\mathbf{r}_{\perp F}, f) = \int_{-\infty}^{\infty} \exp(ik_1 z) \iint_S \Delta\varepsilon_r(\mathbf{r}_{\perp 1}, z_1) Q_\perp(\mathbf{r}_{\perp F} - \mathbf{r}_{\perp 1}, f) d\mathbf{r}_{\perp 1} dz_1, \quad (3.24)$$

where

$$Q_\perp(\mathbf{r}_{\perp F} - \mathbf{r}_{\perp 1}, f) \equiv Q(\mathbf{r}_{\perp F} - \mathbf{r}_{\perp 1}, z_1 = 0, f)$$

$$= \iint_S G_\perp^2(\mathbf{r}_{\perp 1} - \mathbf{r}_{\perp 0}) M(\mathbf{r}_{\perp F} - \mathbf{r}_{\perp 0}, f) d\mathbf{r}_{\perp 0}$$

is the transverse TF of the system operated at frequency f. In Figure 3.5c, the results are shown of the computations of the transverse TF $Q_\perp(\mathbf{r}_\perp)$ for frequency $f = 10$ GHz. It is seen that this function precisely repeats the TF for $n = 3$, until some depth z_{max}; deeper than z_{max}, this approximation becomes incorrect. A degree of localization of the tested inhomogeneity, that is, the TF $Q_\perp(\mathbf{r}_\perp)$ along $\mathbf{r}_\perp(x, y)$, determines a potential efficiency of reconstruction of the system in the horizontal plane.

For transition to the 3-D TF over all operated frequencies used during the test measurements, according to the procedure proposed in [13, 14], the Fourier transform for Equation 3.24 must be used. As a result, we obtain:

$$\tilde{F}(\mathbf{r}_{\perp F}, f) \equiv \int F(\mathbf{r}_{\perp F}, f) \exp\{-2\pi f t\} df$$

$$= \iiint_{V_1} \Delta \varepsilon_r(\mathbf{r}_{\perp 1}, z_1) \tilde{Q}_\perp\left(\mathbf{r}_{\perp F} - \mathbf{r}_{\perp 1}, \frac{ct}{2n} - z_1\right) d\mathbf{r}_{\perp 1} dz_1 \quad (3.25)$$

where

$$\tilde{Q}_\perp\left(\mathbf{r}_{\perp F} - \mathbf{r}_{\perp 1}, \frac{ct}{2n} - z_1\right) = \int Q_\perp(\mathbf{r}_{\perp F} - \mathbf{r}_{\perp 1}, f) \exp\left[-i2\pi f\left(t - \frac{2nz_1}{c}\right)\right] df$$

is the TF of the system in the joint spatiotemporal domain. Numerical computations of TF for a frequency band from 0.5 to 17 GHz showed that the function $\tilde{Q}_\perp$ is localized over all three spatial coordinates (see Figure 3.6). The process of computation was made with the assumption that the depth of location of the point inhomogeneity was $z_1 = 2.5$ cm.

In the frame of the presented approximation for reconstruction of the spatial distribution of inhomogeneities, $\Delta \varepsilon(\mathbf{r}_{\perp 1}, z_1)$, located in the region under investigation, it is necessary to rewrite the convolutional integral (Equation 3.25). Such a problem is well known and can be resolved by use of Wiener filtering with regularization [13, 14]. The fact of satisfied localization of the TF, obtained in the frame of theoretical analysis, allows us, as the first approximation, to suppose that:

$$\Delta \varepsilon(\mathbf{r}_{\perp 1}, z_F) \sim \tilde{F}(\mathbf{r}_{\perp F}, 2nz_F/c)$$

$$= \int_{-\infty}^{\infty} \exp(-i2knz_F) \iint_S E(\mathbf{r}_{\perp 0}, f) M(\mathbf{r}_{\perp F} - \mathbf{r}_{\perp 0}, f) d\mathbf{r}_{\perp 0} df \quad (3.26)$$

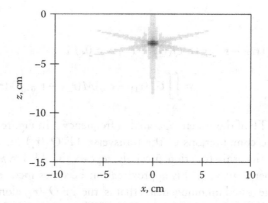

Figure 3.6 3-D TF of system.

Accuracy in resolving of the problem of inhomogeneity (object) localization is defined by a scale of localization of the TF of the system. In the frame of the above-mentioned approximations, the solution of the inverse problem of sub-surface tomography can be obtained by the use of focusing radiation to the point located at the surface of the medium under test and of the inverse Fourier transform over frequencies. Use of the fast Fourier transform (FFT) allows significant acceleration of the data processing procedure.

3.1.6 Two-Step Focusing

The method proposed in Section 3.1.5 of a solution to the tomography problem by the use of fixing the focus at the boundary of two media, as discussed, is good if a second medium is strongly refractive with respect to the first one [4]. In this case, the lower medium (see, e.g., Figure 3.4) works as the refractive prism, which preserves the focusing radiation inside the medium as the same as it is at the boundary surface between the two media.

The situation is changed significantly in the case of a weakly refractive lower medium, or when its refractive index is lower than that of the upper medium. Partially focused radiation starts to spread into the lower medium. To improve this situation, it is necessary to carry out the additional (second) focusing procedure, which is called *two-step focusing*.

We consider that the transmitter and the receiver of the monochromatic wave move parallel to the plate boundary between the two media, at some distance from it. With the defined mesh, the sounding of the surface between two media is achieved where a point of radiation and of detection were designed together, and together they move at the height $h = const$ in the air above the boundary between the two media.

Let us suppose that the first focusing procedure was carried out at an arbitrary point at the boundary of the two media. Now, we will carry out the additional (second) focusing procedure by use of a convolutional-type integral, following the framework proposed in references [13, 14]:

$$F_1(\mathbf{r}_{\perp F}, h_1, f) = \iint_S F(\mathbf{r}_{\perp F}, f) M_1(\mathbf{r}_{\perp F_1} - \mathbf{r}_{\perp F}, h_1, f) d\mathbf{r}_{\perp F} \quad (3.27)$$

where:

$M(\mathbf{r}_{\perp F}, h_1, f) = \exp\left\{-i2kn\sqrt{\mathbf{r}_{\perp F}^2 + h_1^2}\right\}$ is a weighting function of the focusing procedure

$\mathbf{r}_1 = (\mathbf{r}_{\perp F_1}, h_1)$ is a second focusing point in the lower medium

h_1 is the depth of second focusing point along the z-axis (the vertical axis)

S is a plane of the boundary between the two media, which represents a plane of the secondary synthesized aperture

It is also supposed that values of the refractive index for a second medium are known *a priori*. The argument of function M_1 from Equation 3.27 is taken as a difference of arguments, that is, $\mathbf{r}_{\perp F_1} - \mathbf{r}_{\perp F}$, assuming homogeneity of the synthesized aperture. Taking Equation 3.24 into account, the focusing field can be present in the following form:

$$F_1(\mathbf{r}_{\perp F_1}, h_1, f) = \iiint_{V_1} \Delta\varepsilon_r(\mathbf{r}_{\perp 1}, z_1) Q_1(\mathbf{r}_{\perp 1}, \mathbf{r}_{\perp F_1}, h_1, f) d\mathbf{r}_{\perp 1} \quad (3.28)$$

where function

$$Q_1(\mathbf{r}_{\perp 1}, \mathbf{r}_{\perp F_1}, h_1, f) = \iint_S Q(\mathbf{r}_{\perp 1}, z_1, \mathbf{r}_{\perp F}) M_1(\mathbf{r}_{\perp F_1} - \mathbf{r}_{\perp F}, h_1) d\mathbf{r}_{\perp F} \quad (3.29)$$

corresponds to the response of the system from the point inhomogeneity located at the point $\mathbf{r}_1 = (\mathbf{r}_{\perp F_1}, h_1)$, that is, it presents the TF of the system operating at frequency f during focusing at the sub-surface point $\mathbf{r}_{\perp F_1}$.

In the case of a large aperture S, Equation 3.29 is transformed into convolutional-type integrals such as:

$$Q_1(\mathbf{r}_{\perp 1}, \mathbf{r}_{\perp F_1}, h_1, f) = k_1^2 \iint_S G_0^2(\mathbf{r}_{\perp 1} - \mathbf{r}_{\perp 0}) M(\mathbf{r}_{\perp F} - \mathbf{r}_{\perp 0}, f) d\mathbf{r}_{\perp 0}$$
$$\iint_S M_1(\mathbf{r}_{\perp F_1} - \mathbf{r}_{\perp F}, h_1) d\mathbf{r}_{\perp F} \quad (3.30)$$

62 ■ *Electromagnetic and Acoustic Wave Tomography*

Changing $G_0^2(\mathbf{r}_{\perp 1} - \mathbf{r}_{\perp 0})$ to $g(\mathbf{r}_{\perp 1} - \mathbf{r}_{\perp 0})$, we present Equation 3.30 as:

$$Q_1(\mathbf{r}_{\perp 1}, \mathbf{r}_{\perp F_1}, h_1, f) = k_1^2 \iint_S g(\mathbf{r}_{\perp 1} - \mathbf{r}_{\perp 0}) M_1(\mathbf{r}_{\perp F_1} - \mathbf{r}_{\perp F}, f) d\mathbf{r}_{\perp 0} d\mathbf{r}_{\perp F_1} \quad (3.31)$$

As a result, after integration, we finally obtain:

$$Q_1 = \frac{k_1^2}{2\pi} \iint_S \tilde{g}(-\mathbf{\kappa}_{\perp 2}) \tilde{M}(\mathbf{\kappa}_{\perp 3}) \tilde{M}(\mathbf{\kappa}_{\perp 3}) \exp\{i\mathbf{\kappa}_{\perp 3}(\mathbf{r}_{\perp F_1} - \mathbf{r}_{\perp 1})\} d\mathbf{\kappa}_{\perp 3}$$

$$= Q_1(\mathbf{r}_{\perp F_1} - \mathbf{r}_{\perp 1}, h_1, z_1) \quad (3.32)$$

Here, $\tilde{g}(-\mathbf{\kappa}_{\perp 2})$, $\tilde{M}(\mathbf{\kappa}_{\perp 3})$, and $\tilde{M}(\mathbf{\kappa}_{\perp 3})$ are the spatial wave spectra of functions $g(\mathbf{r}_{\perp 1} - \mathbf{r}_{\perp 0})$, $M(\mathbf{r}_{\perp F}, h, f)$, and $M_1(\mathbf{r}_{\perp F_1}, h, f)$, respectively. Based on Equation 3.32, Equation 3.28 transforms into a convolutional-type integral, and can be written:

$$F_1(\mathbf{r}_{\perp F_1}, h_1, f) = \iiint_{V_1} \Delta\varepsilon_r(\mathbf{r}_{\perp 1}, z_1) Q_1(\mathbf{r}_{\perp 1} - \mathbf{r}_{\perp F_1}, h_1, f) d\mathbf{r}_{\perp 1} \quad (3.33)$$

Function $Q_1(\mathbf{r}_{\perp 1}, h_1, z_1)$ plays the role of the TF of the focusing system using the double synthesization procedure. In Figure 3.7, this TF for two-step focusing is presented for a depth of 10 cm under the boundary between two media.

The left-hand side of Figure 3.7 presents the TF for the two-step focusing procedure, while the right-hand side presents its approximate solution.

If we additionally suppose that in the proximity of the focusing point, the waves propagate normally to the boundary surface, this small region will be localized in

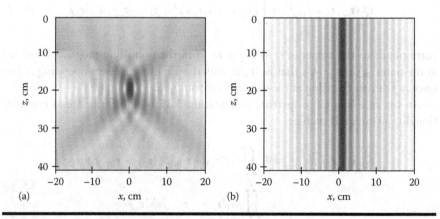

Figure 3.7 TF (a) in two-step focusing; (b) approximate representation.

the horizontal plane (x, y) and elongated along the vertical z-axis for a large height h_1. In such a small region, the TF can be presented as:

$$Q_1 = Q_1(\mathbf{r}_{\perp F_1} - \mathbf{r}_{\perp 1}, z_1 - h_1) \approx Q_1(\mathbf{r}_{\perp F_1} - \mathbf{r}_{\perp 1})\exp\{i2kn(z_1 - h_1)\}$$

and the focusing field will have the following integral form:

$$F(\mathbf{r}_{\perp F}, h_1) = \iint \Delta\varepsilon_r(\mathbf{r}_{\perp 1}, z_1) Q_{1\perp}(\mathbf{r}_{\perp F_1} - \mathbf{r}_{\perp 1})\exp\{i2kn(z_1 - h_1)\}d\mathbf{r}_{\perp 1}dz_1.$$

If we now use a Fourier transform and transfer the integral from the frequency domain to the time domain, we can obtain localization along the depth and recover an approximate image of the scattered inhomogeneities, $\Delta\varepsilon(\mathbf{r}_{\perp F}, z_F)$, in this manner:

$$\int F_1(\mathbf{r}_{\perp F_1}, h_1)\exp\{i2knz_F\}df$$
$$\approx \int \Delta\varepsilon_r(\mathbf{r}_{\perp 1}, z_1)\delta(z_F - [z_1 - h_1])\{2\pi/(2n/c)\}d\mathbf{r}_{\perp 1}dz_1 \approx \Delta\varepsilon_r(\mathbf{r}_F, z_F) \quad (3.34)$$

Here, z_F is the depth at which the focusing procedure was carried out, and $t_F = 2nz_F/c$ is the time of achieving the wave at the focus point and returning back to the point of observation. Equation 3.34 gives the solution to the problem of wave tomography by use of the two-step focusing procedure.

3.1.7 Group Focusing Technique

In some practical applications, the two-step focusing procedure would be not so precise, namely, in a multilayered medium, or even in a homogeneous regular environment. In this case, the focusing procedure should be carried out either in parallel or from one to the other (i.e., by slicing). We will describe this procedure of focusing with an example of the homogeneous regular background medium using the Born approximation (e.g., single scattering [15, 16]); that is, a scattered field will present as:

$$E(\mathbf{r}_{\perp 0}, f) = k_1^2 \iiint_{V_1} \Delta\varepsilon_r(\mathbf{r}_{\perp 1}, z_1) G_0^2(\mathbf{r}_{\perp 1} - \mathbf{r}_{\perp 0}, z_1)d\mathbf{r}_{\perp 1}dz_1 \quad (3.35)$$

Here, as in Section 3.1.4, $k_1 \equiv k = 2\pi f/c$ is the wave number corresponding to the background environment. The Green function in this case is presented as a field of spherical waves:

$$G_0(\mathbf{r}_{\perp 1} - \mathbf{r}_{\perp 0}, z_1) = \frac{\exp\{ik\sqrt{(\mathbf{r}_{\perp 1} - \mathbf{r}_{\perp 0})^2 + z_1^2}\}}{4\pi\sqrt{(\mathbf{r}_{\perp 1} - \mathbf{r}_{\perp 0})^2 + z_1^2}} \quad (3.36)$$

Differentiating Equation 3.35, we obtain:

$$\frac{d}{dk}\left\{\frac{E(\mathbf{r}_{\perp 0}, f)}{k^2}\right\} = \iiint_{V_1} \Delta\varepsilon_r(\mathbf{r}_{\perp 1}, z_1) G_2(\mathbf{r}_{\perp 1} - \mathbf{r}_{\perp 0}, z_1) d\mathbf{r}_{\perp 1} dz_1$$

where

$$G_2(\mathbf{r}_{\perp 1} - \mathbf{r}_{\perp 0}, z_1) = \frac{\exp\left\{i2k\sqrt{(\mathbf{r}_{\perp 1} - \mathbf{r}_{\perp 0})^2 + z_1^2}\right\}}{4\pi\sqrt{(\mathbf{r}_{\perp 1} - \mathbf{r}_{\perp 0})^2 + z_1^2}}$$

It is possible to expand this formula onto the spectrum of plane waves (according to Weyl's expansion of plane waves):

$$G_2(\mathbf{r}_{\perp 1} - \mathbf{r}_{\perp 0}, z_1) = \frac{i}{(2\pi)^2} \iint \frac{\exp\left\{i[\boldsymbol{\kappa}_\perp(\mathbf{r}_{\perp 1} - \mathbf{r}_{\perp 0}) + \kappa_z z_1]\right\}}{2\kappa_z} d\boldsymbol{\kappa}_\perp \quad (3.37)$$

Here, $\kappa_z = \sqrt{(2k)^2 - \kappa_\perp^2}$. Accounting for the latter presentation, the equation can be written:

$$E(\mathbf{u}_\perp, f) = \iint \exp\{i\mathbf{u}_\perp \mathbf{r}_{\perp 0}\} \frac{d}{dk}\left\{\frac{E(\mathbf{r}_{\perp 0}, f)}{k^2}\right\} d\mathbf{r}_{\perp 0}$$
$$= \iiint_{V_1} \Delta\varepsilon_r(\mathbf{r}_{\perp 1}, z_1) \frac{\exp\{i(\mathbf{u}_\perp \mathbf{r}_{\perp 1} + u_z z_1)\}}{2iu_z} d\mathbf{r}_{\perp 1} dz_1 \quad (3.38)$$

where $u_z = \sqrt{(2k)^2 - \mathbf{u}_\perp^2} = \sqrt{(4\pi f/c)^2 - \mathbf{u}_\perp^2}$. This expression describes the spectrum of spatial frequencies of inhomogeneities defined by Equation 3.38 with great accuracy (of some small product value), that is,

$$\Delta\varepsilon_r(\mathbf{u}_\perp, u_z) = \iiint_{V_1} \Delta\varepsilon_r(\mathbf{r}_{\perp 1}, z_1) \exp\{i(\mathbf{u}_\perp \mathbf{r}_{\perp 1} + u_z z_1)\} d\mathbf{r}_{\perp 1} dz_1 =$$
$$\quad (3.39)$$
$$= i2u_z E_1(\mathbf{u}_\perp, f)$$

Finally, for reconstruction of the spatial distribution of inhomogeneities, use of the inverse 3-D Fourier transform should be implied. But, there exists one computing peculiarity in computations of Equation 3.38: transferring from frequencies f to the corresponding spatial frequencies u_z, as defined, which can be realized with the help of simple interpolation. Equation 3.38 realizes the method of focusing over all distances. Moreover, Equation 3.38 is a generalized technique of the Stolt migration, described in Section 2.4, and accounts for the difference between functions $G_2 = (\mathbf{r}_\perp - \mathbf{r}_{\perp 0}, z_1)$ used here and $G_0^2(\mathbf{r}_{\perp 1} - \mathbf{r}_{\perp 0}, z_1)$ used in Stolt's method. But, this does not have such a serious impact on the effect of focusing. The difference is that the Green function plays a role only in selection of the *window function* (also known as the *anodization function* or *tapering function*). Therefore, the results of modeling the method of radiolocation by use of these two approaches gives similar results in reconstruction of the position of point inhomogeneity by use of the Stolt migration and group focusing technique, which is clearly seen from results presented in Chapter 2 by Figure 2.13.

The proposed method of group focusing can be used for the case of a multilayered medium. In this case, in Equation 3.38, the product $\exp\{i(u_z z_1)\}$ should be changed to

$$\exp\left\{i\left(\sum_j u_{zj} z_j + u_z z_1\right)\right\}$$

which accounts for the phase shifts in all previous layers with a width of z_j and a known index of refraction n_j. The normal projection of the wave number in each layer can be found as $u_{zi} = \sqrt{(2kn_j)^2 - \mathbf{u}_\perp^2}$. This simple exchange of the main wave and medium parameters does not dramatically change the time of computation for the direct and the inverse problems.

Finally, we should emphasize that all methods based on the focusing effects allow us significantly decrease the obstructive impact of multiple interactions occurring in the multilayered, inhomogeneous, and irregular media and increase the role of the dominant mechanisms of wave radiation interaction with the material or any object. Within the basis of radio tomography lies the idea of radiation focusing creation as a result of the interference of multiple partial waves. The main algorithms of the inverse problem solution and reconstruction of the objects' images, presented in Sections 2.2 through 2.4, are fast enough and can be realized in real-time timescales.

3.2 Method of Refocusing

A group focusing method allows an increase in the spatial resolution of the test objects or inhomogeneous medium. A further increase in spatial resolution can be undertaken, and we can account for the possibility of transferring from multi-angle measurements of distribution of the locational response of the object (or medium) to the equivalent spectrum of the spatial frequencies. With the help of the Fourier transform, this spectrum is directly related to the spatial distribution of a wave field at some of the equivalent transverse apertures (as seen from Figure 3.8). Reconstructing a spatial field distribution in such a manner, we can take a technology of synthesis of the large aperture and focus the recorded field at the arbitrary given distances located within the limits of Fresnel's diffraction zone. Let us consider, step by step, how this procedure can be done mathematically, based on results discussed in references [9, 13].

We denote by $\Gamma(\mathbf{r}_\perp)$, $\mathbf{r}_\perp = (x, y)$ the transverse distribution of current over the limited aperture of the antenna system, shown in Figure 3.8.

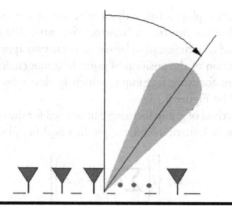

Figure 3.8 Directivity pattern and equivalent aperture of antenna.

The Fourier transform of this distribution gives the corresponding spectrum of spatial frequencies $\mathbf{u}_\perp = (u_x, u_y)$ of the radiated field in the transverse direction:

$$\tilde{\Gamma}(\mathbf{u}_\perp) = \int_{-\infty}^{\infty}\int_{-\infty}^{\infty} \Gamma(\mathbf{r}_\perp)\exp\{i(\mathbf{u}_\perp \cdot \mathbf{r}_\perp)\}d\mathbf{r}_\perp \qquad (3.40a)$$

and, conversely, the inverse Fourier transform gives the field distribution over the antenna's aperture from the spectrum of spatial frequencies:

$$\Gamma(\mathbf{r}_\perp) = \frac{1}{(2\pi)^2}\int_{-\infty}^{\infty}\int_{-\infty}^{\infty} \tilde{\Gamma}(\mathbf{u}_\perp)\exp\{-i(\mathbf{u}_\perp \cdot \mathbf{r}_\perp)\}d\mathbf{u}_\perp \qquad (3.40b)$$

Using radiation with the wave number k, the spatial frequency can be related to the wave direction, defined by angle variables, as follows:

$$u_x = k\sin\theta\cos\varphi, u_y = k\sin\theta\cos\varphi, u_z = k\cos\theta$$

Then we get: $d\mathbf{u}_\perp = k^2\cos\theta(\sin\theta d\theta d\varphi) = ku_z d\Omega$, where $\sin\theta d\theta d\varphi = d\Omega$ is the element of the spatial angle measured in *steradians*. We can then rewrite Equation 3.40b as:

$$\Gamma(\mathbf{r}_\perp) = \int_{-\infty}^{\infty}\int_{-\infty}^{\infty} \tilde{\Gamma}(\mathbf{u}_\perp)ku_z\exp\{-i(\mathbf{u}_\perp \cdot \mathbf{r}_\perp)\}d\Omega$$
$$= \iint R(\theta,\varphi)\exp\{-i(\mathbf{u}_\perp \cdot \mathbf{r}_\perp)\}d\Omega \qquad (3.41)$$

The directivity function of the antenna is defined by:

$$R(\theta,\phi) = \tilde{\Gamma}(\mathbf{u}_\perp)ku_z \qquad (3.42)$$

After the procedure of normalization at the maximum of its module, this function defines the directivity pattern of the antenna, that is, the distribution of the radiated field pattern over angle variables θ and φ in the elevation and azimuth domains, respectively. Equation 3.42 describes the relation between the directivity function of the spatial field distribution in the joint azimuth and elevation domain and the spectrum of its spatial frequencies. By knowledge of the angle field distribution, we can reconstruct the spectrum of spatial frequencies by the following normalization:

$$\tilde{\Gamma}(\mathbf{u}_\perp) = R(\theta,\phi)/ku_z \qquad (3.43)$$

Now, by the use of the inverse Fourier transform, we can immediately obtain the transverse current distribution over the antenna aperture $\Gamma(\mathbf{r}_\perp)$. By knowledge of the current distribution over the aperture, we can focus it into the given point of space **r**, as shown by Figure 3.9.

For this purpose, it is enough to use the in-phase summation of all values of the current with respect to the point **r**. This procedure can be written as:

$$E(\mathbf{r}) = \int_{-\infty}^{\infty}\int_{-\infty}^{\infty} \Gamma(\mathbf{r}_\perp)ku_z \exp\{-ik|\mathbf{r}_\perp - \mathbf{r}|\}d\mathbf{r}_\perp \qquad (3.44)$$

which, for the case of radio-location applications, when the transmitter and receiver are located at the same spatial point, can be rewritten as:

$$E(\mathbf{r}) = \int_{-\infty}^{\infty}\int_{-\infty}^{\infty} \Gamma(\mathbf{r}_\perp)ku_z \exp\{-2ik|\mathbf{r}_\perp - \mathbf{r}|\}d\mathbf{r}_\perp \qquad (3.45)$$

If, at the focusing point, there exists real inhomogeneity, the degree of its response will be larger with an increase of the dimensions of this tested inhomogeneity. If the inhomogeneity is absent at the focusing point, the response will be close to zero. Changing the position of the focusing point, it may be scanned over a whole sounding space and, finally, we obtain the tomogram of the inhomogeneous medium.

The procedure of refocusing described is equivalent to the use of the method of aperture synthesis, and can usually be realized by using computer simulation algorithms [1–14]. The dimensions of the transverse spatial resolution of inhomogeneities can essentially be improved by use of the following procedure. In the first approximation, the spatial (transverse) resolution, achieved in the region of focusing by use of the radio-location method with SAR, can be estimated as:

$$\Delta x = 2\lambda z/2D \qquad (3.46)$$

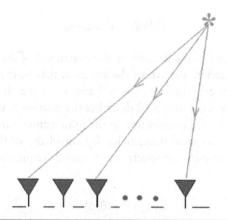

Figure 3.9 Synthetic and focused radiation at a given point.

where:
- λ is the wave length of the radiation
- x is the distance from the antenna aperture to the focusing point
- $2D$ is the size of the synthesis aperture

This estimation is determined by the distance between the first zero of the antenna pattern.

We note that this procedure is valid only within a zone of Fresnel diffraction (i.e., the *near field zone*), which is defined by the following constraint:

$$z \leq z_D \equiv D^2 / \lambda \qquad (3.47)$$

Here, z_D is the diffraction length of the wave beam with radius D, which, in the case under consideration, determines the radius of aperture [13]. In other words, z_D is the distance within the limits of which the phase difference from the center and from the edge of the aperture to the point of partial waves' summation are enough to be determined experimentally. Usually, this zone is called in the literature the *Fresnel zone of diffraction* [15]. For example, for the wave length of radiation $\lambda = 0.3$ m ($f = 1$ GHz) and for a size of aperture $D = 3$ m, the value of z_D, according to Equation 2.66, equals 30 m. In contrast, for $\lambda = 0.1$ m ($f = 3$ GHz) and for the same aperture size, this distance increases to $z_D = 90$ m, and for $\lambda = 0.04$ m ($f = 7.1$ GHz) and for $D = 1.5$ m, we get $z_D = 53$ m. For such an inhomogeneous medium as a forest (see Chapter 10) the latter distance covers the whole region of the exponential attenuation of the radiated field.

3.3 Single-Side Monostatic Tomography of Non-Transparent Objects

Let us consider the situation where sounding of a non-transparent object is carried out at an arbitrary frequency and along some plane S. We will use here the method discussed in Section 3.1.4 (see Equation 3.18). We will rewrite, following reference [14], the result of such a focusing procedure in the following form [13, 14]:

$$U(\mathbf{r}_F) = \sum_m \sum_n M(\mathbf{r}_F, \mathbf{r}_m, \mathbf{r}_{0n}) E(\mathbf{r}_m, r_{0n}) = \iint_S M(\mathbf{r}_F, \mathbf{r}_s) E(\mathbf{r}_s) ds \quad (3.48)$$

where the focusing function is given in the form of

$$M(\mathbf{r}_F, \mathbf{r}_s) = \exp\{-2ik|\mathbf{r}_F - \mathbf{r}_s|\}$$

Now, changing the order of integration of Equation 3.48, we get:

$$U(\mathbf{r}_F) = \iiint_V F(\mathbf{r}') Q(\mathbf{r}', \mathbf{r}_F) dx' dy' dz' \quad (3.49)$$

Here, for a large enough area of focusing synthesis ($k^2 S \gg 1$), the TF of the system can be presented in a simpler form, similar to what was done in Section 3.1.5; that is, in the form of

$$Q(\mathbf{r}', \mathbf{r}_F) \equiv (2k)^2 \iint_S M(\mathbf{r}_F, \mathbf{r}_s) G_0^2(\mathbf{r}' - \mathbf{r}_s) ds \approx Q(\mathbf{r}' - \mathbf{r}_F)$$

$$= Q_\perp(\mathbf{r}'_\perp - \mathbf{r}_{\perp F}, f) \exp\{2ik(z' - z_F)\} \quad (3.50)$$

Here, $Q_\perp(\mathbf{r}'_\perp - \mathbf{r}_{\perp F}, f)$ is the TF in the transverse direction of wave radiation defined by Equation 3.29. This simplified procedure can be proved by the physical fact that in the region of focusing, the wave front of the radiated wave has the form of the collimated beam, as shown in Figure 2.9. In the case where precise localization is achieved, we can assume that the inverse focusing function can be presented in the form of:

$$U(\mathbf{r}_F) \equiv U(\mathbf{r}_{\perp F}, z_F) = \exp\{2ik[Z_s(\mathbf{r}_{\perp F}) - z_F]\} \iint Q_\perp(\mathbf{r}_{\perp F}, f) dx dy \quad (3.51)$$

Here, as in Section 3.1.5, $\mathbf{r}_\perp = (x, y)$ and $|\mathbf{r}_\perp| = \sqrt{x^2 + y^2}$ are the vector and its absolute value in the direction across the wave radiation. Equation 3.51, in fact, is the solution of the inverse problem of reconstruction (imaging) of the form of the illuminated part of a non-transparent object or inhomogeneity. It is given by the following expression:

$$Z_s(\mathbf{r}_{\perp F}) - z_F = \frac{1}{2k} \arg \frac{U(\mathbf{r}_{\perp F}, z_F)}{U(0, z_F)} \qquad (3.52)$$

The much higher the sounding wave frequency, the closer the localization of radiation will be during the focusing procedure in the direction across the wave propagation and a form of the non-transparent object surface will be reconstructed or recovered more precisely.

We should notice two important issues:

1. Because the function arg(x) in Equation 3.52 is a multivalued function, it is necessary to reconstruct a full phase over the neighboring segments of the phase profile.
2. Better results can be obtained after averaging over multi-frequency measurements.

We now present an example of computer simulation of the reconstruction of the surface of the parabolic reflector by use of the Kirchhoff approximation, describing the reflection of the plane wave from the parabolic reflector. In Figure 3.10a, the distribution of one of the quadrature components of the field at the receiving aperture is shown for a frequency of 1 GHz, which was located above the ideal parabolic reflector (without any defect) with a diameter of 5 m. The focus was located at a distance of 1.8 m from the apex of the dish, and the sounding aperture was shifted upward to 40 cm from the top.

As follows from the computer simulation of the ideal parabolic reflector (see Figure 3.10a), the field reflected from such a dish has good axis symmetry. This symmetry is totally broken when any deformation exists (see Figure 3.11).

The solution of the inverse problem (Equation 3.52) allows us to reconstruct the dish with the defect and to localize its position and recover its form (shown by the arrow in Figure 3.11), as clearly seen from the results of the simulation shown in Figure 3.10b. Hence, the proposed methodology based on the Kirchhoff approximation allows us to reconstruct precisely enough the arbitrary inhomogeneity or any object located in the tested medium, whether homogeneous or not.

3.4 Diffraction Tomography Techniques

The research in this section, which is based on results obtained and discussed in references [18–29], describes the problem of reconstructing an unknown object on the basis of the field scattered and diffracted by this object. In the literature, this problem is known as *inverse scattering* and *diffraction imaging* or the *diffraction tomography technique* [18–29]. As shown in Sections 3.1.1 through 3.1.7, the problem of reconstruction or recovering any target or foreign object can be exactly solvable by use of the Born approximation [18–23], according to which, due to the

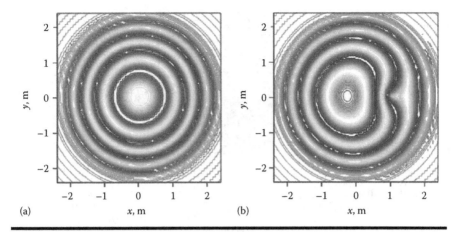

Figure 3.10 Distribution of one of the quadrature components of a reflected wave field in aperture of parabolic reflector antenna (a) in absence of and (b) in presence of deformation.

weak response of the target on the incident electromagnetic wave, an accurate image, localization, and form of the object can be obtained [24–27]. At the same time, as was shown and mentioned in Sections 3.1.1 through 3.1.7, in the multidimensional case and in complicated clutter conditions, the solution of the Born approximation (or single scattering, without and with diffraction) is not unique enough; even the uniqueness of the solution does not help us to find it in practice.

The key problem that arises is that even the direct problem (evaluation of the field scattered by a known scatterer) is essentially non-linear; to obtain a reasonable analytical solution for the direct problem, it is usually confined to the first (linear) term in the whole series of expansions of the first-, second-, and higher-order terms of the main scattered field perturbations introduced by a target under detection. This leads to the Born approximation, when the perturbation approach is applied to the field itself or to the Rytov approximation, when this is performed for the complex phase.

These linearized approximations are valid for small (smooth) weak scatterers, and fail when strong scattering is observed. The Rytov approximation, being non-linear with respect to the wave field, accounts for multiple scattering effects and, in this sense, is more advantageous compared with the Born approximation.

However, this gain is not sufficient and, moreover, approximations are utilized, and for the far field measurements (the evanescent waves are neglected), unwrapping appears in the case when the frequency domain measurements are performed. To improve the Rytov approximation and overcome, at least partially, the above-mentioned difficulties, it was proposed to evaluate the wave intensity of the scattered field by using an essentially non-linear propagator [28, 29]. This would allow the reconstruction of strong scatterers with better accuracy, and the removal of the problem of phase estimation.

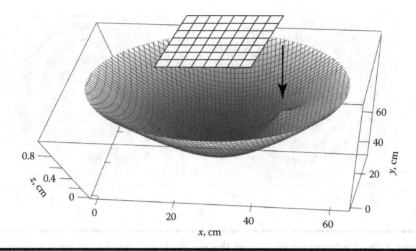

Figure 3.11 Reconstructed shape of parabolic reflector with local deformation (indicated by arrow) and aperture of sounding antenna.

Additionally, we suppose that the proposed approach, called *diffraction tomography*, will allow us to estimate high spatial frequencies, which are related to the evanescent waves under single scattering conditions; that is, to attack the super-resolution problem. The drawback of the approach is that the result of reconstruction is not the scattering potential itself, but the static structure function of the object. However, this seems to be not such a serious problem, since in most practical cases it is possible to transfer from the structure function to the scattered function, and to that responsible for diffraction rays also.

3.4.1 Diffraction Tomography Method Based on Rytov Approximation

The basic equation of diffraction tomography can be obtained as a result of the solution of a scattering inverse problem of a plane electromagnetic wave diffracted by the object under investigation with a radius of $2k$, where $k = 2\pi/\lambda$ is the wave number. Following references [18, 24–27], we consider an object characterized by a refractive index $n(\mathbf{r}) = 1 + n_\delta(\mathbf{r}) \equiv 1 + f$, where f is equal to zero outside the refracting object. The incident plane harmonic $(\exp(-i\omega t))$ wave $U_I(\mathbf{r}) = \exp[ik(\boldsymbol{\theta} \cdot \mathbf{r})]$ is scattered by the object; $\boldsymbol{\theta}$ is a unit vector pointing in the direction of the wave propagation; $n_\delta(\mathbf{r})$ is the deviations of the refractive index; $k = \omega/c$; ω is the radiation frequency; c is the speed of light. In the case of direct scattering, the total field $U = U_I + U_S$ (where $U_S(\mathbf{r})$ is a scattered wave) satisfies the given wave equation

$$\Delta U + k^2(1+f)^2 U = 0 \qquad (3.53)$$

and the boundary condition, that is, the condition of radiation, is at infinity. Here, Δ is the Laplace operator. The scattered field U_S may be found using Equation 3.53 in the first-order Born approximation. So, for the scattered field, $U_S^{(1)}(\mathbf{r})$ may be written in an integral form [18, 24–27]:

$$U_S^{(1)}(\mathbf{r}) = \int G(\mathbf{r} - \mathbf{r}') Q(\mathbf{r}') U_I(\mathbf{r}') d\mathbf{r}' \qquad (3.54)$$

where:
$G(\mathbf{r} - \mathbf{r}') = (i/4) H_0^{(1)}(k\,|\mathbf{r} - \mathbf{r}'|)$ is a Green function
$H_0^{(1)}$ is a Hankel function of the first kind of zero order; $Q(r) = k^2(2f + f^2)$

The integral representation (Equation 3.54) of the scattered field $U_S^{(1)}(\mathbf{r})$ is permitted under the condition $U_S \ll U_I$.

In the inverse scattering problem, function f should be found with a known scattered field U_S. The solution of such a problem using Equation 3.54 allows us to obtain the main equation of diffraction tomography [18, 24–27]. After that, we may find functions $Q(\mathbf{r})$ and $f(\mathbf{r})$, if we know (after calculations or experimental results) $U_S^{(1)}$ or $U_I(k\Phi_R)$. In our case, we are dealing with the Rytov approximation, which can describe the situation where both $U_S \ll U_I$ and $U_S \geq U_I$; that is, it can describe both weak and strong inhomogeneities and allow the designers to obtain information about the amplitude and phase of waves scattered and diffracted from the test objects.

3.4.2 Method of Diffraction Tomography Based on Feynman's Path Integral

In this section, we present the non-linear approach of diffraction tomography based on Feynman's path integral evaluation, which allows us to obtain better imaging of the shape and dimensions of any foreign object embedded in clutter conditions. To show the efficiency of such a technique, in references [18, 29] it was shown experimentally that when using the proposed approach of reconstruction of objects, better imaging of any object hidden in the clutter could be obtained, which finally allows the shape and dimensions of the test object to be clearly seen.

Now, to enter deeper into the subject, we start with the reduced Helmholtz equation describing the propagation and scattering of scalar time-harmonic waves in an inhomogeneous medium. Following references [28, 29], we can then define the Green function as follows:

$$\nabla^2 G(\mathbf{R} \,|\, \mathbf{R}_0) + k^2 \left[1 + \tilde{\varepsilon}(\mathbf{R})\right] G(\mathbf{R} \,|\, \mathbf{R}_0) = -\delta(\mathbf{R} - \mathbf{R}_0) \qquad (3.55)$$

where:
- **R** is the position vector in m-dimensional space ($m = 2$ or 3)
- k is the wave number associated with a homogeneous medium
- $\varepsilon(\mathbf{R}) = 1 + \tilde{\varepsilon}(\mathbf{R})$ is the random permittivity distribution

We suppose that while ε is a real function, k contains an infinitesimally small positive imaginary part that provides the convergence of some integrals appearing in the course of the work. Equation 3.55 is known to serve as a reasonable model for acoustic wave propagation, and also for some electromagnetic problems in which the polarization effects can be neglected.

Next, we introduce, according to references [28, 29], an auxiliary parabolic

$$2ik\partial_\tau g + \nabla^2 g + k^2 \tilde{\varepsilon}(\mathbf{R}) g(\mathbf{R},\tau|\mathbf{R}_0,\tau_0) = 0, \quad \tau > \tau_0 \quad (3.56)$$

with the initial condition:

$$g(\mathbf{R},\tau_0|\mathbf{R}_0,\tau_0) = \delta(\mathbf{R}-\mathbf{R}_0)$$

Then, the Green function $G(\mathbf{R}|\mathbf{R}_0)$ is defined through the solution of the latter equation as

$$G(\mathbf{R}|\mathbf{R}_0) = \frac{i}{2k}\int_{\tau_0}^{\infty} d\tau \exp\left[i\frac{k}{2}(\tau-\tau_0)\right] g(\mathbf{R},\tau|\mathbf{R}_0,\tau_0). \quad (3.57)$$

The generalized parabolic equation (Equation 3.56) for the normalized Green function $g(\mathbf{R},\tau|\mathbf{R}_0,\tau_0)$ coincides with the non-stationary Schrödinger equation in quantum mechanics. Using this analogy, the solution of Equation 3.57 can be presented via Feynman's path integral according to references [18, 28, 29]:

$$g(\mathbf{R},\tau|\mathbf{R}_0,\tau_0) = \int_{\mathbf{R}(\tau_0)=\mathbf{R}_0}^{\mathbf{R}(\tau)=\mathbf{R}} D\mathbf{R}(t) \exp\{iS[\mathbf{R}(t)]\} \quad (3.58)$$

where the integration $\int D\mathbf{R}(t)$ in the continuum of possible trajectories is interpreted as a sum of contributions of arbitrary paths over which a wave propagates from the point $\mathbf{R}_0$ at the moment τ_0 to the point R at the moment τ, and the functional

$$S[\mathbf{R}(t)] = \frac{k}{2}\int_{\tau_0}^{\tau} dt \left\{ \left[\dot{\mathbf{R}}(t)\right]^2 + \tilde{\varepsilon}[\mathbf{R}(t)] \right\} \quad (3.59)$$

can be related to the phase accumulated along the corresponding path. To simplify the propagator obtained, we use a self-consistent two-step procedure as follows. First, to dispose of the integral over τ, which is difficult to invert, we present the

Green function in Equation 3.57 as a product of two factors, the first corresponding to free space and the second related to the effects of the scatterer. Equation 3.57 is then presented in the following form [18, 28, 29]:

$$G(\mathbf{R}|\mathbf{R}_0) = \frac{i}{2k}\int_{\tau_0}^{\infty} d\tau \exp\left[i\frac{k}{2}(\tau-\tau_0)\right] g_0(\mathbf{R},\tau|\mathbf{R}_0,\tau_0)$$
$$g_\varepsilon(\mathbf{R},\tau|\mathbf{R}_0,\tau_0). \tag{3.60}$$

Evaluating the latter integral asymptotically allows us to obtain

$$G(\mathbf{R}|\mathbf{R}_0) = G_0(\mathbf{R}|\mathbf{R}_0) G_\varepsilon(\mathbf{R}|\mathbf{R}_0), \tag{3.61}$$

where the inhomogeneous factor $G_\varepsilon(\mathbf{R}|\mathbf{R}_0)$ in the 3-D case is given by the series in derivatives of the generalized parabolic equation solution:

$$G_\varepsilon(\mathbf{R}|\mathbf{R}_0) = g_\varepsilon(\mathbf{R},\tau_0+L|\mathbf{R}_0,\tau_0) + i(L/2k) g_\varepsilon''(\mathbf{R},\tau_0+L|\mathbf{R}_0,\tau_0)$$
$$+ (L/2k^2) g_\varepsilon'''(\mathbf{R},\tau_0+L|\mathbf{R}_0,\tau_0) + \ldots \tag{3.62}$$

Keeping only the first term in this series and neglecting all the derivatives we obtain

$$G_\varepsilon(\mathbf{R}|\mathbf{R}_0) \approx g_\varepsilon(\mathbf{R},\tau_0+L|\mathbf{R}_0,\tau_0) \tag{3.63}$$

Though the path integral cannot be evaluated exactly, we may propose a perturbation approach, which will allow us to obtain its value with reasonable accuracy. A propagator obtained in such a way was tested recently in problems of wave propagation and localization in random media [18, 28, 29]. The results have been compared with those obtained in the framework of the naive Rytov approximation. In particular, it was found that the improved propagator applied in references [18, 28, 29] to the mean (coherent) field leads to the result coinciding exactly with that given by the Bourret approximation for the Dyson equation [30]. At the same time, if the Rytov approximation is applied directly to the Helmholtz equation without any embedding procedure, this leads to a result that not only differs from that of the Bourret approximation, but, being divergent with the distance from the source, contradicts the energy conservation principle [18, 28, 29].

References

1. Yakubov, V. P. and M. L. Masharuev, Method of double focusing for microwave tomography, *Microwave Optical Technologies Letters*, vol. 13, no. 4, 1996, pp. 187–189.
2. Pimchuk, R., V. P. Yakubov, D. Ya. Sukhanov, et al. Spatial-temporal synthesis in microwave tomography, *Izv. Vuzov, Physics*, vol. 33, no. 9/2, 2010, pp. 108–109 (in Russian).

3. Sukhanov, D. Ya. and V. P. Yakubov, Usage of linear-frequency-modulated signals in the three-dimensional radio tomography, *Journal of Technical Physics (JTP)*, vol. 80. no. 4, 2010, pp. 115–119 (in Russian).
4. Yakubov, V. P., D. Y. Suhanov, A. S., Omar, V. P. Kutov,and N. G. Spiliotis, New fast SAR method for 3-D subsurface radiotomography, *Proceedings of the Tenth International Conference Ground Penetrating Radar, GPR 2004*, Eds, E. Slob, A. Yarovoy, J. B. Rhebergen, Delft, the Netherlands: Delft University of Technology, pp. 103–106.
5. Yakubov, V. P., E. D. Tel'pukhovskii, and G. M. Tsepelev, Radio locational tomography, *Atmospheric and Oceanic Optics*, vol. 49, no. 12, 2006, pp. 1081–1086 (in Russian).
6. Yakubov, V. P., K. G. Skolyarchk, R. V. Pinchuk, et al., Radiowave tomography of hidden objects for security systems, *Izv. Vuzov, Physics*, vol. 51, no. 10, 2008, pp. 63–79 (in Russian).
7. Yakubova, O. V., E. D. Tel'pukhovskii, and V. P. Yakubov, Focusing of ultra-wideband radiation by lens from artificial dielectric, *Izv. Vuzov, Physics*, vol. 48, no. 6, 2005, pp. 153–154 (in Russian).
8. Yakubov, V. P., E. D. Tel'pukhovskii, and G. M. Tsepelev, Radio locational tomography of inhomogeneous media, *Izv. Vuzov, Physics*, vol. 49, no. 9, 2006, pp. 20–24.
9. Klokov, A., V. Yakubov, and M. Sato, Ultra-wideband radiolocation of a forest, *Proceedings of the 2010 IEICE General Conference*, March 16–19, 2010, Sendai, Japan, Institute of Electronics, Information and Communication Engineers, USA, pp. 298–299.
10. Cloude, S. R., P. D. Smith, A. Milne, D. Parkes, and K. Trafford, Analysis of time domain UWB radar signals, *SPIE, Pulse Engineering*, vol. 1631, 1992, pp. 111–122.
11. Colton, D. and R. Kress, *Inverse Acoustic and Electromagnetic Scattering Theory*, New York: Springer, 1998.
12. Hamran, S., D. T. Gjessing, J. Hjelmstad, and E. Aarholt, Ground penetrating synthetic pulse radar: dynamic range and modes of operation, *J. Appl. Geophys.*, vol. 33, 1995, pp. 7–14.
13. Yakubov, V. P., S. E. Shipilov, D. Ya. Sukhanov, and A. V. Klokov, *Radiowave Tomography – Achievements and Perspectives*, Tomsk: NTL, 2016 (in Russian).
14. Yakubov, V. P. and S. E. Shipilov, *Inverse Problems of Radiophysics*, Tomsk: NTL, 2016 (in Russian).
15. Blaunstein, N. Sh., M. B. Sergeev, and A. P. Shepeta, *Applied Aspects of Electrodynamics*, St. Petersburg: Agaf, 2016 (in Russian).
16. Meincke, P., Efficient calculation of Born scattering for fixed-offset ground-penetrating radar surveys, *IEEE Geoscience and Remote Sensing Letters*, vol. 4, no. 1, pp. 88–92, 2007.
17. Sukhanov, D. Ya. and E. S. Berzina, Recovering of current distribution in the plate objects via distance measurements of the vector of magnetic induction, *Izv. Vuzov, Physics*, vol. 55, no. 8/2, 2012, pp. 163–167 (in Russian).
18. Blaunstein, N., Ch. Christodoulou, and M. Sergeev, *Introduction to Radio Engineering*, Boca Raton, FL: CRC Press, Taylor & Francis, 2017.
19. Tie Jun Cui and Weng Cho Chew, Diffraction tomographic algorithm for detection of three-dimensional objects buried in a lossy half-space, *IEEE Transactions on Antennas and Propagation*, vol. 50, no. 1, pp. 42–49, 2002.

20. Lin-Ping Song, Qing Muo Liu, Fenghua Li, and Zhong Qing Zhang, Reconstruction of three-dimensional objects in layered media: Numerical experiment, *IEEE Transactions on Antennas and Propagation*, vol. 53, no. 4, 2005, pp. 1556–1561.
21. Meincke, P., Efficient calculation of Born scattering for fixed-offset ground-penetrating radar surveys, *IEEE Geoscience and Remote Sensing Letters*, vol. 4, no. 1, 2007, pp. 88–92.
22. Chan, L. C., D. L. Moffat, and L. Peters, Subsurface radar target imaging estimates, *Proc. IEEE*, vol. 67, 1979, pp. 991–1000.
23. Lo Monte, L., D. Erricolo, F. Soldovieri, and M. C. Wicks, Radio frequency tomography for tunnel detection, *IEEE Transactions on Geoscience and Remote Sensing*, vol. 48, no. 3, 2010, pp. 1128–1137.
24. Gavrilov, S. P. and A. A. Vertiy, Detection of the cylindrical object buried in dielectric half-space by using wave interference of the two different frequencies, *4th International Conference on Millimeter and Submillimeter Waves and Applications*, July 20–23, 1998, San Diego, Calfornia, USA, 5 pages.
25. Vertiy, A. A., S. P. Gavrilov, and G. Gençay, Microwave tomography systems for investigation of the ware structure, *4th International Conference on Millimeter and Submillimeter Waves and Applications*. July 20–23, 1998, San Diego, California, USA, 5 pages.
26. Gavrilov, S. P. and A. A. Vertiy, Application of tomography method in millimeter wavelengths band, I. Theoretical, *International Journal of Infrared and Millimeter Waves*, vol. 18, no. 9, 1997, pp. 1739–1760.
27. Vertiy, A. A. and S. P. Gavrilov, Application of tomography method in millimeter wavelengths band, II. Experimental, *International Journal of Infrared and Millimeter Waves*, vol. 18, no. 9, 1997, 1761–1781.
28. Samelsohn, G. and R. Mazar, Path-integral analysis of scalar wave propagation in multiple-scattering random media, *Physical Review*, vol. E 54, 1996, pp. 5697–5706.
29. Blaunstein, N., Recognition of foreign objects hidden in clutter by novel method of diffraction tomography, *Proceedings of IEEE Conference on Radar Applications*, Kiev, Ukraine, 2010, 5 pages.
30. Blaunstein, N. and C. Christodoulou, *Radio Propagation and Adaptive Antennas in Wireless Communication Links – Terrestrial, Atmospheric and Ionospheric*, Chapter 3, Wiley Interscience, Hoboken, New Jersey, USA, 614 pages.

Chapter 4

Low-Frequency Magnetic and Electrostatic Tomography

Vladimir Yakubov, Sergey Shipilov,
Dmitry Sukhanov, and Andrey Klokov

Contents
4.1   Low-Frequency Magnetic Tomography 79
4.2   Impedance Electrostatic Tomography Methods 84
References 86

4.1 Low-Frequency Magnetic Tomography

The method of magnetic tomography deals with low-frequency magnetic fields and is based on its high penetration of electrically conductive media. Therefore, this method is usually used for diagnostics of electrically conductive objects or for detecting fully conductive objects located behind dielectric obstructions. Moreover, weakly varying magnetic fields pass through metallic obstructions, allowing definition of the objects behind them [1–6]. However, the components of magnetic fields have a narrow spatial spectrum far from the source and, therefore, have poor localization of the magnetic field at large ranges from the source, leading to a low resolution of the recovered tomographic images.

Let us consider the method of reconstruction of the current density distribution according to the measurements of the only z-component of the vector of the magnetic field (the so-called *1-D problem* [3, 4]). If $j(\mathbf{r}_1)$ is the vector of spatial distribution of the eddy current (Foucault current) density in volume V_1, then, according to the well-known Stratton–Chu formula, the magnetic field strength vector, created by this current, is defined by the following integral:

$$\mathbf{H}(\mathbf{r}) = \iiint_{V_1} \left[\mathbf{j}(\mathbf{r}_1) \times \nabla_1 G_0(\mathbf{r}_1 - \mathbf{r}) \right] d^3 \mathbf{r}_1 \qquad (4.1)$$

where, as in Section 2.3, $G_0(\mathbf{r}) = \exp(ik|\mathbf{r}|)/(4\pi|\mathbf{r}|)$ is a Green function of the point source in the background medium, which, within the frame of the quasi-static approximation, can be rewritten as $G_0(\mathbf{r}) = 1/(4\pi|\mathbf{r}|)$. This equation is similar to Equation 2.7 (see Chapter 2), introduced to describe wave tomography. If measurements of the magnetic field vector $\mathbf{H}(\mathbf{r})$ are carried out in any plane perpendicular to the axis Oz (see Figure 4.1), the normal component of the magnetic field vector equals $H(\mathbf{r}) = \mathbf{e}_z \cdot \mathbf{H}(\mathbf{r})$.

Here, $\mathbf{e}_z$ is the unit vector along the normal axis Oz, and can be defined via the components of the eddy currents that are parallel to the plane of measurements, that is,

$$H(\mathbf{r}) = \mathbf{e}_z \iiint_{V_1} \left[\mathbf{j}_\perp(\mathbf{r}_1) \times \nabla_1 G_0(\mathbf{r}_1 - \mathbf{r}) \right] d^3 \mathbf{r}_1 \qquad (4.2)$$

Here, it is assumed that the normal components of the current do not contribute to the normal components of the magnetic field. Selection in such a manner of the normal components of the magnetic field is assumed due to its easier measurements.

Equation 4.2 can be rewritten as:

$$H(\mathbf{r}) = \mathbf{e}_z \nabla \times \iiint_{V_1} \mathbf{j}_\perp(\mathbf{r}_1) \nabla_1 G_0(\mathbf{r}_1 - \mathbf{r}) d^3 \mathbf{r}_1 \qquad (4.3)$$

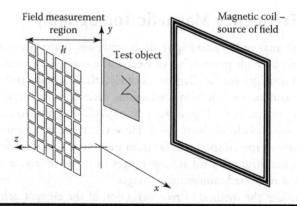

Figure 4.1 Measurement scheme.

In the particular case of the constant current I, floating along the circular contour L, we can obtain a simple expression:

$$H(\mathbf{r}) = I \oint_L \mathbf{e}_z \frac{[(\mathbf{r}_1 - \mathbf{r}) \times d\mathbf{r}_1]}{4\pi(\mathbf{r}_1 - \mathbf{r})^3} \qquad (4.4)$$

The closed contour L can be divided into the sum of straight linear conductors. Then, Equation 4.4 yields

$$H(\mathbf{r}) = I \sum_j \oint_{L_j} \mathbf{e}_z \frac{[(\mathbf{r}_1 - \mathbf{r}) \times d\mathbf{r}_1]}{4\pi(\mathbf{r}_1 - \mathbf{r})^3} \qquad (4.5)$$

Each of these integrals can be calculated analytically. In Figure 4.2, the result of the computations of the magnetic field distribution is shown for the case of two current rectangular frames shifting diagonally with respect to each other.

It is seen that the magnetic field created by such current frames has a distributed character, but the weight center of this distribution is co-planed with that of the frame system. This result was proved by experimental data.

Let us now consider an *inverse problem*—reconstruction of the eddy currents' distribution based on the knowledge of the measured magnetic field distribution. We return to the geometry presented in Figure 4.1, where the point of observation lies at distance z from the plane xOy. For $\mathbf{r} = (x, y, z)$ and $\mathbf{r}_\perp = (x, y)$, the 2-D spectrum of the spatial frequencies can be presented as:

$$H(\mathbf{\kappa}_\perp, z) = \iiint_{V_1} H(\mathbf{r}_\perp, z) \exp(-i\mathbf{\kappa}_\perp \cdot \mathbf{r}_\perp) d^3 r_\perp \qquad (4.6)$$

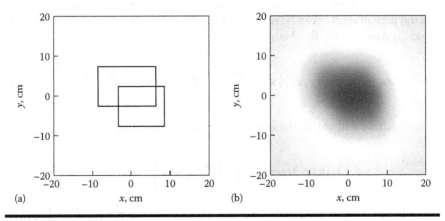

Figure 4.2 Simulation of the magnetic field of two current loops: (a) simulated object; (b) z-component of the magnetic field at a distance of 5 cm.

In accordance with Weyl's formula, introduced in Section 2.4, Equation 4.6 can be rewritten as

$$H(\kappa_\perp, z) = -\frac{\exp(i\kappa_z z)}{2\kappa_z} \mathbf{e}_z \cdot [\kappa \times \mathbf{j}_\perp(\kappa)] \quad (4.7)$$

Here, $\mathbf{j}_\perp(\kappa) = \iiint_{V_1} \mathbf{j}_\perp(\mathbf{r}_1) \exp(-i\kappa \cdot \mathbf{r}_1) d^3 \mathbf{r}_1$ is the spectrum of spatial frequencies of the vector of the eddy current density within volume V_1. The vector of spatial frequencies has the following components: $\kappa = (\kappa_x, \kappa_y, \kappa_z)$, $\kappa_\perp = (\kappa_x, \kappa_y)$, $\kappa_z = \sqrt{k^2 - \kappa_x^2 - \kappa_y^2} = \sqrt{k^2 - \kappa_\perp^2}$. If we now suppose that $\kappa_z = \sqrt{k^2 - \kappa_\perp^2} \approx i|\kappa_\perp|$, we finally obtain:

$$H(\kappa_\perp, z) = i \frac{\exp(i\kappa_z z)}{2} \mathbf{e}_z \cdot \left(\frac{\kappa_\perp}{|\kappa_\perp|} \times \mathbf{j}_\perp(\kappa) \right) \quad (4.8)$$

Timing both parts of Equation 4.8 on $\mathbf{e}_z$, we can reconstruct the z-component of the magnetic field vector. Then, using the vector product operation, we obtain:

$$i(\kappa \times \mathbf{e}_z) H(\kappa_\perp, z) \equiv i(\kappa_\perp \times \mathbf{e}_z) H(\kappa_\perp, z)$$
$$= -\frac{\exp(i\kappa_z z)}{2} \mathbf{e}_z \cdot \left[\kappa_\perp \times \left(\frac{\kappa_\perp}{|\kappa_\perp|} \times \mathbf{j}_\perp(\kappa) \right) \right] \quad (4.9)$$

Let us discover the double vector product, while taking note that in the quasi-stationary approximation, the law of charge conservation $\nabla \cdot \mathbf{j}_\perp(\mathbf{r}) = 0$ is valid. In the spectral presentation, it can be rewritten as $i[\kappa \times \mathbf{j}_\perp(\kappa)] \equiv i[\kappa_\perp \times \mathbf{j}_\perp(\kappa)] = 0$. In such an approximation, we finally obtain:

$$i(\kappa_\perp \times \mathbf{e}_z) H(\kappa_\perp, z) = |\kappa_\perp| \frac{\exp(i\kappa_z z)}{2} \mathbf{j}_\perp(\kappa) = |\kappa_\perp| \frac{\exp(-|\kappa_\perp| z)}{2} \quad (4.10)$$

from which we will obtain the solution for a spectrum of the eddy currents:

$$\mathbf{j}_\perp(\kappa) = 2i \exp(|\kappa_\perp| z) \left(\frac{\kappa_\perp}{|\kappa_\perp|} \times \mathbf{e}_z \right) H(\kappa_\perp, z) \quad (4.11)$$

Formally speaking, Equation 4.11 is the solution of the inverse problem. It is only necessary to take from Equation 4.11 the inverse Fourier transform over the spatial frequencies.

However, two circumstances should be taken into account here. First, in the presence of measured noises that always exist, the exponential product in Equation 4.10 leads to non-convergence of the expression in Equation 4.11. Overcoming this drawback can be achieved using Wiener filtering with the regularization procedure, mentioned in Section 1.8, by exchanging $\exp(|\kappa_\perp|z)$ for

$$\frac{\exp(-|\kappa_\perp|z)}{\exp(-2|\kappa_\perp|z)+\alpha}$$

where α is the parameter of regularization, which is usually selected experimentally. Secondly, if the eddy current runs in the plane xOy, $\mathbf{j}_\perp(\mathbf{r}_1)$ should be exchanged for $\mathbf{j}_S(\mathbf{r}_{\perp1})\delta(z_1)$, and then $\mathbf{j}_\perp(\kappa) = \mathbf{j}_S(\kappa_\perp)$. Accounting for all these comments, the solution of the inverse problem can finally be written as:

$$\mathbf{j}_S(\mathbf{r}_\perp) = \frac{1}{(2\pi)^2}\iint \mathbf{j}_S(\kappa_\perp)\exp(i\kappa_\perp \cdot \mathbf{r}_\perp)d\kappa_\perp \qquad (4.12)$$

where

$$\mathbf{j}_S(\kappa_\perp) = 2i\left(\frac{\kappa_\perp}{|\kappa_\perp|}\times \mathbf{e}_z\right)H(\kappa_\perp,z)\frac{\exp(-|\kappa_\perp|z)}{\exp(-2|\kappa_\perp|z)+\alpha} \qquad (4.13)$$

The solution of Equation 4.13 is written in the form of the quasi-static approximation, and is a full analogue of the solution obtained in Section 2.4 by use of the method of tomographic synthesis. It is important to notice that all transformations can be performed by the use of the fast and inverse Fourier transforms (FFT and IFT, respectively). Use of Equation 2.23 from Section 2.4 for the model shown in Figure 4.2 and repeated in Figure 4.3a allows us to reconstruct the given

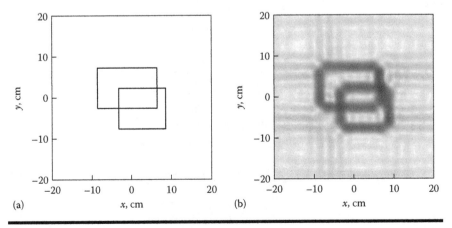

Figure 4.3 (a) Predefined and (b) reconstructed current distribution in the form of rectangular loops: numerical simulation.

distribution of the eddy currents. The distribution of the absolute value of the currents $|\mathbf{j}_S(\mathbf{\kappa}_\perp)|$, taken according to Equation 4.13, is shown in Figure 4.3b.

Taking account of the spreading of the initial distribution of the magnetic field, shown in Figure 4.2b, the obtained distribution of the corresponding eddy currents, presented in Figure 4.3b, is sufficiently correct.

4.2 Impedance Electrostatic Tomography Methods

Another methodology, existing separately in the literature, *impedance tomography* is an example of a method to solve the direct and inverse problems by use of stationary currents in electrostatics; namely, when it is important to recover or reconstruct the distribution of currents or electrical charges in various materials. Impedance tomography is based on measurements of electrical conductivity between a pair of points located at the surface of the current conductive body, which is sounded under multiple angles, as shown in Figure 4.4. These points can be stationary or moving. The sounding instrument is the current passing inside the body under measurements. This current can be considered as a sum of large amounts of elementary currents, that is, $I = \iint_S \mathbf{j} \cdot d\mathbf{s}$, where $\mathbf{j}$ is the vector of current density, and integration is made along any surface surrounding one of two electrodes located at either point 1 or point 2.

According to the differential Ohm's law, the vector $\mathbf{j}$ relates (at each of the electric strength lines) to the surface distribution of the normalized resistivity ρ, or to the normalized conductivity $\sigma = 1/\rho$, as $\mathbf{j} = \sigma \mathbf{E} = \mathbf{E}/\rho$.

At the same time, the vector of the electric field relates to the potential difference (i.e., the voltage U) between the points of placement of electrodes 1 and 2, as [1]:

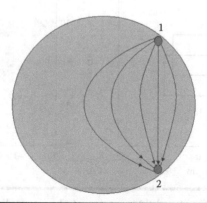

Figure 4.4 Current tubes inside impedance object.

$$U = \int_L \mathbf{E} \cdot d\mathbf{l}$$

Here, integration is made along the arbitrary strength line or a curve connecting point 1 with point 2, including a straight line, as shown in Figure 4.4.

The relations presented here allow us to write, for a measured resistance of the body under testing, the following integral presentation [1]:

$$R_{12} = \frac{U}{I} = \frac{\int_L \mathbf{E} \cdot d\mathbf{l}}{\iint_S \mathbf{j} \cdot d\mathbf{s}} = \int_L \frac{\rho(l)}{\bar{S}(L)} dl \qquad (4.14)$$

Here, $\bar{S}(l)$ is the effective cross-section area (in square meters) of the current tube, which is defined as

$$\bar{S}(L) = \frac{\iint_S \mathbf{j} \cdot d\mathbf{s}}{j(L)} \qquad (4.15)$$

Now, using a theorem of averaging, the integral (Equation 4.14) can be rewritten as

$$R_{12} = \frac{1}{S_{12}} \int_L \rho(l) dl \qquad (4.16)$$

where $S_{12} \equiv \bar{S}(L)$ is the effective cross-section of the current tube, taken at an arbitrary point along the strength lines. This integral equation finally gives

$$F_{12} = R_{12} S_{12} = \int_L \rho(l) dl \qquad (4.17)$$

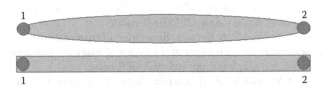

Figure 4.5 Schematic explanation of idea of equalization of cross-section of current tube.

We notice that the task of reconstruction of the non-known function F_{12}, which defines distribution of the resistivity density ρ, relates to the task of shadowing projections described in Sections 2.1 and 2.2.

Only one problem exists here: the dependence of the effective cross-section of the current tube, S_{12}, on the distance d between electrodes. This characteristic plays the role of the equalization function, which allows us to use the task of shadowing projection in solving this problem. By use of this function, the ellipsoidal current tube with a changing cross-section area within it becomes cylindrical with a constant cross-section area. Such a transformation is clearly illustrated by Figure 4.5.

Computations made while accounting for a special conforming transformation show that for a plane homogeneous medium, the effective cross-section area of the current tube changes according to the following law:

$$S_{12} = \frac{\pi \cdot d}{\ln\left[\frac{d}{2} + r^{-1}\sqrt{\left(\frac{d}{2}\right)^2 - r^2}\right]} \approx \frac{\pi \cdot d}{\ln\left[\frac{d}{r}\right]} \qquad (4.18)$$

for $d \gg r$, where r is the radius of the electrodes.

Hence, the equalizing function S_{12} is a monotonic function and its presentation view for different configurations of the measured system can be established with the help of calibration measurements carried out in the mean homogeneous medium.

The above-mentioned theoretical frameworks have found a lot of significant limitations in practical applications. Therefore, the eddy current tomography techniques that have reflected more general physical applications will be the subject of the next chapter.

References

1. Yakubov, V. P. and S. E. Shipilov, *Inverse Problems in Radiophysics*, Tomsk, Russia: Scientific-Technical Literature (STL) Edition Group, 2016.
2. Yakubov, V. P., S. E. Shipilov, D. Ya. Sukhanov, and A. V. Klokov, *Radiowave Tomography: Achievements and Perspectives*, Tomsk, Russia: Scientific-Technical Literature (STL) Edition Group, 2016 (in Russian).
3. Sukhanov, D. Ya., and E. S. Berzina, Recovering of current distribution in the plate objects via distance measurements of the vector of magnetic induction, *Izv. Vuzov, Physics*, vol. 55, no. 8/2, 2012, pp. 163–167 (in Russian).
4. Sukhanov, D. Ya., and M. A. Goncharik, Distance managing by induction currents with help of the system of magnetic coils, *Izv. Vuzov, Physics*, vol. 55, no. 8/2, 2012, pp. 158–162 (in Russian).

5. Sukhanov, D. Ya., and E. S. Berzina, Magnetic introscopy with usage of the grating of sensors of the magnetic field, *Izv. Vuzov, Physics*, vol. 56, no. 8/2, 2013, pp. 23–26 (in Russian).
6. Sukhanov, D. Ya., and M. A. Goncharik, Definition of the form of electro-conductive objects via distance perturbations of the varied magnetic field, *Izv. Vuzov, Physics*, vol. 56, no. 8/2, 2013, pp. 41–43 (in Russian).

Chapter 5
Eddy Current Tomography

Nathan Blaunstein and Alexey Vertiy

Contents

5.1 Overview of Methodology 89
5.2 Methods of Inverse Problem Solution Using Eddy Current Testing 94
5.3 Theoretical Consideration of Diffraction Tomography Approach...............102
5.4 Conclusion..115
References 116

5.1 Overview of Methodology

The non-destructive testing of highly stressed devices is becoming more and more important for new industrial applications. Therefore, new methods have been developed to identify and locate cracks and other material defects.

The effect of eddy currents follows from results described in references [1–19], because the effect used in eddy current testing is the detection of a distorted magnetic field caused by an eddy current displacement due to faults in the material. This can be done by measuring the magnetic flux density on the surface. The methods of eddy current analysis have been considered in many works, namely, in references [1–23].

Thus, in 1968, analytical solutions to eddy current probe-coil problems were developed [1]. Solutions were obtained for axially symmetrical eddy current problems regarding two configurations of wide applicability. In both cases, the eddy currents were assumed to be produced by a circular coil of rectangular cross-section, driven by a constant amplitude alternating current. One solution is for a coil above a semi-infinite conducting slab with a plane surface, covered with a uniform layer of another conductor. The other solution is for a coil surrounding an infinitely long circular conducting rod with a uniformly thick coating of another conductor. The solutions were presented in the form of integrals of first-order Bessel function giving the vector potential, from which the other electromagnetic quantities of interest could be obtained.

The use of boundary conditions in eddy current problems is proposed in references [3–5]. The use of boundary integral equation methods for the solution of some electromagnetic field problems were discussed in reference [3]. Boundary integral formulations using simple-layer potentials, and the Green function, are explained there. This section concludes with the corresponding discussions that attempt to relate boundary integrals to partial differential methods.

The hybrid method, which tries to use both the finite element method (FEM) and the boundary element method (BEM) to resolve the same problem, was proposed in reference [4]. The relative advantages and disadvantages of the hybrid method were explained there. The basic formulation of the given method was stated, and corresponding examples were presented for the 2-D axisymmetric vector and scalar problems. The extension of the hybrid method to 3-D problems was also discussed.

Reference [5] examines the use of impedance boundary conditions (IBC) for the reduction of the field problem encountered in the computation of eddy currents in non-magnetic and magnetic conductors with small penetration depths in a simpler exterior problem. The formulation of the original field problem and the approximate IBC problem in terms of boundary integral equations are developed for 2-D and 3-D linear problems. These formulations are used for analysis of the eddy currents in conducting magnetic cylinders of circular or rectangular cross-section when the object is directed to transverse magnetic time-harmonic fields.

Non-linear methods of analysis in eddy current problems were considered in references [6–9]. Methods of estimating the eddy current losses induced in ferromagnetic materials have been described in reference [6]. A simple 1-D cylinder has been used as the object under investigation. It was shown that a time-harmonic solution of the eddy current problem, in which the effective magnetic permeability is a function of the field strength H, provides an estimate of the total disturbed and non-disturbed losses that differ from each other on more than 10% of the values predicted by the actual time-domain solution of the problem.

Reference [7] presents an analytical method for magnetic field calculation in a non-linear ferromagnetic semi-infinite plate in the presence of eddy currents, magnetized by a short coil wound around the plate. The current through the short coil is sinusoidal in time.

It has been shown that the results obtained by the method described are better than those evaluated by assuming a constant permeability in the material.

The authors of reference [7] described an application of the BEM technique to 2-D and axisymmetric non-linear magnetic shielding. It was noted that because of the highly accurate vector fields attainable by the BEM, this technique is particularly applicable to studying and detecting the small perturbations produced by the shield on the otherwise very homogenous fields required within the magnet coils for the imaging applications.

The volume integral equation method (VIM) [9] is adopted to analyze 3-D non-linear eddy current problems in induction heating. The derived integral equation is solved effectively by adopting the relation between the surface magnetic field and the magnetic fluxes passing through the metal. Computed values are compared with measured values. Both are in reasonably good agreement.

The FEM was used in references [3, 4]. A FEM based on an exponential shape function was also introduced for eddy current problems in reference [10]. It was shown that for 1-D problems in a Cartesian geometry, the resulting FEM equations are identical to difference equations derived previously using the singular perturbation theory, as well as application to several 1-D problems, showing that the technique is strongly skin depth independent. The results obtained by the method are uniformly accurate for small and large skin depth problems. The finite element solution of the Helmholtz equation,

$$\left(\nabla^2 - \gamma^2\right)u = 0, \gamma^2 = i\mu\omega\sigma \tag{5.1}$$

used in eddy current problems, presents serious difficulties when the skin depth is very small. These difficulties follow the rapid changes of the field variable in the skin (boundary) layer. One has to take a very fine mesh to obtain an accurate solution.

A skin depth independent FEM solution of Equation 5.1 is constructed using exponential basis functions:

$$u = D_1 e^{\frac{x}{\gamma}} + D_2 e^{\frac{-x}{\gamma}} \tag{5.2}$$

If $u(x_{k-1}) = u_1$ and $u(x_k) = u_2$, then Equation 5.2 yields

$$u = N_1 U_1 + N_2 U_2 \tag{5.3}$$

Functions N_1 and N_2 are the shape functions and given by

$$N_1 = \frac{\sinh \frac{x_k - x}{\gamma}}{\sinh \frac{x_k - x_{k-1}}{\gamma}}; N_2 = \frac{\sinh \frac{x - x_{k-1}}{\gamma}}{\sinh \frac{x_k - x_{k-1}}{\gamma}} \tag{5.4}$$

Performing a standard FEM system of the assembled equations for a model of dimension k, we finally got:

$$AU_{k-1} - BU_k + CU_{k+1} = 0 \quad 0 < k \leq N-1, \tag{5.5}$$

where

$$A = \frac{1}{\sinh\frac{h_1}{\gamma}}; \quad B = \coth\frac{h_1}{\gamma} + \coth\frac{h_2}{\gamma}; \quad C = \frac{1}{\sinh\frac{h_2}{\gamma}}$$

$$h_1 = x_k - x_{k-1}; \quad h_2 = x_{k+1} - x_k$$

The FEM solution constructed decay exponential basis functions have properties of uniform convergence and skin depth independence. The same shape functions provide an accurate FEM solution for the 1-D cylindrical test problem (consideration of the eddy current distribution in 1-D conducting cylinder) with skin depth independence being observed for a/δ ratio (ratio of the region size to skin depth) greater than 7.

The authors of reference [10] applied a FEM that uses a hinged weighting function for 1-D and 2-D eddy current problems. A method of choosing the hinge parameters such that the resulting piecewise linear function approximates an exponential weighting function is developed. A special purpose quadrilateral element that incorporates the hinged weighting function is derived. The hinged elements were used to solve 1-D and 2-D eddy current test problems. Computed results that illustrate the advantages of the hinged FEM are presented in Section 5.2.

Numerical computations were used in reference [11]. The paper deals with the computation of global parameters such as forces, torques, and power dissipation from numerical models based on the computation of the vector potential for 2-D (plane or axisymmetric) eddy current problems, including linear or non-linear magnetic materials. The numerical method is a coupling between the FEM and the BEM. It provides a precise value of the potential and tangential fields on the boundary of the subdomains. The values of normal induction and of the tangential electric field on the boundary are accurately computed from the values of the vector potential on the boundary and the terminal voltage. All these quantities allow the efficient computation of global parameters as line integrals on the boundary of 2-D domains. The forces and torques are computed using the Maxwell stress tensor. The power dissipation is obtained using the Poynting theorem.

The computation of time-dependent electromagnetic fields in eddy current problems was considered in references [12–14]. Thus, in reference [12], a method was described for the computation of time-dependent electromagnetic fields of 3-D coil conducting plane arrangements. The method is based on the idea of solving available 2-D formulas for eddy current solution problems and to employ them in a 3-D context. Time-dependent field calculations for 3-D coil arrangements above a conducting plane have been carried out with this approach. The procedure employed by the authors starts with a current time function $i(t)$ that was sampled at regular intervals to obtain a discrete function $I(t)$. The time-domain values of $I(t)$ then were converted into frequency-domain values $I(\omega)$ by the inverse Fourier integral transformation. The function $I(t)$ is approximated as a set of linear interpolations between the sampling points and the corresponding analytic Fourier transforms were evaluated and superposed. With the frequency-domain values $I(\omega)$ obtained in such a manner, the corresponding values of the B-field vector components were evaluated from integral formulations that are based on the assumption of a filamentary source current [13].

It was suggested to conduct an eddy current test in the transient mode [14] by using a pulsed eddy current technique. Eddy current tests in the sinusoidal mode are of great interest for detecting flaws in metallic structures. A limitation of this classical method concerns the detection of deep defects in very conductive materials. Pulsed eddy current techniques require a particular signal processing that differs from the usual amplitude and phase analysis. The temporal and spectral analysis of pulsed signals is very informative for characterizing and discriminating different flaws. The purpose of the study was to determine, in a particular and significant application, the parameters characterizing discontinuities in simulated structures, to detect and size flaws.

The null field integral (NFI) equations in the solution of eddy current magnetic field problems are discussed in reference [15]. Their usage is approached through a development of the classical boundary integral equation technique as applied to irregularly shaped 2-D objects.

A transformation from diffusion fields to wave fields was examined in reference [16] as an approach to inverting eddy current non-destructive evaluation (NDE) data. An analytic inversion to the transformation was used as a meaning to gain insight into robustness issues associated with the method. A discretized version of the transformation was utilized to solve a 1-D inverse problem with a direct wave-based approach. Two regularization schemes were used to stabilize the inversion, which were themselves enhanced by an averaging algorithm.

In reference [17], an original method was proposed for the physical analysis of electromagnetic interactions between an inductive sensor and any homogeneous conducting plane target. From the solution of Maxwell's equations, an analytical model was obtained for the relationship between the target properties and the electrical signals for a U-shaped sensor. Analytical relations can be interpreted as an extension of the electrical image method, even in the most difficult case of conducting and magnetic targets.

The remote-field eddy current (RFEC) technique was used to non-destructively inspect both ferrous and nonferrous metal tubes [18]. To elucidate the physics of the RFEC measurements, the mutual impedance between two induction coils placed inside a long metal pipe was examined in detail. Cases where the coils are placed inside the innermost of two metal pipes were also considered. The impedance was decomposed into terms that represent waveguide modes (associated with the poles of its Fourier transform) and a radiation term (associated with the branch point singularity of its Fourier transform). The terms associated with the pole and branch point singularity were computed separately and compared with the total mutual impedance. It was shown that RFEC measurements can be made when the branch cut term (also known as a lateral or head wave) is dominant. A comparison between theoretically computed results and experimental measurements was carried out in reference [18]. The corresponding effective eddy current or permeability-contrast detectors were reported in reference [19].

The method for the 3-D analysis of magnetic fields around a thin magnetic conductive layer using vector potential was described in references [20–23]. The present method did not require numerical analysis of the electromagnetic field inside the layer, which is formidable due to its extreme spatial structure. The method gives reasonable solutions to numerical examples.

It should be emphasized that this problem is one of the main problems in the non-destructive testing area for the calculation of the perturbation of the magnetic field due to a slot or a crack in a plate with induced eddy currents. Eddy currents analysis by the integral equation method, utilizing loop electric and surface magnetic currents as unknowns, was proposed in references [21–25] using different frameworks to solve a problem. The surface integral equations whose unknowns are the surface electric and magnetic currents were widely used in eddy current analysis.

The eddy current distortion is three-dimensional and, therefore, has to be calculated with 3-D field computation methods. An accurate and fast 3-D field computation is therefore required that can resolve very small field perturbations generated by a sub-surface crack. In references [19–22], the implementation of a boundary element formulation with four variables on linear or quadratic isoperimetric elements that is able to fulfill these requirements was presented.

In references [23–25], the identification of the crack depth using signals obtained from eddy current testing is demonstrated. The identification method is based on finite elements with the pre-computed unflawed database approach and a meshless crack representation technique, and parameter estimation in non-linear problems. Some different cracks are estimated by using measured data.

5.2 Methods of Inverse Problem Solution Using Eddy Current Testing

As noted in Section 5.1, an eddy current model for 3-D inversion was considered in reference [1]. A model was presented for the inversion of eddy current data to be used for the detection of flaws. This model is based on rigorous electromagnetic theory and uses a multi-frequency approach to make it truly 3-D. The resulting integral equations were discretized and solved using the least squares technique. The numerical problems involved in this algorithm were discretized and solved using the same least squares technique. This means that the numerical problems involved in this algorithm were discussed and the corresponding solution of reconstructions of the flows generated by computer will be presented here. The flow can be considered as an anomalous region embedded in a homogeneous material with known properties. Maxwell's equations for the two rigorous regions can be written as follows:

For a known region

$$\nabla \times \mathbf{E}_0 = -i\omega\mu_0 \mathbf{H}_0 \qquad (5.6)$$

$$\nabla \times \mathbf{H}_0 = (\sigma_0 + i\omega\mu_0)\mathbf{E}_0, \qquad (5.7)$$

For a flawed region

$$\nabla \times \mathbf{E}_f = -i\omega\mu_0 \mathbf{H}_f \qquad (5.8)$$

$$\nabla \times \mathbf{H}_f = (\sigma_f + i\omega\varepsilon_0)\mathbf{E}_f, \qquad (5.9)$$

The slowed problem is explained by Figure 5.1.

From Figure 5.1, it can be seen that the region of interest (a tube) consists of three parts: the interior of the tube (1), the tube wall (2), and the region exterior to the tube (3). The dyadic Green function, $G_{ij}(\mathbf{r}|\mathbf{r}')$, can be introduced for the ith region (i = 1, 2, 3). The Green function is symmetrical, that is,

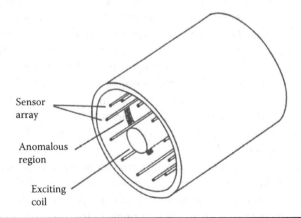

Figure 5.1 Three-dimensional tube with anomalous region, sensor array, and exciting coil.

$G_{ij}(\mathbf{r}|\mathbf{r}') = G_{ji}(\mathbf{r}'|\mathbf{r})$. It is assumed that the Green function $G_{ij}(\vec{r}|\vec{r}')$ or $G_{ji}(\mathbf{r}'|\mathbf{r})$ of the problem is known.

The set of base integral equations used for the solution of the inverse problem [definition of $\sigma_f(\mathbf{r})$ from the measurement data obtained by receivers in region 2], are as follows:

$$\mathbf{E}_2(\mathbf{r}) + i\omega\mu_0\sigma_0(\mathbf{r}) \iiint_{\text{flaw}} G_{22}(\mathbf{r}|\mathbf{r}')\mathbf{E}_2(\mathbf{r}')\left(\frac{\sigma_f(\mathbf{r}')}{\sigma_0(\mathbf{r}')} - 1\right) d\mathbf{r}' = \mathbf{E}_0(\mathbf{r}) \quad (5.10)$$

$$\mathbf{E}_0(\mathbf{r}) = -i\omega\mu_0 \iiint_{\substack{\text{exciting}\\\text{coil}}} G_{22}(\mathbf{r}|\mathbf{r}')\mathbf{J}_0(\mathbf{r}') d\mathbf{r}' \quad (5.11)$$

where $\mathbf{J}_0$ is the exciting coil current density. Then, the field difference between the two regions equals

$$E_1(\mathbf{r}) - E_f(\mathbf{r}) = i\omega\mu_0 \iiint_{\text{flaw}} G_{12}(\mathbf{r}|\mathbf{r}') E_f(\mathbf{r}') [\sigma_f(\vec{r}) - \sigma_0(\mathbf{r}')] d\mathbf{r}' \quad (5.12)$$

Equation 5.12 was obtained from Maxwell's equations (Equations 5.6 and 5.7). Field $E_0(\mathbf{r})$ can be considered as the incident (known) wave propagating into region 2 from the exciting source. The second term on the right-hand side of Equation 5.12 is the field scattered by flaw and the first term is the total field in region 2.

The calculated field $E_1(\mathbf{r})$ can be then compared with the measured field. Calculations start with an assumed distribution for $\sigma_f(\vec{r})$. An iterative procedure can be used in the solution of the inverse-scattering non-linear problem considered. But, the authors have found that if the flaw is relatively small, then field $E_f(\mathbf{r})$ in Equation 5.10 can be approximated by the known field $E_0(\mathbf{r})$ produced by the exciting coil. This linearizes the problem and decouples the inverse problem from the direct scattering problem. The perturbed electric field results of the solution to the linearized problem are obtained after substitution of expressions for $E_0(\mathbf{r})$ and the Green function in Equation 5.10. On taking the operation *curl* $\equiv \nabla \times$ of the perturbed electric field in accordance with the first of Maxwell's equations, the perturbed magnetic induction field measured by the sensor can be obtained.

The fundamental integral equations give the components of the perturbed magnetic induction field in terms of the 2-D Fourier transform of the conductivity distribution of the anomalous region at each of the N_r layers within the tube wall.

The field component can be sensed at a single radius r, but at many different z and φ locations. The following transform equation can be obtained when the 2-D Fourier transform of the field is taken with respect to φ and z:

$$\tilde{B}(m,h,\omega) = \sum_{k=1}^{N_r} \tilde{H}_k(m,h\omega)\,\tilde{\sigma}_k(m,h) \qquad (5.13)$$

where:

$\tilde{B}(m,h,\omega)$ is the 2-D transform of the z-component of the perturbed induction field

$\tilde{\sigma}_k(m,h)$ is the 2-D transform of the conductivity distribution at the kth layer of the anomalous region

$\tilde{H}_k(m,h,\omega)$ is the transfer function from the kth layer to the sensor

m is the spatial frequency in the φ direction and is an integer, being a simple Fourier series harmonic

h is the spatial frequency in the z-direction

$\tilde{B}(m,h,\omega)$ is obtained from measurement data; $\tilde{H}_k(m,h,\omega)$ is calculated by using results of described analyses. The unknown function $\tilde{\sigma}_k(m,h)$ is not dependent on frequency ω, and this allows us to acquire data to solve Equation 5.13 for $\tilde{\sigma}_k$ as a function of (m, h). This can be done by measuring the perturbed induction field at several frequencies and then writing an equation such as Equation 5.13 for each frequency. This system of equations is solved for each (m, h), which gives the Fourier transform of the unknown conductivity at each layer. The inverse Fourier transform of this solution vector gives the actual distribution of conductivity within the anomalous region.

A numerical experiment was carried out in accordance with the described theory. The measured magnetic induction field data were used for calculation of the linearized Equation 5.13, and then the latter was used for inversion. The authors of reference [7] have pointed out that the results of the reconstruction, with 5% random noise, showed that the method is reasonably robust relative to small modeling errors. But, the described inversion algorithm is computationally intensive. From the results of applying the multi-frequency model and its algorithm to the pyramid flaw (see Figure 5.2a and b) in the wall of a cylinder tube, one can see that the reconstruction is quite faithful, though dispersion in the z-direction (along the axis of the tube) reduces the resolution. In the model, 10 layers in the radial direction and 10 frequencies $\omega_1, \ldots, \omega_{10}$ were used.

The major problem in eddy current flaw reconstruction is that the error functional for general 3-D situations may have several local minimums. A first-order optimization strategy may thus give different reconstructed flaws, depending on the initial estimate. To overcome this problem, a stochastic global optimization algorithm may be used [9].

Most widely, eddy current flaw reconstruction strategies are based on the minimization of the non-linear least squares error function. The theory of eddy current inversion was considered in reference [3]. Then, it was developed for a problem accounting for an arbitrary material, type of eddy-current problem, and the materials

98 ■ Electromagnetic and Acoustic Wave Tomography

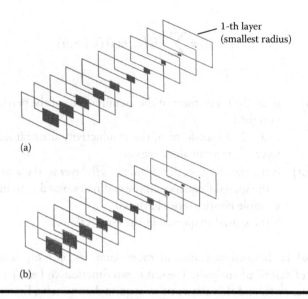

Figure 5.2 (a) Original inverted pyramid flaw; (b) reconstructed.

defects' shapes. A fast 3-D forward solver is created to rapidly predict eddy current signals in the inversion shell. The high speed of the signal evaluation results from utilizing a reaction data set constructed before performing the inversion by a finite element electromagnetic field simulator. The same pre-calculated reaction data set supports the quick evaluation of sensitivity information, thereby ensuring the efficient implementation of an optimization algorithm. Two general types of inverse problems are considered: the reconstruction of a conductivity distribution and of the shape of an inclusion or crack whose conductivity is known or assumed to be zero.

The use of the least squares criterion was proposed as the basis of an iterative scheme for reconstruction of the flaw function. The iterative scheme proceeds by updating the flaw function at each iteration in such a way that the mean square difference or error between the measured and predicted data is driven to zero or to a minimum in the presence of noise and model inaccuracies. The mean square error may be regarded as a function over the function space comprised of all possible flaw functions, and the main purpose to use this parameter is to select such a flaw function from all spatial flaw functions that minimizes the ordinary function using some suitable algorithm.

Let us consider the direct problem, assuming that impedance data may be recorded as a function of frequency or probe position, or both, whichever is appropriate.

Assuming time-harmonic excitation [exp($-i\omega t$)] and neglecting the displacement current, Maxwell's equations read

$$\nabla \times \mathbf{E}(\mathbf{r}) = i\omega\mu_0(\mathbf{r}) \qquad (5.14)$$

$$\nabla \times \mathbf{H}(\mathbf{r}) = \sigma(\mathbf{r})\mathbf{E}(\mathbf{r}) = \sigma_0 \mathbf{E}(\mathbf{r}) + \mathbf{p}(\mathbf{r}) \tag{5.15}$$

where

$$\mathbf{p}(\mathbf{r}) = [\sigma(\mathbf{r}) - \sigma_0]\mathbf{E}(\mathbf{r}) = \sigma_0 g(\mathbf{r})\mathbf{E}(\mathbf{r}) \tag{5.16}$$

The function $\mathbf{p}(\mathbf{r})$ is an effective current dipole density at the flaw due to the variation $(\sigma(\mathbf{r}) - \sigma_0)$ of the conductivity from that of the host σ_0. The solution of Maxwell's equations in integral form may be expressed as

$$\mathbf{E}(\mathbf{r}) = \mathbf{E}^{(i)}(\mathbf{r}) + i\omega\mu_0 \int G(\mathbf{r}|\mathbf{r}') \cdot \mathbf{p}(\mathbf{r}') d|\mathbf{r}' \tag{5.17}$$

where the integration is over the volume of the flaw. $\mathbf{E}^{(i)}(\mathbf{r})$ is the incident field produced by a primary source in the absence of the flaw, and $G(\mathbf{r}|\mathbf{r}')$ is the dyadic Green function obeying

$$\nabla \times \nabla \times \mathbf{G}(\mathbf{r}|\mathbf{r}') - k^2 \mathbf{G}(\mathbf{r}|\mathbf{r}') = \delta(|\mathbf{r} - \mathbf{r}'|)\mathbf{I} \tag{5.18}$$

where $k^2 = i\omega\mu_0\sigma_0$, $\delta(|\mathbf{r} - \mathbf{r}'|)$ is the 3-D Dirac delta function and $\mathbf{I}$ is the unit dyad. From Equations 5.14 and 5.15, the following integral equation can be obtained:

$$\mathbf{P}(\mathbf{r}) = \mathbf{P}^{(i)}(\mathbf{r}) + g(\mathbf{r})k^2 \int \vec{G}(\mathbf{r}|\mathbf{r}') \cdot \mathbf{p}(\mathbf{r}') d\mathbf{r}' \tag{5.19}$$

where $\mathbf{P}^{(i)}(\mathbf{r}) = \sigma_0 g(\mathbf{r})\mathbf{E}^{(i)}(\mathbf{r})$.

On the other hand, the impedance Z due to the flaw is given by

$$Z = -\int_V \mathbf{E}^{(i)}(\mathbf{r}) \cdot \mathbf{P}^{(i)}(\mathbf{r}') d\mathbf{r}' \tag{5.20}$$

where the integration is over the volume of the flaw.

Generally speaking and based on similar methodology, as in reference [3], the direct problem may be defined as follows [6–13]: given the flaw function $g(\mathbf{r})$ and incident field $\mathbf{E}^{(i)}(\mathbf{r})$ produced by the primary source, compute the impedance Z. To accomplish this, the linear integral Equation 5.17 is solved for the polarization $\mathbf{p}(\mathbf{r}')$, which, in turn, is substituted into Equation 5.19.

Now, for reconstruction of an arbitrary conductivity distribution, it is supposed that $g(\mathbf{r})$ is an estimate of the time flaw function $g_{true}(\mathbf{r})$. Then, on the basis of the estimate $g(\mathbf{r})$, the direct problem can be solved to obtain the predicted measurement. Impedance Z can be detected by a probe sensor at the probe position $\mathbf{r}_0$ at an excitation frequency ω.

So, the inverse eddy current problem can be described as the task of reconstructing an unknown distribution of electrical conductivity from eddy current

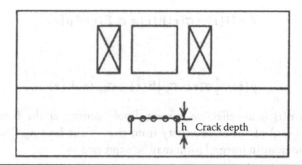

Figure 5.3 Assumed shape estimation.

probe impedance measurements recorded as a function of probe position, excitation frequency, or both. In eddy current NDE, this is widely recognized as a central theoretical problem whose solution is likely to have a significant impact on the characterization of flaws in conducting materials [4, 20–25].

Inverse problems in which the position or shape of a crack is estimated have become very important for many industrial systems; for example, nuclear power generation systems. Reference [4] presents inverse analysis by using BEM with Laplace transform including the terminal voltage method. The method is applied to a simple problem in eddy current testing (ECT) [23]. Some crack shapes in a conductive specimen are estimated from distributions of the transient eddy current on its sensing surface and magnetic flux density in the liftoff space. Because the transient behavior includes information on various frequency components, the method is applicable to the shape estimation of a comparatively small crack. We will not enter into a complicated mathematical description of the inverse problem described in reference [4], giving some physically vivid results of numerical computations carried out in the proposed theoretical framework.

Thus, Figures 5.3 and 5.4 show the assumed crack shape and the initial shape, respectively, for the estimation following the framework proposed in reference [4]. The exciting voltage was given a winding of 300 turns as a step function at $t = t_0$ (resistance $R = 5.0\ \Omega$, exciting voltage $V = 5.0$ V). The conditions used in this

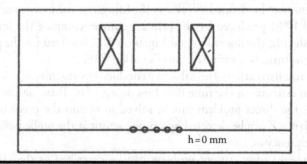

Figure 5.4 Initial shape for estimation.

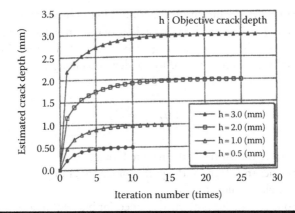

Figure 5.5 Convergence of crack shape with eddy current density distribution.

analysis were the relative permeability of the conductor $\mu_r = 1$ and the resistivity $\rho = 1.73 \times 10^{-8}$ Ωm.

Figure 5.5 shows the obtained results and the convergence of this method with the eddy current density distribution when the complex normalized wavelength parameter λ equals 512 (values of λ varied in a range from 1.0 to 16,384).

The points of convergence were: the maximum relative error of eddy current at all the estimating points was 0.01%; the relative error of the objective function at the kth iteration was 0.001%.

It is clearly seen from the presented illustrations of the objective crack depths and those estimated during theoretical calculations that the obtained crack shapes were in good agreement with the numerically predicted shapes. It was concluded by the authors of reference [10] that the method considered is effective in obtaining the inner detail information in conductive materials and more useful in developing a non-destructive testing method in comparison with the conventional methods.

The deconvolution algorithms applied to the measured data were also described in references [5, 6]. Thus, in reference [5], it was shown that the connectivity of surface-breaking structures whose overall size is comparable to that of the transducer can be extracted from eddy current data by the application of image processing techniques. Specially designed probes and image processing can be combined for scanning at increased rates to cover greater areas for high-resolution scanning for automatic extraction of quantitative detail, or for enhanced probability of flaw detection.

The interaction of the eddy current probe with the surface structure can be represented as a convolution of the point spread function with the structure. Processing steps described in reference [5] involved: background removal to improve signal to noise; smoothing or tapering of the image edges to reduce the Gibbs phenomenon; insertion of both the point spread function and the blurred image into large zero images to eliminate aliasing; and, then, restoring the shape of the surface structure with a Wiener filter. The point spread function is found by scanning the probe over a small symmetrical artifact, such as a deep hole of 0.25 mm diameter. The data was obtained at 500 kHz. The restored image was constructed via the inverse Fourier transform of the Wiener filter applied to the degraded image:

$$R \frac{H^*}{HH^* + K} \cdot G$$

where:
  R is the Fourier transform of the restored image
  H is the Fourier transform of the point spread function
  H^* is the complex conjugate of H
  G is the Fourier transform of the degraded (or blurred) image
  K is a constant

If K defines the ratio of the noise power spectral density, this formula directly describes a classic Wiener filter. The constant may be varied and should (usually) be of the order of the variance of the noise in the frame background.

As mentioned in Section 5.1, this approach describes a digital signal processing method called *post-focus processing*. The method reduces the transducer blurring effect and improves the resolution of images by simultaneously performing 3-D (2-D spatial and temporal) deconvolution using the monostatic point reflector spreading function (MPSF) of the transducer. Experiments were conducted using post-focus processing and three known image resolution improving methods, including axial-Wiener deconvolution, 2-D Wiener deconvolution, and pseudo 3-D Wiener deconvolution. Such processing was partly or fully used to recover obstructions, such as cracks or flashes, by other authors (see, e.g., references [19–25]). All technical details described there are not part of the scope of this book. The reader who is interested in the technical and computational details of the proposed methods of the direct and inverse problems in eddy current tomography can find them in all of the papers mentioned in this section.

Now, based on the diffraction tomography approach mentioned in Section 3.4.1 and the Born approximation, which follows in the next section, we will consider the eddy current image processing that we use for the reconstruction of defects in various materials.

5.3 Theoretical Consideration of Diffraction Tomography Approach

For image reconstruction of defects, a plane wave spectrum of the scattered field is used. This means that the scattered field $\psi(x, y_1)$ at line $y = y_1$ (1-D case: Figure 5.6) is represented in the form of the Fourier integral

Eddy Current Tomography

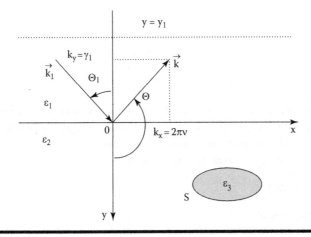

Figure 5.6 Region under consideration in the *x, y*-plane and projections of plane wave vector $\vec{k}$ from plane wave spectrum of scattered field on axes of reference.

$$\psi(x, y_1) = \int_{-\infty}^{\infty} \hat{\psi}(\nu, y_1) \exp(2\pi i x \nu) d\nu \qquad (5.21)$$

where $\hat{\psi}(\nu, y_1)$ is the Fourier image of $\psi(x, y_1)$ and is defined as

$$\hat{\psi}(\nu, y_1) = \varphi(\nu)\exp\left[-iy_1\sqrt{k^2-(2\pi\nu)^2}\right] = \varphi(\nu)\exp(-i\gamma_1 y_1); \qquad (5.22)$$

where:

$\gamma_1 = \sqrt{k^2-(2\pi\nu)^2}$
$k = (\omega/c)$ is the wave number of the plane wave in free space
ω is the cyclic frequency
c is the velocity of light

The function $\varphi(\nu)$ is the angle spectrum of the scattered field. The variable ν is the space frequency in Equation 5.21 and it also defines a direction of propagation of the plane wave in the expansion of the scattered field in terms of the plane waves. This is seen from Figure 5.6. The following relation connects value ν and angle θ:

$$2\pi\nu = (\omega/c)\sin\theta \qquad (5.23)$$

The function $\varphi(\nu)$ may be written as

$$\varphi(\nu) = c_1(\nu)c_2(\nu) \qquad (5.24)$$

where

$$c_1(\nu) = \frac{ik_2^2 T}{\gamma_1 + \gamma_2}. \quad (5.25)$$

where:
- T is the Fresnel transmittance of the boundary between medium 1 and 2 with dielectric permittivity $\varepsilon_1 = \varepsilon_0$ (air) and $\varepsilon_2 = \varepsilon_{r2}\varepsilon_0$, respectively
- ε_0 is the dielectric permittivity of the vacuum
- ε_{r2} is the relative dielectric permittivity of medium 2

$\gamma_2 = \sqrt{k_2^2 - (2\pi\nu)^2}$
$k_2 = \omega^2 \varepsilon_2 \mu_0 + i\omega\mu_0\sigma_2$
- k_2 and σ_2 are the wave number and conductivity, respectively, of medium 2
- μ_0 is the magnetic permeability of the vacuum

Function $c_2(\nu)$ may be written in an integral form:

$$c_2(\nu) = \iint_S K(x', y') \exp\left[-2\pi i(\alpha x' + \beta y')\right] dx' dy' \quad (5.26)$$

where:
- S denotes that integration is over the cross-section S of the object under investigation
- $K(x', y')$ represents the normalized polarization current that is sought

$$\alpha = \nu - \frac{\omega}{c}\frac{1}{2\pi}\sin\theta_1 \quad (5.27)$$

$$-2\pi\beta = \sqrt{\left[\left(\frac{\omega}{c}\right)^2 \varepsilon_{r2} - (2\pi\nu)^2\right] + i\frac{\omega}{c}120\pi\sigma_2} \\ + \frac{\omega}{c}\sqrt{(\varepsilon_{r2} - \sin^2\theta_1) + i\frac{c}{\omega}120\pi\sigma_2} \quad (5.28)$$

where:
- θ_1 is the angle of incidence

Note that to obtain Equations 5.5 through 5.8, the results of reference [26] were used. From Equations 5.27 and 5.28, we can see that α is a real value for all ν, and β is a complex value, if $\sigma_2 \neq 0$ or if $\sigma_2 = 0$; $|2\pi\nu| > \frac{\omega}{c}\sqrt{\varepsilon_{r2}}$.

Let us now consider function $c_2(v)$. This function can be rewritten in the following form:

$$c_2(v) = \int_{-\infty}^{\infty}\left(\int_{-\infty}^{\infty} K(x,y)\exp[-2\pi i\alpha(v)x]dx\right)\exp[-2\pi i\beta(v)y]dy$$

$$= \int_{-\infty}^{\infty} p[\alpha(v),y]\exp[-2\pi i\beta(v)y]dy \tag{5.29}$$

Let us consider the following integral also:

$$F(\beta) = \int_{-\infty}^{\infty} f(y)\exp(-2\pi i\beta y)dy \tag{5.30}$$

where $f(y)$ is an arbitrary function. The function $F(\beta)$ is defined for those complex values of β for which the integral exists. If the integral in Equation 5.30 exists for values of $2\pi\beta$ in $A < 2\pi\mathrm{Im}(\beta) < B$, then the function $f(y)$ can be defined by the inversion integral as

$$f(y) = \int_{\frac{-i\gamma}{2\pi}-\infty}^{\frac{-i\gamma}{2\pi}+\infty} F(\beta)\exp(2\pi i\beta y)d\beta; \quad A < \gamma < B \tag{5.31}$$

From this result, the function $p(\alpha, y)$ in Equation 5.29 can be written in the form of a contour integral in the complex β-plane along a contour L:

$$p(\alpha, y) = \int_L c_2(\alpha,\beta)\exp(2\pi i\beta y)d\beta \tag{5.32}$$

The contour is a direct line in plane $\alpha = \mathit{const}$ of complex α; the β-space is parallel to axis β' at $\beta'' = \dfrac{-\gamma}{2\pi}$. The polarization current distribution $K(x, y)$ is defined as

$$K(x,y) = \int_{-\infty}^{\infty}\int_{\frac{-i\gamma}{2\pi}-\infty}^{\frac{-i\gamma}{2\pi}+\infty} c_2(\alpha,\beta)\exp[2\pi i(\alpha x + \beta y)]d\beta d\alpha \tag{5.33}$$

Function $K(x, y)$ is calculated for the dielectric half-space with ε_2. If variables x, y are taken to be in region S, then the function $K(x, y) \cong K(x', y')$ will give the polarization current distribution of the object under investigation. We take the image function in the form of $|K(x,y)|$. Let $y=y_0$ be constant in Equation 5.33. We can find function $K(x, y_0)$ at different values of z, since in the 3-D case the scattered field is also changed along this axis. So, we can obtain the polarization current distribution at an object in the plane of x, z at a constant depth y_0, measuring the scattered field $\psi(x, z, y_1)$ at a constant y_1 in this plane:

$$K(x,z) = \int_{-\infty}^{\infty} \int_{-\frac{i\gamma}{2\pi}-\infty}^{\frac{-i\gamma}{2\pi}+\infty} \hat{\psi}(\alpha,\beta,z,y_1) c(\alpha,\beta,y_1) \exp\left[2\pi i(\alpha x + \beta y_0)\right] d\beta d\alpha \quad (5.34)$$

where $c(\alpha, \beta, y) = \exp[i\gamma_1(\alpha, \beta)y_1]/c_1(\alpha, \beta)$. Functions $K(x, y)$ or $K(x, z)$ depend on frequency f, and can be calculated for a set of frequencies $f_1, f_2, ..., f_N$ from frequency band Δf. In this case, the image functions are defined as

$$|K(x,y)| = \left|\sum_{i=1}^{N} K_{f_i}(x,y)\right| \text{ or } |K(x,z)| = \left|\sum_{i=1}^{N} K_{f_i}(x,z)\right|.$$

In the direct problem, the scattered field in the Born approximation was modeled and calculated by the method described in reference [27], based on the parameters' values used in the inverse problem.

The solving of the direct problem has special interest, as it allows us to study possibilities of the given approach for obtaining the current function $K(x, y)$ of a cylindrical object of arbitrary shape. On the other hand, weakly scattering objects are usually objects of practical investigation.

According to the Born approximation, defined by the exciting field $\psi(x', y') \ll E_z^i(x', y')$, the expression for the normalized polarization current $K(x', y')$ reduces to

$$K(x', y') = \left[\frac{k_3^2(x', y')}{k_2^2} - 1\right] \quad (5.35)$$

where:
  $k_3(x', y')$ is the wave number for the medium of the object
  k_2 is the wave number for medium 2, and does not depend on incident and diffracted fields

If the electromagnetic parameters of the object under investigation do not depend on coordinates x', y', then, according to Equation 5.35, function $K(x', y')$ also does not depend on these coordinates and it may be removed from the integral in Equation 5.26. Then, the expression for the Fourier image of field $\psi(x', y')$ has the form

$$\hat{\psi}(\nu, y_1) = C(\nu, y_1) K \iint_S \exp\left[-2\pi i(\alpha x' + \beta y')\right] dx' dy' \qquad (5.36)$$

where K is a value taking on a constant value in the region S according to reference [15], and which is equal to 0 outside of the region S. The region S has a rectangular cross-section with borders along the x-axis: $x_1 = a$; $x_2 = b$ ($b > a$); and along the y-axis: $y_1 = c_1$; $y_2 = d$ ($d > c_1$). After analytical integration, the following expression for function $\hat{\psi}(\nu, y_1)$ describing the Fourier image of the scattered field is obtained:

$$\hat{\psi}(\nu, y_1) = (1/4\pi^2 \alpha \beta) C(\nu, y_1) K I_1 I_2 \qquad (5.37)$$

where

$$I_1 = (\sin 2\pi\alpha b - \sin 2\pi\alpha a) + i(\cos 2\pi\alpha b - \cos 2\pi\alpha a) \qquad (5.38)$$

$$I_2 = (\sin 2\pi\beta d - \sin 2\pi\beta c_1) + i(\cos 2\pi\beta d - \cos 2\pi\beta c_1) \qquad (5.39)$$

The field $\psi(x, y_1)$ scattered by the object is found at the line $y = y_1$ by the inverse Fourier transformation of function $\hat{\psi}(\nu, y_1)$.

Equation 5.37 allows us to study the scattered field in the Born approximation and use the inverse problem solutions in Equation 5.33 or 5.34 to obtain function $K(x, y)$ to image the object, if the parameters of the problem are close to the experimental conclusions. In this case, the obtained function $K(x, y)$ is different from the one given initially, having a constant value in a given rectangular region and a value of 0 outside of this region. It follows from this that in the inverse problem, the integrand function is known only in some region of the plane of variables α, β in the result of the angle of incidence and frequency variation of the plane wave, illuminating the boundary of the media.

In a numerical experiment, a reconstruction of imaging function $\tilde{K}(x, y) \sim |K(x, y)|$ at normal angle of incidence $\theta \cong 0$ was carried out in the frequency band $\Delta f = 20$–200 KHz at 19 frequencies with step of $\Delta f = 0.1$ KHz. Relative dielectric permittivities and conductivities of the media (air–metal) are equal to: $\varepsilon_{r1} = 1$; $\varepsilon_{r1} = 0.8$; $\sigma_2 = 3.54 \times 10^7$ S/m; $\varepsilon_{r3} = 0.9$; $\sigma_3 = 1.0 \times 10^7$ S/m (object). The relative magnetic permeability of the media was taken as 1.0. The electric field amplitude of incident plane wave is equal to 1 V/m and the angle of incidence $\theta_1 \cong 0$. The scattered field $\psi(x, y_1)$ was calculated using Equation 5.21 in the bounded limits of integration by $v = -1000$; 1000 m^{-1} with the constant step $\Delta v = 5.0$ m^{-1} at values of $x = -9.1666... \times 10^{-6}\lambda_0$; $9.1666... \times 10^{-6}\lambda_0$ ($\lambda_0 = 3.0 \times 10^3$ m) with the constant step of computations $\Delta x = 0.1833... \times 10^{-3}$ m. These values of v and x are also used in the inverse problem. The corresponding images were reconstructed at values of $y \cong 0$; $0.6666... \times 10^{-6} \lambda_0$.

The magnitude β assumes complex values, but only the real part $\text{Re}(\beta)$ of β was used in reconstruction, because the inverse is generally an ill-posed problem and the calculation of the integrals (Equations 5.33 and 5.34) at complex values of β [imaginary part $\text{Im}(\beta) < 0$] can be executed only at sufficiently small values of $y > 0$ if $|\text{Im}(\beta)|$ is large or at large values of $y > 0$ if $|\text{Im}(\beta)|$ is taken to be small enough. In our calculation, $\text{Im}(\beta) < 0$ and $|\text{Im}(\beta)|$ is large, as the metal conductivities are $\sim 10^7$ S/m.

In the modeling, we also varied the dimensions of objects investigated (cylinders of rectangular cross-section) and the depths of their occurrence under the surface of the media division (positions of objects studied are shown in Figures 5.7 and 5.8).

It is assumed in the reconstruction algorithm that the back-scattered field is known at points on the direct line L, which is in the air at distance $y_1 = -0.01 \times 10^{-3}$ m above the metal surface.

Figure 5.9 shows the modulus of complex amplitude of the back-scattered field $|\psi(x, y_1)|$ on the direct line L in the case of metal with narrow slots 1–3 (Figure 5.7). One can see in Figure 5.10 the dependence on v of the complex Fourier transform modulus $|\hat{\psi}(v, y_1)|$ (the modulus of the complex space spectrum) of function $\psi(x, y_1)$

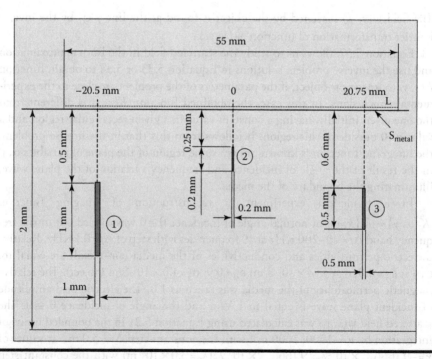

Figure 5.7 Positions and sizes of objects (narrow rectangular slots) studied in modeling.

Eddy Current Tomography ■ 109

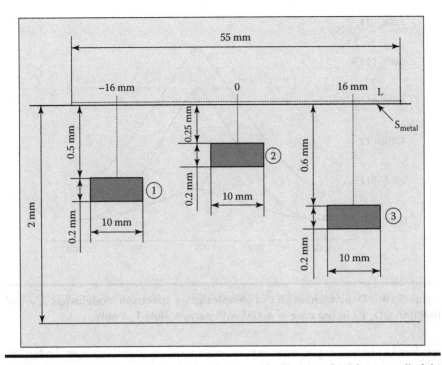

Figure 5.8 Positions and sizes of objects (wide rectangular slots) studied in modeling.

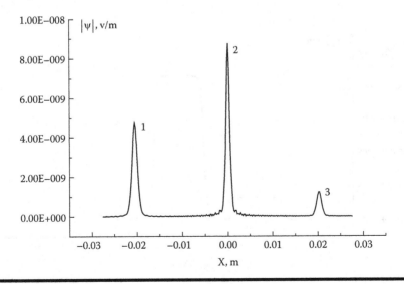

Figure 5.9 Distribution $|\psi(x,y_1)|$ on direct line L at distance $y_1 = -0.01 \times 10^{-3}$ m in air above surface of metal with narrow slots.

110 ■ *Electromagnetic and Acoustic Wave Tomography*

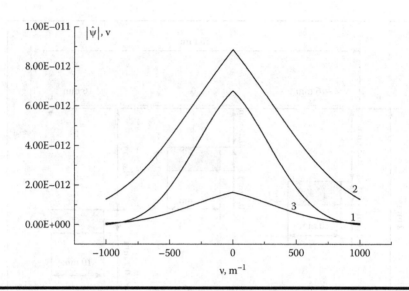

Figure 5.10 Dependences on ν of complex space spectrum modulus $|\hat{\psi}(\nu, y_1)|$ of function $\psi(x, y_1)$ in the case of metal with narrow slots 1–3 only.

for slots 1–3 only. The presented dependencies were calculated by using the formulas in Equations 5.21 and 5.37.

The maxima in Figure 5.9 denoted by numbers 1–3 and the numbers 1, 2, 3 near to the curves in Figure 5.10 are adequate for slots 1–3 in Figure 5.7.

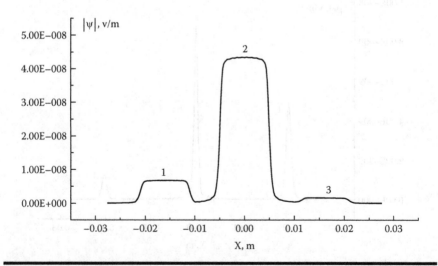

Figure 5.11 Distribution $\psi(x, y_1)$ on the direct line L at distance $y_1 = -0.01 \times 10^{-3}$ m in air above surface of metal with wide and narrow slots.

Eddy Current Tomography ■ 111

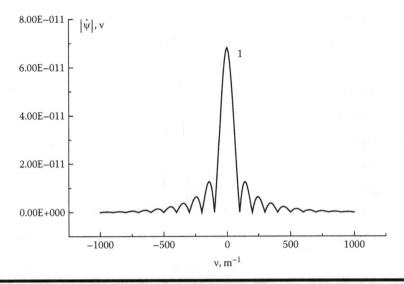

Figure 5.12 Dependences on v of complex space spectrum modulus $|\hat{\psi}(v, y_1)|$ of function $\psi(x, y_1)$ in the case of metal with narrow slot 1 only.

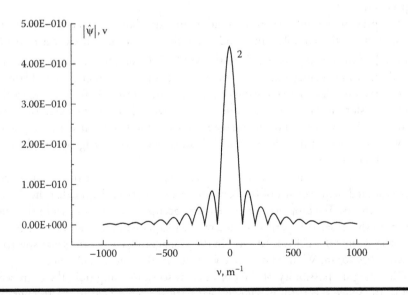

Figure 5.13 Dependences on v of complex space spectrum modulus $|\hat{\psi}(v, y_1)|$ of function $\psi(x, y_1)$ in the case of metal with narrow slot 2 only.

112 ■ *Electromagnetic and Acoustic Wave Tomography*

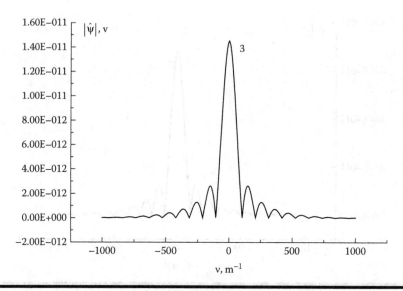

Figure 5.14 Dependences on v of complex space spectrum modulus $|\hat{\psi}(v, y_1)|$ of function $\psi(x, y_1)$ in the case of metal with narrow slot 3 only.

Figures 5.11 through 5.14 present similar functions in the case of metal with wide slots. Figure 5.11 shows the distribution $|\psi(x, y_1)|$.

Figures 5.12 through 5.14 illustrate the function $|\hat{\psi}(v, y_1)|$ for slots 1–3, respectively.

The presented results show us that in the air above the metal surface, the distribution of the complex amplitude modulus of the near back-scattered field $|\psi(v, y_1)|$ coincides approximately to the width of the slot. So, this function can be used directly as the image function for the detection of different slots and laminations in the non-destructive testing of metals. But, information about the vertical length of a slot (in the y-direction) and its depth will be lost in this case. The developed tomography method of the image reconstruction (eddy current tomography) allows us to obtain this information and construct a 3-D image of an object by using 2-D scanning.

It is necessary to note that the calculated space spectra of the required near back-scattered fields consist of the basic evanescent part as the incident fields have low frequencies. The propagating plane waves occupy a very small part of the space spectra in this case. This part is defined by the inequality $|2\pi v| < \omega/c$, and so v takes values between $\pm 0.3333\ldots \times 10^{-3}$ m^{-1} in the propagating part of the space spectrum for $\lambda_0 = 3.0 \times 10^3$ m. We used only one value for $v = 0$ in our calculation.

The imaging possibility of the object cross-sections using only the evanescent part of the space spectrum of known near back-scattered field in the tomography reconstruction algorithm was shown in references [26, 28]. Here, the modeling

Eddy Current Tomography ■ 113

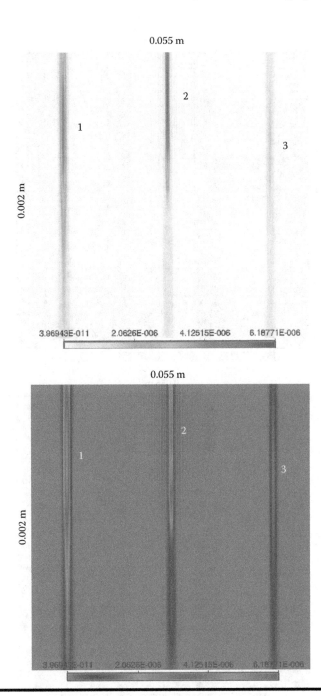

Figure 5.15 Cross-section images of narrow slots in metal (in different colors) obtained by eddy current tomography method.

114 ■ *Electromagnetic and Acoustic Wave Tomography*

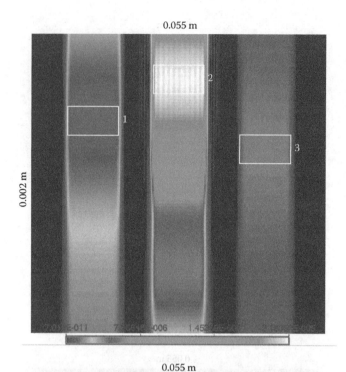

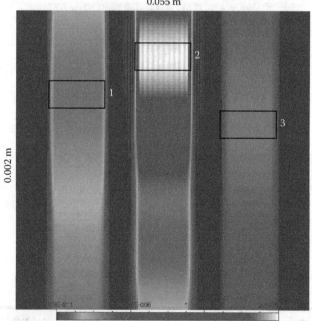

Figure 5.16 Cross-section images of wide slots in metal (in different colors) obtained by eddy current tomography method.

results of imaging the object cross-sections with the help of eddy current tomography are presented. The objects were the slots in the metal shown in Figures 5.7 and 5.8, as described. Images of these inhomogeneities are presented in Figure 5.15 (for narrow slots in metal) and in Figure 5.16 (for wide slots in metal) and were reconstructed in accordance with Equation 5.33 by using the approximation of complex β by its real part Re(β). One can see a very sharp image of slot 2 in Figure 5.15. The size and sub-surface depth of this slot are in accordance with the position of slot 2 in Figure 5.7. Slots 1 and 3 in Figure 5.15 can also be clearly observed. The depths of these slots are evaluated from the position of the maximum values of the image function on the slot cross-sections and the sizes in the sub-surface direction can be found from the sizes of the cross-sections. Similar behavior of the reconstructed images is observed in Figure 5.16.

So, the modeling conducted showed that the eddy current tomography method described allows us to reconstruct the cross-section images of slightly scattering inhomogeneities placed under the surface of a metal. The basic evanescent part of the near back-scattered field was used in the reconstruction algorithm. As the spatial period (the distance between zero-phase lines in the space spectrum component) is $P=|v|^{-1}$ ($P=1$ mm for the maximum values of v), the reconstructed images have very high space frequencies, although the length of the incident wave is very large at $\sim 3 \times 10^3$ m.

5.4 Conclusion

The results of modeling foreign objects hidden inside different materials and constructions showed that the eddy current tomography method described in this chapter, based on different techniques including the diffraction tomography approach proposed by the authors, allows us to reconstruct the cross-section images of the slightly scattered inhomogeneities placed under the surface of a metal. The basic evanescent part of the near back-scattered field was used in the reconstruction algorithm. As the spatial period (the distance between zero-phase lines in the space spectrum component) is $P=|v|^{-1}$ ($P=1$ mm for the maximum values of v), the reconstructed images have very high space frequencies, although the length of incident wave is very large at $\sim 3 \times 10^3$ m.

Calculations for the near field for the constant current thin circular loop antenna (vertical magnetic dipole) near to a plane interface have been conducted. Due to the very complicated mathematical operation that describes the near field generated by such a circular loop antenna, we will only will give the reader the main results regarding the basic phenomena, referring him to the excellent papers where this technique was fully described [29, 30]. The principal results obtained are the following. Components of the vector electromagnetic fields of the vertical circular loop above and below a planar interface between two media (air–plastic, air–composite material or air–metal) were expressed as a spectrum of plane waves.

The spectral analysis was based on knowledge of the incident electric field (field of the isolated loop) on the planes enclosing the loop. It was assumed that the uniform current excites in a loop of radius a placed in air. Analysis was undertaken of the electrically small loops provided that $k_0 a \ll 1$ where k_0 was the wave number in the air. The operating frequency varied from 1.0×10^4 to 1.0×10^5 Hz, the radius a of the loops was 0.005 m, and the height H of the loops above the surface varied between 0.005 m and 0.01 m. The resulting integrals were evaluated numerically according to the methodology described in references [29, 30]. Based on the methodology presented in references [29, 30], the authors of reference [28] found components of the spectral density function and showed that the plane wave spectra of the electromagnetic field of a vertical isolated loop consist of spectral components that represent evanescent waves in the normal direction to the surface. Investigating the properties of the reflection and transmittance coefficients in the spectral domain, the authors have found a new phenomenon concerning the distribution of the vector components of the electromagnetic fields of the circular loops under consideration placed near to the planar interface of two media. The results obtained by the authors of reference [28] could be considered to be a continuation of investigations of circular loop antennas placed near to a planar interface.

At the same time, we should emphasize that the proposed diffraction tomography method, described in Section 5.2, was then used vertically to the material surface plane (the image reconstruction plane) from the measured data for the cross-section reconstruction of objects. It was supposed that the measured data are created by the E_y component of the electromagnetic field (the image reconstruction plane is the x, z-plane) and that the scattered field inside of the sub-surface object is formed by only one component from the plane wave spectrum of the transmitted field, which propagates in the direction of the z-axis (normally to the surface of the sample) [28]. The images of the variant slots and holes placed in the test samples made from different materials were obtained. Finally, it was shown by the authors that the obtained theoretical results can be used in practice in all cases when low frequencies and circular loops of small electrical size are used. This could be, for example, the non-destructive testing of materials. In Chapter 14, some of the most interesting applications of the proposed eddy current tomography techniques for non-destructive testing of defects, cracks, and holes in different materials will be presented.

References

1. Dodd, C. V. and W. E. Deeds, Analytical solutions to eddy-current probe-coil problems, *Journal of Applied Physics*, vol. 39, no. 6, 1968, pp. 2829–2838.
2. Lean, M. H., Application of boundary integral equation methods to electromagnetics, *IEEE Transactions on Magnetics*, vol. MAG-21, no. 5, 1985, pp. 1823–1828.
3. Salon, S. J., The hybrid finite element-boundary element method in electromagnetics, *IEEE Transactions on Magnetics*, vol. MAG-21, no. 5, 1985, pp. 1829–1834.

4. Fawzi, T. H., M. Taher Ahmed, and P. E. Burke, On the use of the impedance boundary conditions in eddy current problems, *IEEE Transactions on Magnetics*, vol. MAG-21, no. 5, 1985, pp. 1835–1840.
5. Lavers, I. D., M. R. Ahmed, M. Cao, and S. Kaloichelvan, An evaluation of loss models for nonlinear eddy current problems, *IEEE Transactions on Magnetics*, vol. MAG-21, no. 5, 1985, pp. 1850–1852.
6. Rabinovici, R. and B. Z. Kaplan, Magnetization of a nonlinear ferromagnetic semi-infinitive plate by a short coil, in the presence of eddy-currents, *IEEE Transactions on Magnetics*, vol. MAG-21, no. 5, 1985, pp. 1859–1861.
7. Ma, X. and A. Wexler, BEM calculation for magnetic shielding with steel sheets, *IEEE Transactions on Magnetics*, vol. MAG-21, no. 6, 1985, pp. 2153–2156.
8. Ishibashi, K., Nonlinear eddy current analysis by volume integral equation method, *IEEE Transactions on Magnetics*, vol. MAG-23, no. 5, 1987, pp. 3038–3040.
9. Keran, S. and J. D. Lavers, A skin depth-independent finite element method for eddy current problems, *IEEE Transactions on Magnetics*, vol. MAG-22, no. 5, 1986, pp. 1248–1250.
10. Keran, S. and J. D. Lavers, On the application of hinged finite elements to eddy current problems, *IEEE Transactions on Magnetics*, vol. MAG-23, no. 5, 1987, pp. 3059–3061.
11. Bamps, N., A. Genon, H. Hedia, W. Legros, and A. Nicolet, Line integrals for the computation of global parameters from 2D eddy current numerical computations, *IEEE Transactions on Magnetics*, vol. 27, no. 6, 1991, pp. 4987–4997.
12. Konrad, A., A. J. Roeth, G. Bedrosian, and M. V. K. Chari, Time domain analysis of eddy current effects in conducting structures with pulsed excitation, *IEEE Transactions on Magnetics*, vol. MAG-22, no. 5, 1986, pp. 1251–1253.
13. Chari, M. V. K. and P. Reece, Magnetic field distribution in solid metallic structures in the vicinity of current carrying conductors and associated eddy-current losses, *IEEE Transactions on Power Apparatus and Systems*, vol. PAS-93, no. 1, 1974, pp. 45–56.
14. Lebrun, B., Y. Jayet, and J.-C. Baboux, Pulsed eddy current signal analysis: Application to the experimental detection and characterization of deep flaws in highly conductive materials, *NDT&E International*, vol. 30, no. 3, 1997, pp. 163–170.
15. Davey, K. R. and Y. Kanai, The use of null field integral equations in magnetic-field problems, *IEEE Transactions on Magnetics*, vol. MAG-22, no. 4, 1986, pp. 292–298.
16. Ross, S., M. Lusk, and W. Lord, Application of a diffusion-to-wave transformation for inverting eddy current nondestructive evaluation data, *IEEE Transactions on Magnetics*, vol. 32, no. 2, 1996, pp. 535–546.
17. Dufour, I. and D. Placko, An original approach to eddy current problems through a complex electrical image concept, *IEEE Transactions on Magnetics*, vol. 32, no. 2, 1996, pp. 2893–2896.
18. Haugland, S. M., Fundamental analysis of the remote-field eddy-current effect, *IEEE Transactions on Magnetics*, vol. 32, no. 4, 1996, pp. 3195–3211.
19. Michelsson, O. and F. H. Uhlmann, On the use of the 3-D H- ϕ- formulation for the forward solution for eddy current nondestructive testing, *IEEE Transactions on Magnetics*, vol. 34, no. 5, 1998, pp. 2672–2675.
20. Sabbagh, H. A. and L. David Sabbagh, An eddy-current model for three-dimensional inversion, *IEEE Transactions on Magnetics*, vol. MAG-22, no. 4, 1986, pp. 282–291.

21. Badics, Z. and J. Pavo, Fast flaw reconstruction from 3D eddy current data, *IEEE Transactions on Magnetics*, vol. 34, no. 5, 1988, pp. 2823–2828.
22. Norton, S. J. and J. R. Bowler, Theory of eddy current inversion, *Journal of Applied Physics*, vol. 73, no. 2, 1993, pp. 501–512.
23. Enokizono, M., T. Todaka, and K. Shibao, Numerical approach for ECT by using boundary element method with Laplace transform, *IEEE Transactions on Magnetics*, vol. 33, no. 2, 1997, pp. 2135–2138.
24. Joynson, R. F., R. O. McCary, D. N. Oliver, K. M. Silverstein-Medengren, and L. L. Thumhart, Eddy current imaging of surface-breaking structures, *IEEE Transactions on Magnetics*, vol. MAG-22, no. 5, 1986, pp. 1260–1262.
25. Doi, T., S. Hayano, and Y. Saito, Wavelet solution of the inverse source problems, *IEEE Transactions on Magnetics*, vol. 33, no. 2, 1997, pp. 1935–1938.
26. Chommeloux, L., Ch. Pichot, and J. Ch. Bolomey, Electromagnetic modeling for microwave imaging of cylindrical buried inhomogeneities, *IEEE Transactions on Microwave Theory Techniques*, vol. MTT-34, no. 10, 1986, pp. 1064–1076.
27. Vertiy, A. A. and S. P. Gavrilov, Modelling of microwave images of buried cylindrical objects, *International Journal of Infrared and Millimeter Waves*, vol. 19, no. 9, 1998, pp. 1201–1220.
28. Vertiy, A. A, S. P. Gavrilov, S. Aksoy, I. V. Voynovskyy, A. M. Kudelya, V. N. Stepanyuk, Reconstruction of microwave images of the subsurface objects by diffraction tomography and stepped-frequency radar methods, *Zarubejnaya Radioelektronika. Uspehi Sovremennoy Radioelektroniki [Foreign Radioelectronics. Success in Modern Radio Electronics]*, no. 7, 2001, pp. 17–52 (in Russian).
29. Smith, G. S., Directive properties of antennas for transmission into a material halfspace, *IEEE Transactions on Antennas and Propagation*, vol. AP-32, no. 3, 1984, pp. 232–246.
30. Li, L.-W., M.-S. Leong, P.-S. Kooi, and T.-S. Yeo, Exact solutions of electromagnetic fields in both near and far zones radiated by thin circular-loop antennas: A general representation, *IEEE Transactions on Antennas and Propagation*, vol. 45, no. 12, 1997, pp. 1741–1748.

II

EXPERIMENTAL VERIFICATION OF WAVE TOMOGRAPHY THEORETICAL FRAMEWORK

Chapter 6

Radio Tomography of Various Objects Hidden in Clutter Conditions

Vladimir Yakubov, Sergey Shipilov,
Dmitry Sukhanov, and Andrey Klokov

Contents

6.1 Special Experimental Setup and Antennas 122
6.2 Radio Tomography Based on UWB Tomographic Synthesis 130
    6.2.1 UWB Tomographic Synthesis of Building Constructions 130
    6.2.2 UWB Tomographic Synthesis of Bags and Luggage 138
    6.2.3 UWB Tomographic Synthesis of Object with Metallic Inclusions 138
6.3 Tomography Based on Linear Frequency Modulation Radiation 146
6.4 Transmission Tomography of Semitransparent Objects 147
    6.4.1 Experimental Setup 148
    6.4.2 Tomography of Semitransparent Objects 151
6.5 Radio Tomography of Non-Transparent Objects 152
    6.5.1 Transmission Tomography of Form of Radio-Opaque Object 153
    6.5.2 Multiple-Angle Tomography of Opaque Objects' Shapes 156
    6.5.3 Unilateral Location Tomography of the Shapes of
          Radiopaque Objects 158
    6.5.4 Recovery of the Focusing Properties of Combined
          Reflector Antennas 158
References 163

The methods and techniques described and discussed in Chapter 2 were proven there by the results of simulation or numerical modeling. But, more important for practical applications is to prove these results by real experiments. The aim of this chapter is to estimate objectively the accuracy of object recovery and reconstruction by use of the methods proposed previously.

At the same time, working experimental (laboratory) systems constructed for these purposes were taken as basic prototypes for future industrial radiograph and tomography image production. In this chapter, the results of works described in references [1–29] were analyzed.

6.1 Special Experimental Setup and Antennas

The basic equipment for ultra-wide band (UWB) radio-wave sounding includes: 2-D scanner, as a two-coordinate positioning frame (see Figure 6.1); scanner management unit, stroboscopic oscilloscope, and bipolar pulse generator (Figure 6.2), personal computer (PC), and transmitting–receiving (transceiver) antenna module (Figure 6.3).

The latter consists of the original construction of a UWB antenna, the form of which is shown in Figure 6.4a.

This antenna is a combined antenna consisting of two electric antennas—a transverse electromagnetic (TEM) horn and asymmetric dipole, and also a magnetic antenna in the form of a loop. Overlap of near-field zone different (electric and magnetic) antennas essentially allows a decrease of reactive fields inside the antenna and, therefore, extension of the operative frequency band [1–6]. In Figure 6.4b are

Figure 6.1 Two-dimensional scanner for UWB measurements.

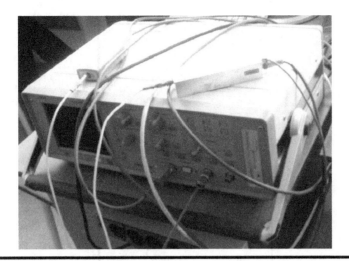

Figure 6.2 TMR8140 stroboscopic oscilloscope and pulse generator with a duration of 0.1 ns.

shown the frequency characteristics of a smaller antenna presented in Figure 6.4a at the far-right position. It is seen from Figure 6.4b that the frequency band for such an antenna is wide enough and is in the range 1–18 GHz. Here, the voltage standing wave ratio (VSWR) does not exceed 2, that is, VSWR < 2, for the tested antenna. Movement of the phase center of the antenna over the frequency spectrum does not exceed its diameter. The radiation pattern of the antenna, due to its small dimensions, is large enough (~30°–70°) in both the azimuthal and elevational

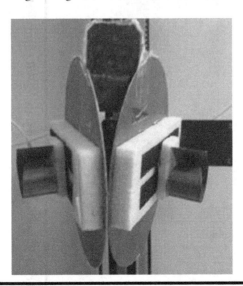

Figure 6.3 Transceiver module.

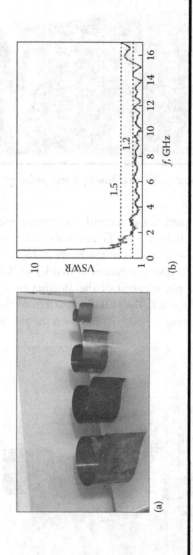

Figure 6.4 (a) View of antennas. (b) VSWR of smallest antenna.

planes, as seen from Figure 6.5a, where the smallest UWB antenna pattern is presented. Figure 6.5b shows a view and geometry of such an antenna in the joint 2-D azimuth–elevation domain [1–6].

In Figure 6.6, the radiation pattern (RP) of the antenna is shown for two radiated frequencies, 3 GHz and 9 GHz. For radio sounding, signals with periods of 0.1, 0.2, and 1.5 ns were tested. The special generator and the receiver of UWB signals were used during measurement. The frequency of pulse repetition was taken to equal 100 kHz.

The analysis of the signals' form and their spectra is presented in Figures 6.7a and b, where the first figure shows UWB pulses with different periods of (1) 0.1 ns, (2) 0.2 ns, and (3) 1.5 ns. The results presented here show that for radio tomography, the more effective impulses are those of the periods from 0.1 to 0.2 ns. In Figure 6.7b, the spectral dependence of such signals is shown. It is also seen that pulses of (3) are very poor with respect to those of (1) and (2).

Consequently, the latter signals should have good spatial resolution, and their spectra are in good agreement with the selected construction of the transmitting–receiving antenna [1–6].

In Figure 6.8, a view of one of typical signal with time period of 0.1 ns is shown after its reflection from the tested object.

It is seen that the signal is influenced by a strong fade, and its main maximum is shifted in the time domain with a time delay of 0.8 ns. This time delay is related to the given distance between the tested object and the antenna module. In Figure 6.9, the full raster picture of the recording signals, obtained during sounding of the tested object by use of a scanner, is shown. Here, each raster obtained for each line is placed in the following manner: the current raster is positioned after the previous raster.

The reverse movement of the scanner along the line begins immediately after scanning the previous line in the forward direction. This allows a decrease in scanning time. It is important to note that, during each pass of the line, the characteristic diffraction hyperbola is clearly seen, informing us about the localization of the sounding object.

The scanner control and management unit consists of a USB-to-COM converter and allows acquisition of the position in two directions—vertical and horizontal. Generally speaking, the existing experimental setup allows the movement of the receiving–transmitting antenna module within the limits of the plane ~84×84 cm, with accuracy of positioning of an arbitrary foreign object in the order of 2 mm and with speed of movement up to 3 cm/s.

The use of monostatic methods of radiometry to resolve tomography problems requires, in many cases, the use of one of the antennas as both the transmitter and the receiver, which achieves much better reconstruction of the radio image of the tested object. However, in UWB tomography, the usage of only one antenna is very difficult to achieve and there is usually a minimum of two frequency tracks, which should be separated both spatially and spectrally, let us say, by separate use of the

126 ■ *Electromagnetic and Acoustic Wave Tomography*

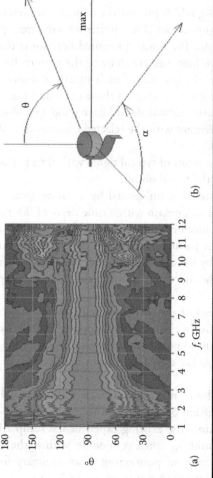

Figure 6.5 (a) UWB antenna 2-D pattern; (b) geometry of UWB antenna.

Radio Tomography of Hidden Objects ■ 127

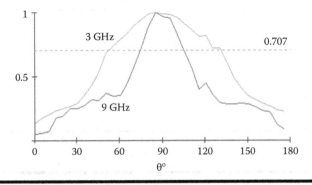

Figure 6.6 Radiation pattern of antenna shown in Figure 6.5b.

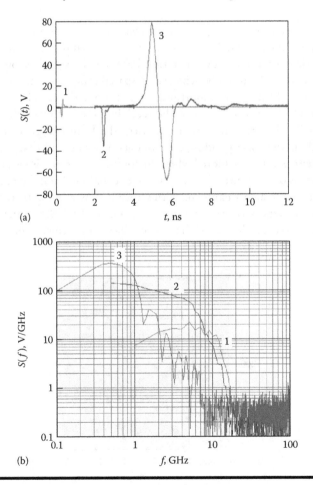

Figure 6.7 (a) UWB pulses of different duration used for sounding, and (b) their spectra. Curve 1: 100 ps; curve 2: 200 ps; curve 3: 1500 ps.

128 ■ *Electromagnetic and Acoustic Wave Tomography*

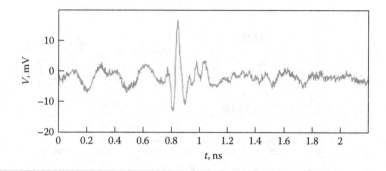

Figure 6.8 UWB signal reflected from the test object.

decimeter and meter wave length bands, as was proposed in references [2–6]. As stated, based on experimental data, to obtain good isolation between the receivers and the transmitter, the energy separation between the signals should be in the order of 25–30 dB, which can be achieved by spatial ranging of the antennas at distances exceeding 20 cm. This, unfortunately, does not coincide with UWB tomography. To increase isolation between the receiving and the transmitting antenna in references [1–3], a metallic screen was used, the form and geometry of which were defined via the corresponding experiments. It was found that good isolation between antennas can be achieved when the form of the screen is close to elliptical with the lengths of its large and small semi-axes equaling 9 and 5 cm, respectively. The arrangement of such an antenna with screen is shown in Figure 6.3.

Recently, references [1–6] mentioned the use of the UWB antenna array to increase the spatial resolution of the tested objects or elements of such objects.

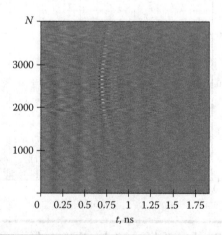

Figure 6.9 Raw data of test section.

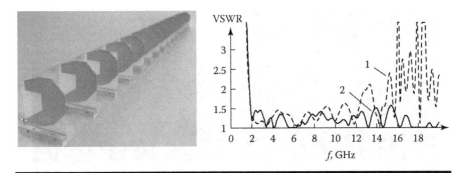

Figure 6.10 View of UWB antenna array (left), and its VSWR spectral dependence in the frequency domain (right).

Figure 6.10 (left) presents a view of such an array of small antennas, while the VSWR of each individual antenna element is shown in Figure 6.10 (right).

What is interesting to emphasize is that the form of the radiated pulse sent by one UWB antenna (denoted *1* in Figure 6.11a and b) and by an array of antennas (denoted *2* in Figure 6.11a and b) are similar. The same tendency was observed during analysis of the peak amplitude at (1) the *E*-plane and (2) the *H*-plane presented in Figure 6.11b.

The arrangement of such antenna elements into a linear array for the purpose of increasing the efficiency of radio-location and positioning and imaging the test objects or their elements is shown in Figure 6.12a and b.

As an example, we present such an antenna consisting of 22 UWB elements. The extremely short pulse (ESP) has a period of $T = 0.2$ ns, and the frequency band is spread over the wide band of ~1–10 GHz. The observation time of each foreign object was about 1.2 s, and the resolution of objects does not exceed 3 cm.

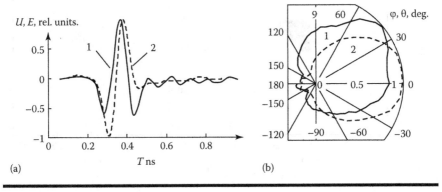

Figure 6.11 (a) Radiated pulse pattern in the time domain; and (b) in the frequency domain.

130 ■ *Electromagnetic and Acoustic Wave Tomography*

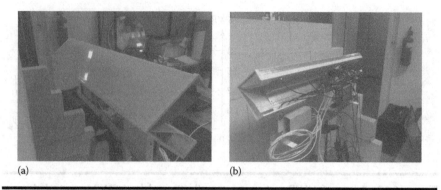

Figure 6.12 The front (a) and back (b) view s of the antenna array.

6.2 Radio Tomography Based on UWB Tomographic Synthesis

6.2.1 UWB Tomographic Synthesis of Building Constructions

Usage of the method of inverse focusing described in Section 2.4 was adapted for signal processing after experimental radio scanning of objects hidden in clutter environments, surrounded by barriers, and in structures of walls and other architectural elements. We will follow the results of experimental measurements and the results of the corresponding signal processing described in detail in references [7–16]. Thus, in Figure 6.13a is shown the experimental data for the plate test object in the form of a stepped triangular pistol with a scale of 5 cm for each step of its real form. Its reconstructed image, as the result of single focusing at the real distance to the object, is shown in Figure 6.13b.

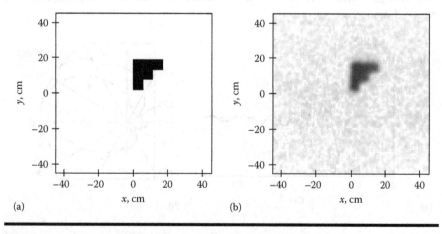

Figure 6.13 Stepped triangle test object: (a) true shape; (b) reconstructed radio tomogram.

Radio Tomography of Hidden Objects ■ 131

Figure 6.14 Experiment on tomography of thin metallic object.

It is seen that the form of the tested object is fully recovered, but there exists a slight blur of its boundaries, which was estimated to be roughly 2 cm. Another example is shown in Figure 6.14, where scanning was carried out of an object in the form of a metallic triangle, which was arranged on a simple piece of paper between two foam concrete blocks with a thickness of 10 cm.

The blocks with the embedded object were ranged at 80 cm from the scanner. An example of the signal reflected from the wall and from the hidden object is shown in Figure 6.15.

As seen from Figure 6.15, the observed locational signals were strongly affected by fading, which occurred due to multiple reflections from the two blocks and from the triangular test object. Figure 6.15 shows the results of measurements of the radio locational scanner for a scene with the object hidden in a wall. The reconstructed image of such an object with focus realized at the hall wall is presented in

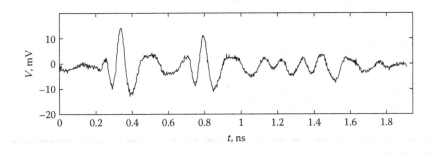

Figure 6.15 Signal reflected from the wall and hidden object inside it.

Figure 6.16a, whereas Figure 6.16b shows the same type of imaging, but for focus realized at the hidden object.

As follows from the illustrations presented, the hidden object was clearly reconstructed. This allowed the researchers to state that longitudinal (distance) resolution exists. An experiment on the usage of the elements of 3-G tomography was undertaken to search the inner structure of a brick wall, and its tomogram is shown in Figure 6.17a. Figure 6.17b presents its pulse signal distribution in the time domain. The details of imaging at different depths of radar penetration inside the wall is shown in Figure 6.17 for focusing carried out at the surface of the wall (left, $z = 0$), at a depth of penetration inside the wall of 2 cm (middle), and at a penetration depth of 15.5 cm (right). A water pipe is clearly seen from the illustrations presented in Figures 6.16a and 6.17a from the beginning (near the wall's outer surface), as well as other inhomogeneous structures existing inside the wall (see Figure 6.17c).

Additional tests of various walls via UWB tomography and their imaging reconstruction at different depths were carried out in parallel in several countries. In all of the experiments and the corresponding reconstruction algorithms described in Section 2.4, similar results were obtained to those shown in Figures 6.16 and 6.17.

Thus, Figure 6.18 shows radio imaging of the wall of a width about 16 cm. The left panel presents results of scanning at a depth of 2 cm, and the right panel at a depth of 9 cm. The layer-after-layer scanning shows that the intrinsic structure of the wall is very inhomogeneous (see Figure 6.18b). Here, an electric cable was found inside the brick wall after 2 cm, which was visualized as a vertical cylindrical inhomogeneity elongated along the y-axis. Along this cable, the corresponding electric current flows. When entering the scan more deeply (at 9 cm), the inhomogeneous structure of the brick wall is clearly seen. These examples allow us to point out that radio tomography algorithms are applicable for the visualization of hidden objects

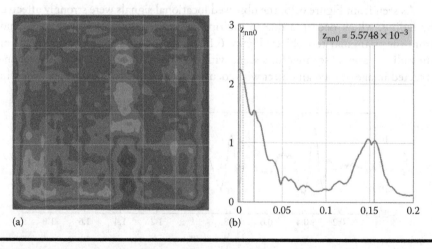

Figure 6.16 Reconstructed image of part of a foam concrete wall: (a) focusing at hall wall; (b) focusing on hidden object.

Radio Tomography of Hidden Objects ■ 133

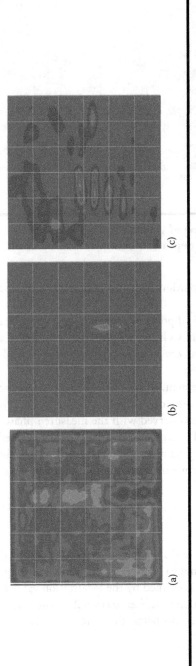

Figure 6.17 Imaging of intrinsic structure of brick wall: (a) at outer surface of wall ($z=0$); (b) at a depth of 2 cm; and (c) at a depth of 15.5 cm.

134 ■ *Electromagnetic and Acoustic Wave Tomography*

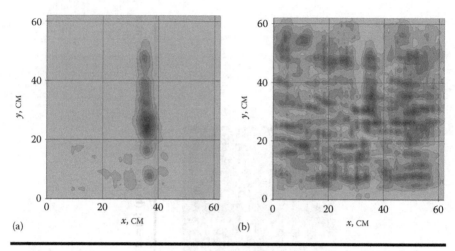

Figure 6.18 Radio imaging of wall with width of 16 cm: (a) scanning at a depth of 2 cm; (b) scanning at a depth of 9 cm.

inside walls, but their resolution becomes poorer close to a scanning distance of 2–3 cm.

From the general physical point of view, it is clear that the spatial resolution of wave tomographic images has a limit, which is defined, first of all, by the size of the focusing region; in the *transverse direction*, by the spatial spreading of testing pulses in the longitudinal direction.

Let us explain this statement. In the transverse direction, focusing is possible up to the limit of the depth of the diffraction of Fresnel's zone, that is, a zone within which the radiated point is observed with the measured phase difference at different points of the registered aperture. If the size of the registered aperture is D, and $\Delta\varphi = \pi/16$ is the meaningful value of the phase difference of the aperture, then the depth of Fresnel's diffraction zone for radiation with a wave length of λ is defined by the distance:

$$H_F = 4D^2 / \lambda \qquad (6.1)$$

The focusing regresses if the distance r becomes close to the depth H_F. The value of this regress is proportional to $\gamma = r/H_F$. If the better focusing size in the focusing center of a closed aperture is defined by the localization value $\delta_k(r_\perp)$ and equals the wave length λ as its diffraction limit (see Section 2.3), then, achieving the boundary of Fresnel's diffraction zone, this better focusing size increases in value proportionally to γ:

$$\rho_\perp = \lambda(1+\gamma) = \frac{\lambda r}{4D^2} \qquad (6.2)$$

Despite the fact that this estimation is approximate, it is close to experimental observations. Namely, for $f = 12$ GHz, $r = 90$ cm, and $D = 84$ cm, resolution in the transverse direction during the focusing procedure can be estimated as $\rho_\perp = 2.52$ cm.

As for the longitudinal resolution, it can be defined by the effective period τ of the operative pulse or by the bandwidth of the corresponding frequencies $\Delta f = 1/\tau$, in such a manner that the full longitudinal discrimination of the focusing procedure can be estimated by the following relation:

$$\rho_\parallel = \frac{c\tau}{2} = \frac{c}{2\Delta f} \qquad (6.3)$$

For a pulse with a bandwidth of 10 GHz, Equation 6.3 gives the estimated value of the longitudinal resolution size $\rho_\parallel = 1.5$ cm.

To estimate the losses in spatial resolution with increase of the distance r to the tested object, a special experiment was carried out with two metallic cylinders ranged at a distance of 6 cm in the transverse direction to the test system, as shown in Figure 6.19.

The test objects were placed at distances of $r = 90$, 135, and 180 cm. Figure 6.20 presents the tomograms obtained. The positions of the cylinders are indicated by *1* and *2*; the occurrence of an artifact (e.g., noisy image) is denoted by *3*. Each cell (mesh) has a size of 5×5 cm. For these distances, by use of Equation 6.2, the

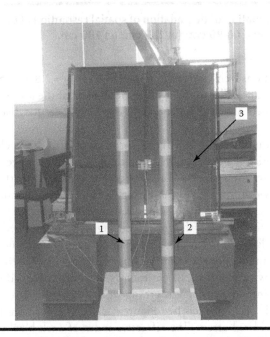

Figure 6.19 Experiment with metallized cylinders. 1, 2: cylinders, 3: scanner.

136 ■ *Electromagnetic and Acoustic Wave Tomography*

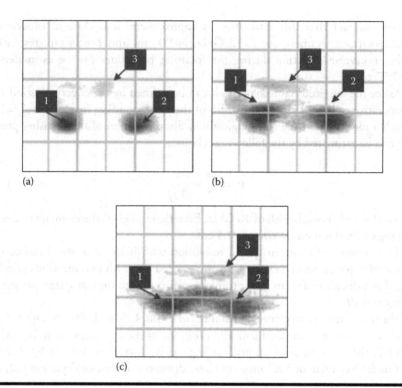

Figure 6.20 Estimation of degradation of spatial resolution at different distances to the test cylinders: (a) 90 cm; (b) 135 cm; (c) 180 cm.

transverse resolutions were estimated, respectively, as $\rho_\perp = 2.8, 4.3, 5.7$ cm. It is seen that for small distances (let us say, $r < 1$ m) resolution of the test objects (cylinders) is sufficient for their observation, and at an upper limit to this distance, resolution can be lost ($r = 180$ cm).

Resolution was fully lost and the occurrence of artifacts is clearly seen. It is interesting to notice that longitudinal resolution does not change with increase of the distance r to the test objects. This fact can be explained by the following: the frequency bandwidth is not changed during the experiment, because the period of sounding pulses is the same during measurements.

For experimental testing of the group focusing method described in Section 3.1.7, and to estimate also the discrimination possibilities of the constructed radio-tomographic system, a model of the medium was constructed consisting of three foam concrete blocks with width of 10 cm each (see Figure 6.21a). Between blocks 2 and 3, a testing inhomogeneity was arranged, as a system of five parallel metallic layers with width of 2 cm each (see Figure 6.21b). The position of this system is shown by the bold red arrow. The refractive index of such blocks was measured,

Radio Tomography of Hidden Objects ■ 137

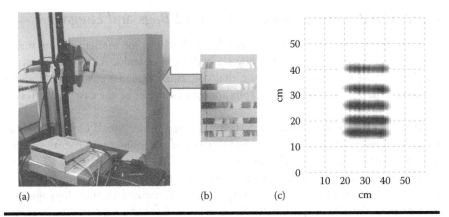

Figure 6.21 (a) Experimental setup; (b) the tested five-layer object; (c) its tomogram.

and equals $n = 1.44$. The result of reconstruction of the image by use of the group focusing method is shown in Figure 6.21c in the form of inverse grades of lightness. Sounding was carried out with the use of UWB pulses with duration s of 200 ns.

Enough precise spatial resolution was observed with all five layers, placed below each other, with accuracy of 2 cm, as was predicted by the use of Equations 6.2 and 6.3.

Similar measurements carried out in the inhomogeneous background medium give much poorer spatial resolution in the presence of a lot of artifacts (noisy shadow images: see Figure 6.22), which produce deformation of real layers' images in both the vertical and horizontal directions. Thus, layer 4 was shifted leftward at 2 cm, but layer 5 was totally removed from the plane of visualization of the tomogram at 16 cm.

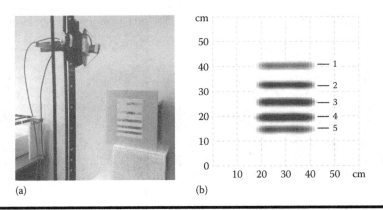

Figure 6.22 (a) Experimental setup; (b) result of UWB sounding.

6.2.2 UWB Tomographic Synthesis of Bags and Luggage

Another application of UWB tomography is checking for foreign objects inside bags and luggage. Figure 6.23a shows a radio-tomographic UWB wave setup, checking a case made of hard plastic that contained a model plastic pistol and a bottle of water. The UWB tomography image of content of the case under test is shown in Figure 6.23b.

A more complicated experiment was arranged to test small objects, such as coins in luggage. A plastic case was taken as the test object (see Figure 6.24a). The corresponding antenna array (see Figure 6.24b), consisting of 22 small UWB antennas, each of the form presented in Figure 6.24c, was used for precise separation of each small foreign object contained in such luggage (see Section 6.1). An ESP of period $\tau = 0.2$ ns was used for these purposes. The time of observation of the luggage was $T = 5$ s. The objects' resolution sides were around 3 cm.

For this purpose, the UWB tomographic setup, which is shown in Figure 6.25a with the luggage under test, was used, and as the test objects, coins were taken, distributed inside the luggage as shown in Figure 6.25b.

The results of the UWB tomography are shown in Figure 6.26.

In Figure 6.26, very precise separation of coins inside the luggage is clearly seen, with a small resolution of 2–3 cm and with weak artifacts.

6.2.3 UWB Tomographic Synthesis of Object with Metallic Inclusions

The experimental setup shown in Figure 6.27 was also used for tomography imaging of foreign objects hidden in closed areas with metallic inclusions. At the first

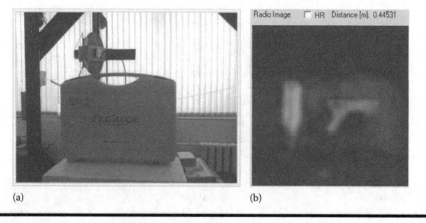

Figure 6.23 (a) Tomographic setup with tested luggage; (b) observed UWB tomogram.

Figure 6.24 (a) Tested plastic luggage; (b) array of UWB antennas; (c) separate view of antenna with its dimensions indicated as diameter of 5–6 cm.

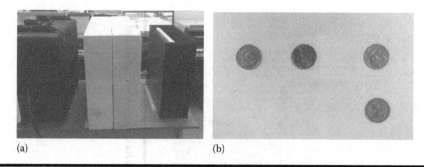

Figure 6.25 (a) UWB setup with luggage; (b) distribution of four coins inside luggage.

step, sounding of the wooden box with metallic edges and with dimensions of $47 \times 41 \times 19$ cm was carried out by use of UWB radiation (see Figure 6.27).

A stepped metallic triangle was positioned inside the box; it had dimensions of each step around 5 cm and a square hole in its center with dimensions of 2×2 cm. Scanning of the transceiver antennas was done with a step of 1 cm in the plane xOy normal to the box. The area of scanning was 60×50 cm, and the test triangle was arranged inside the box, as shown in Figure 6.28, at a range of 39 cm from the radiated scanning antenna. The range was measured along the axis Oz from the transceiver antenna unit.

As the result of UWB scanning and the data processing of the raw data, a 3-D radio tomogram of the inside of the box was performed with the test triangle hidden inside it.

Thus, in Figure 6.29a (top), which corresponds to testing of the top side of the box at a of 18.5 cm range between the scanner and the box, the metallic edges surrounding the top of the box and the top handle in the middle are clearly seen.

Figure 6.29b corresponds to a range of 21.5 cm between the scanning antenna and the object, that is, a deeper penetration of the radiated wave within the box. The wooden plates embedded in the box sides are clearly seen. Figure 6.29c corresponds to a range of 22.5 cm.

Here, the same inhomogeneities as in Figure 6.29b are seen. At a range of 39 cm, which is much deeper inside the box, the triangle starts to be identified (see Figure 6.29d). Figure 6.29e corresponds to the image obtained at a range of 42 cm and shows the rear side of the box. Figure 6.29f shows the shadowing image of the bottom of the triangle, which was expected because the radiation could not penetrate it.

Similar tests have been carried with half-open pure metallic containers, one of which is shown in Figure 6.30, as an outer view of the container's rear side. The stepped metallic object was placed at its middle, as seen from Figure 6.30. As a result of scanning, a tomogram was obtained of the distribution of inhomogeneities inside the test space, which had dimensions of $64 \times 48 \times 41$ cm.

Radio Tomography of Hidden Objects ■ 141

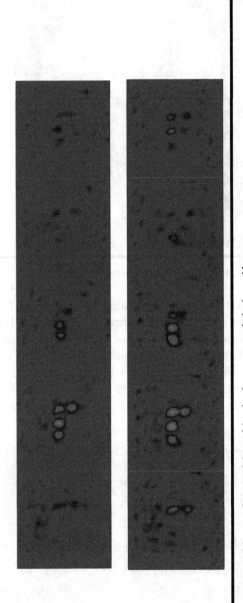

Figure 6.26 Images of four coins positioned in the luggage and their artifacts.

142 ■ *Electromagnetic and Acoustic Wave Tomography*

Figure 6.27 UWB scanner and carrying box.

Figure 6.31 shows the radio images resulting from different 2-D slices of scanning inside a container (i.e., for different depths of scanning). Thus, Figure 6.31a shows the slice that corresponds to a range of 18.5 cm between the scanning transceiver antenna and the rear side (which does not open) of the container. As was expected, only the outline of the container is clearly seen here. In Figure 6.31b, the 2-D slice corresponding to a range of 19 cm is shown. The frame of the door, taken out before testing, is seen. Figure 6.31c shows the 2-D slice corresponding to

Figure 6.28 Test object inside box.

Radio Tomography of Hidden Objects ■ 143

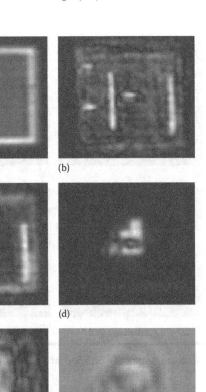

Figure 6.29   Tomogram of carrying box.

Figure 6.30   Test object inside metal safe.

144 ■ *Electromagnetic and Acoustic Wave Tomography*

Figure 6.31 Radio images of sounded area at different distances.

a range of 53 cm. The test object and the hole within it are clearly seen. Finally, in Figure 6.31d, the 2-D slice is shown that corresponds to a range of 62 cm penetration of radiation within the container. This slice corresponds to the rear side of the container.

These few examples of numerous experimental tests allow us to emphasize that use of techniques of tomographic synthesis, described in Section 2.6, permits us to eliminate noises caused by signals reflected from clutter, such as walls, metallic elements, and inclusions. Moreover, the proposed technique allows us to determine the depth of objects hidden or embedded in clutter and, finally, to visualize their shape and form.

Recently, based on the theoretical framework of tomographic synthesis, another application of the proposed methodology was found, namely, testing of big transport vehicles, such as trucks, cars, and so on, for detection and identification of foreign objects hidden inside them, such as weapons, explosives, and so forth. Even though, nowadays, X-ray systems are a more effective scanning system, they are very expensive to use for each side of the object, particularly along the long boundaries. We propose to use a relatively cheaper radio-wave scanning system for detection and visualization of hidden and embedded objects inside transport vehicles, based on usage of UWB tomography and multi-beam antenna grids, a view of which is shown in Figure 6.32. This is an array of 22 UWB antenna elements, with

Radio Tomography of Hidden Objects ■ 145

Figure 6.32 UWB/ESP antenna array with 12 elements.

ESP of 0.2 ns, operating in the frequency band of 2–10 GHz [1–6]. The observation time is around 1.2 s and the resolution of each foreign object hidden in clutter is no more than 3 cm. In other words, each object with dimensions exceeding 3 cm can be detected and identified successfully.

The corresponding experimental setup is sketched in Figure 6.33. Such a system regulates the switching of each element of the UWB antenna array by use of special electro-mechanical commutators.

A vehicle passing through the testing setup with low speed, as shown from Figure 6.33, will allow sounding of its whole volume and discovery of any foreign object hidden in sand, soil, or other construction material.

Figure 6.33 Advanced system for detecting and visualizing hidden objects.

6.3 Tomography Based on Linear Frequency Modulation Radiation

As mentioned in Chapters 2 and 3, to obtain better resolution in radio bandwidth applications with greater perspective, the millimeter-wave bandwidth is used. Their small wave length allows separation during radio sounding and recovery of small elements of the test structure and object or small defects of each detail or element. Moreover, the penetration properties of millimeter waves are satisfactory. To analyze such properties, let us present some corresponding experiments carried out by use of the linear frequency modulation (LFM) signal of the millimeter-wave bandwidth. Here, we will follow the results of measurements and the corresponding signal processing described in details in reference [11]. Thus, Figure 6.34 presents a scheme of measurements, consisting of the transceiver operating at an average frequency of 94 GHz ($\lambda \approx 3.02$ mm) with a corresponding frequency band of 5 GHz. The antenna moves in the xOy plane (across the wave radiation to the object placed at a distance of 50 cm from the transceiver) (Tx/Rx).

The test objects, plastic models of a pistol and a knife, were hidden in a jacket. Then the antenna was moved through the crossing xOy plane with a step of 1.5 mm. Figure 6.35a shows the image of the plastic pistol (gun) after pre-processing of the radio image, while Figure 6.35b shows the reconstruction by use radio tomography technique (i.e., post-processing) together with an actual photograph of the pistol (bottom left).

An example of a physically vivid result of radio tomography using millimeter-wave active LFM radiation is shown in Figure 6.36, where panel (a) presents the modified experimental setup with the knife and plastic pistol positioned behind the non-transparent screen, and the corresponding image obtained after knife reconstruction is shown in panel (b), compared with a simple photograph of the knife

Figure 6.34 Scheme of experimental setup with LFM transceiver and object.

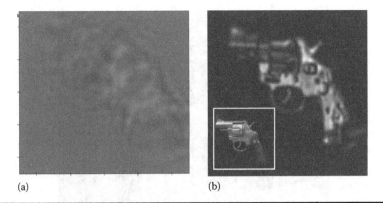

Figure 6.35 Radio tomography of plastic pistol hidden in jacket: (a) initial radio image (after pre-processing); (b) small photo (inset, bottom left) and its reconstruction by use of the radio tomography technique.

(right). In panel (c), we again present the tomogram of the plastic pistol with its photograph at bottom right.

As follows from Figure 6.35a, the radio image of the object was too "blurred" after pre-processing. As expected and following from the theoretical framework presented in Section 2.5, from the spatial distribution of one of the quadrature components of the recording signal reflected from the object at the moment of its maximum contrast, obtained with a step of 1.5 mm, the aureole of the interference picture that was predicted theoretically in Section 2.5 is clearly seen. At the same time, after reconstruction of the object, as is clearly seen from Figures 6.35a and 6.36b and c, the tomogram of the objects obtained after post-processing is very close to their real photos, which proves the theoretical framework proposed in Section 2.5.

6.4 Transmission Tomography of Semitransparent Objects

Transmission radio tomography is related to cross-lighting of the tested areas. As for particular cases, they may be based on shadowing projections or on the Fourier synthesis that was described in Section 2.2. We should again notice that radio tomography differs from X-ray or nuclear magnetic resonance (NMR) tomography, because for radio waves the effects of diffraction and interference of waves and the subsequent multiple interactions lead to a very complicated interference picture with many artifacts. Use of the double focusing method (see Section 2.3) can only decrease these noisy effects, but not definitely overcome them. To obey such problems, in Section 2.4 the phase approximation of the Kirchhoff method was proposed as the simplest and most effective method to solve inverse problems on

148 ■ *Electromagnetic and Acoustic Wave Tomography*

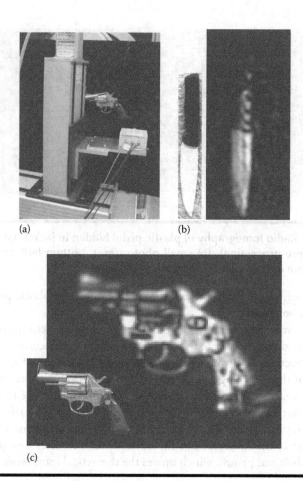

Figure 6.36 (a) Setup with test objects behind screens; (b) radio tomography image of knife with its photo; (c) radio tomography image of plastic pistol with its photo.

recovery in transmission wave tomography of semitransparent media and objects. The subject is based on the results obtained in references [17–22].

For descriptions of transmission tomography of semitransparent and non-transparent media, the approximation of the physical theory of diffraction was briefly given in Sections 2.5 and 2.6, respectively. Here, we will not explain it again, but will describe experiments and the obtained experimental data, comparing them with the theoretical prediction.

6.4.1 Experimental Setup

The setup provides direct focusing as well as inverse focusing at the hardware level. The basic optical circuit and a photograph showing the external appearance of

Radio Tomography of Hidden Objects ■ 149

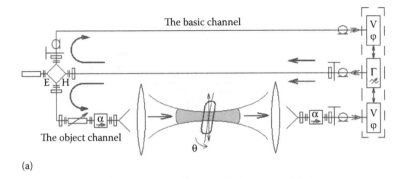

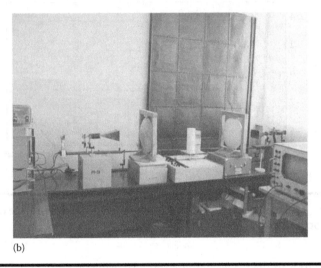

Figure 6.37 (a) Basic optical circuit and (b) photograph of transmission tomography setup.

the transmission tomography setup are shown in Figure 6.37. The current system makes it possible to measure the complex transmission coefficient of the wave channel. Measurements of the amplitude and the phase of the radiation penetrating the object were performed using a P4–36 panoramic transmission coefficient meter. This meter is outlined by the dashed rectangle in Figure 6.37a.

To improve the accuracy of measurement of the phase values of the recorded field (i.e., of the scanned area), a reference channel and an information channel are included in the circuit. The setup operates in the range 8–12 GHz, with a midfrequency of 10 GHz. Vertical wave polarization was applied. The radiation was focused with two lenses made of gypsum. Each lens was 32 cm in diameter, and the focal distance was 16 cm. The focusing system and amplitude-phase distribution of the field in the focusing area are depicted in Figure 6.38. Similar figures, 2.14 and 2.15, were presented in Chapter 2. We show them again here as a matter of

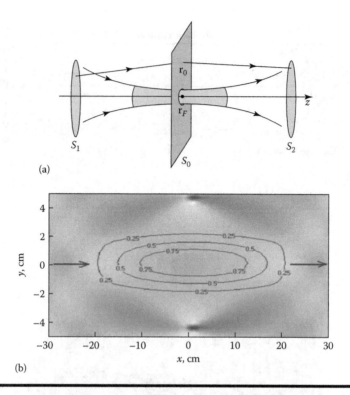

Figure 6.38 (a) Focusing system and (b) computed amplitude-phase distribution of field in focusing area.

convenience to help the reader to understand transmission tomography. The point of combined focusing of both lenses is denoted as r_F.

The region of localization of focusing has the shape of a revolving solid 3 cm in diameter and 30 cm in length. Figure 6.38b depicts the distribution of the calculated values of the phase and amplitude of the focusing field for the case where the transmitting and receiving apertures are separated by 90 cm. Intensity contours, which define the active zone of the wave channel, are displayed in the middle of the figure. The active zone, which is important for propagation of radiation, is the region of maximum interaction of the field with the propagation medium. Of course, this is true for relatively weak interactions. The distribution of phase values is represented in gray-scale. The field in the focusing region has an approximately planar phase front and small phase shifts across the wave channel. In other words, an approximately collimated wave beam is realized.

The amplitude-phase distribution of the wave projections was recorded with the P4–36 meter and controlled with a PC. The true accuracy of the measurements was estimated as ± 1 dB for the amplitude and $\pm 5°$ for the phase. Linear motion of the test object and its rotation about the active zone of the wave channel were used to obtain multi-angle wave projections of the object. The platform was constructed on the base of an x-y flatbed recorder. The control voltages were assigned using two digital–analogue converters, which were controlled by the PC.

6.4.2 Tomography of Semitransparent Objects

Tomographic scanning of an inhomogeneous test scene was performed by means of the setup described in Section 6.4.1. The test scene consisted of two semitransparent figures—two cylinders, one of circular cross-section and the other of square cross-section—filled with polystyrene beads.

Full-angle images of the distribution of the amplitude attenuation and phase disturbance of the wave field that were obtained experimentally are displayed in Figure 6.39. Their behavior is similar to that presented in Figure 2.16 (see Chapter 2). Variation of the measured values are displayed in gray-scale. The areas that are darker gray correspond to more intense amplitude attenuation and greater phase disturbance. The axial symmetry of the two figures reflects the double redundancy of the obtained projections because rotation of the object by 180° should not change these projections. The observed differences of opposite-side projections indicate a good level of measurement accuracy. The differences can be averaged during processing.

The shape of the scanned objects is depicted in Figure 6.39a. It should be noticed that an abrupt (non-monotonic) attenuation of the amplitude of the transmitted radiation is observed when a focused ray is transmitted through the side boundaries of the object, and then the amplitude increases. This is shown in Figure 6.39b and can be explained by the ray refraction effect at the side boundaries. Everything varies monotonically in the phase image; the ray transmission through the side boundary does not cause non-monotonic changes in the phase (Figure 6.39b). Note the smaller phase disturbance due to the rectangular cylinder, in contrast to the circular cylinder. A distinctive characteristic here is a clearly observed gap in the images, which is explained by passage of the focused ray through the space between the cylinders.

By the way, the obtained wave projections (Figure 6.39) do not resemble the shapes of the test objects (see also Figure 2.16a). For adequate reconstruction of the tomogram, the dominant effects of the interaction of radiation with the propagation medium must be taken into account. Such a reconstruction was based on the analytical model of wave projection described in Section 2.1.

Finally, we can outline that the transmission radio-wave tomography is a direct development of X-ray and NMR tomography, but, in contrast to these methods, it takes into account the wave nature of radiation, which is manifested in multiple diffraction and interference interactions.

It has been shown that double counter focusing provides effective localization of the interaction of radiation with matter and reduces the impact of multiple interactions. It has been established that phase relations are of particular importance

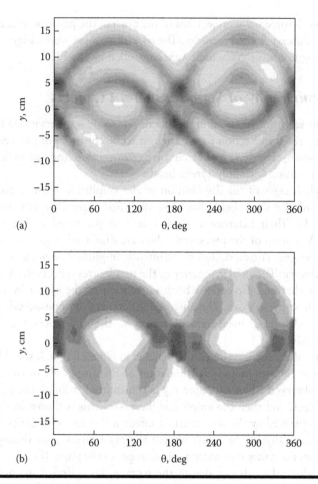

Figure 6.39 (a) Full-angle images of distribution of amplitude attenuation and (b) phase disturbance of wave field when test scene is scanned.

here, and that the phase approximation of the Kirchhoff method, introduced in Chapter 2, provides an efficient means with which to describe them. It has been demonstrated that obtaining the phase portrait of the wave projections in this approximation reduces the inverse problem of shadow projection tomography. All of the results, presented here and in Chapter 2, have been validated with numerical simulations and confirmed experimentally.

6.5 Radio Tomography of Non-Transparent Objects

When sounding non-transparent (opaque) objects, the radiation does not penetrate them, so the main information provided consists of their shapes. Of course, surface reflectivity is also an important factor, but the recovery of object shapes provides far more information about the objects. This information is essential, for example, for identification of hidden objects and for estimation of their integrity and dynamics. All metallic objects and living beings come under the heading of such objects. The water content of dielectric objects renders them opaque. Radio-location (radio-detection) makes it possible to specify types of aircraft, their purpose and weapon systems, and also to detect underground cables and other underground structures. The range of possible applications is extremely wide. Section 2.6 summarizes the working results of references [23–29].

6.5.1 Transmission Tomography of Form of Radio-Opaque Object

All possible terminators create a set of shadow projections, whose inversion is possible by means of the Fourier synthesis technique described in Section 2.2. The characteristic function of the terminator $F(\mathbf{r}_s)$ can be reconstructed by the radio-wave tomographic synthesis method described in Section 2.4. The inverse focusing method discussed there is an equivalent technique for reconstructing the characteristic function of an object.

The following modeling experiment (see Figure 6.40) was performed to illustrate the proposed approach. An opaque object was placed on a turntable (3).

This makes it possible to perform multi-angle measurements of the diffraction field. The object was illuminated by transmitting antenna 2, which was moved behind the object along a straight line. Simulation of the diffraction field was based on calculation of the Kirchhoff integral (see Section 2.4 and the results of numerical simulation obtained there). The numerical simulation results were completely

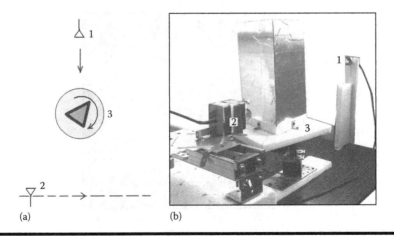

Figure 6.40 (a) Scheme and (b) photograph of experimental setup for transmission tomography of opaque object.

confirmed by experiment. The experiment was performed according to the scheme depicted in Figure 6.40a. The experimental setup included: a fixed radiating antenna (*1*), a mobile receiving antenna (*2*), and a turntable (*3*) on which the test object was placed.

The receiving antenna was mounted on the line scanner and could be moved in a straight line. The range of motion of the receiving antenna was $L = 60$ cm, the distance from the center of rotation of the object to the line of antenna movement was $R = 23$ cm, and the signal frequency was 18 GHz. The widths of the sides of the triangular cylinder (the side lengths of its triangular face) were 195, 185, and 155 mm. UWB antennas (described in Section 6.1) were used as the receiving and transmitting antennas. The object was rotated on the turntable to 32 different positions in the angular range from 0° to 180°. The tomogram of the object displayed in Figure 6.41 was reconstructed by processing the measured field using the methodology described in Chapter 2.

It can be seen that the shape of the object was reconstructed, however, with some distortions that were associated with measurement noise and the finite value of the wave length ($\lambda = 1.7$ cm). Increasing the frequency (decreasing the wave length) of the sounding radiation should increase the resolution of the image of the object and thereby present its shape more clearly. This can be checked most easily with ultrasound. Ultrasonic transceiver sensors operating at a frequency of 40 kHz are widely available. Noting that the speed of ultrasound in air is $v = 330$ m/s, we obtain the value $\lambda = v/f = 8.25$ mm for the wave length. We decreased the wave length by about a factor of two. The experimental setup used was the same, but we shortened the scanning interval to $L = 37.5$ cm.

The results of measurements of one of the quadratures of the wave projection for the triangular cylinder are presented in Figure 6.42, and the results of

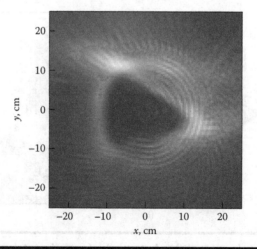

Figure 6.41 **Reconstructed tomogram of triangular cylinder.**

Radio Tomography of Hidden Objects ■ 155

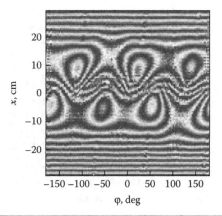

Figure 6.42 Measurement results for wave projection of triangular cylinder.

reconstruction of its shape are shown in Figure 6.44. It can be seen that the spatial resolution has been substantially improved, even though the scanning interval was shortened somewhat. For comparison of radio and acoustic methods of tomography, the experiment with ultrasonic transmission tomography was repeated with two test objects of different shapes: a square cylinder with a side length of 57 mm and a round cylinder 28 mm in diameter. Corresponding results are presented in Figures 6.43 and 6.45.

It can be seen that the objects differ from each other, the object separation is about 7 cm, and one object is twice as large as the other one. However, the shape of the square cylinder is hardly discernible, and the spatial resolution achieved is no better than 3 cm. Arguing against the proposed methodology, it can be outlined

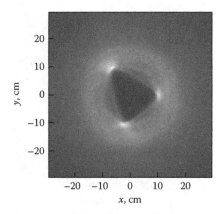

Figure 6.43 Ultrasonic tomogram of triangular cylinder.

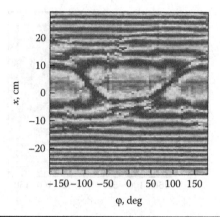

Figure 6.44 Measurement results for wave projection of a square and a round cylinder.

that the shadow transmission wave projection method is not efficient at reconstructing the shapes of objects that produce self-shadowing and mutual-shadowing effects.

6.5.2 Multiple-Angle Tomography of Opaque Objects' Shapes

The location sounding scheme that is the most efficient at reconstructing the shapes of opaque objects is the one in which the radiating and transmitting antennas are located on the same side of the object. First, this scheme is the easiest to realize, and second, there is no shadowing at most of the sounding angles. As was pointed out in Section 6.5.1, this last circumstance has a significant effect on the accuracy of reconstruction of the shape of an opaque object.

The main mathematical methods of inverting wave projections for radio-location purposes were considered in Chapter 3. As mentioned there, a number of

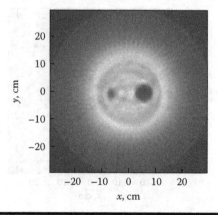

Figure 6.45 Ultrasonic tomogram of a square and a round cylinder.

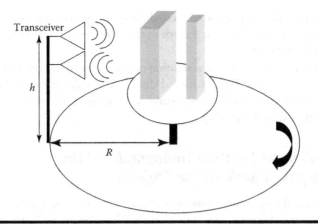

Figure 6.46 Object viewed from all sides.

peculiarities in the scanning of opaque objects was observed that were considered *a priori*. As was shown, in the case of partial *scanning* of the object by the sounding point, the form of the characteristic function is uniquely determined only for the illuminated areas, so the shape of the object can be reconstructed for those areas only. Panoramic scanning of the object from all sides is required for complete reconstruction of the shape of an opaque object (see Figure 6.46).

Experiments were performed on the setup shown in Figure 6.40 with the triangular cylindrical object used previously. The measurements were performed using the R4M–18 vector network analyzer produced by MICRAN (Russia). The scanning radius was $R = 67$ cm, and the frequency bandwidth was 1–18 GHz. The distribution of one of the quadratures of the radar response is presented in Figure 6.47 as a function of the scanning angle and scanning frequency. It can be clearly seen that the distribution is sharper in the high-frequency range. The corresponding spatial

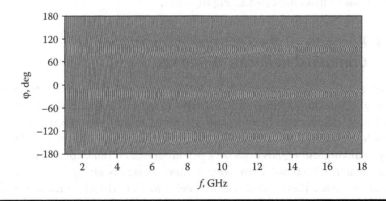

Figure 6.47 Measured distribution of radar response of triangular cylindrical object.

spectrum of the radar responses, converted into Cartesian coordinates, is displayed in Figure 6.48a. The reconstructed shape of the characteristic function of the object is shown in Figure 6.48b.

It can be seen that the spatial resolution of the obtained images is of the order of 1.5 cm. This sharp resolution is achieved thanks to a wide frequency bandwidth of 1–18 GHz. In contrast to the transmission scheme, there are no shadowing effects in multi-angle location sounding.

6.5.3 Unilateral Location Tomography of the Shapes of Radiopaque Objects

Let us consider the situation where location (detection) of an opaque object is performed at a certain frequency in some plane S. We make use of the inverse focusing method described in Chapter 3. As was shown there (see Figures 3.10 and 3.11), a good agreement between the predefined and reconstructed shapes of the parabolic reflector, without and with deformation, was observed.

A dish antenna of 66 cm in diameter with a focal length of 30 cm was investigated to validate the method experimentally (Figure 6.49a). The sounding distance was 56 cm. The antenna was scanned by ultrasound. The signal frequency was 40 kHz, which corresponded to a wave length of 0.82 cm and electromagnetic radiation with a frequency of 36 GHz. From the scattered field data, the shape of this parabolic reflector was reconstructed. It was found that the shape was nearly ideal. The distribution of detected deviations over the reflector surface is shown in Figure 6.49b. The lighter areas correspond to downward deviations from the regular shape, and the darker areas to upward deviations.

It can be seen that the parabolic reflector has smooth distortions of its shape extending over its entire surface and reaching 1 cm at its periphery, and local dimple-shaped non-uniformities up to 2 mm in depth. This is in good agreement with the positions of flaws indicated in Figure 6.49a.

6.5.4 Recovery of the Focusing Properties of Combined Reflector Antennas

The same method as in Section 6.5.3 was used for calculating the distortions of the surface of a reflector antenna, which showed it to be of practical significance. The point here is that systems in which surface distortions of the reflector antenna are present lose their ability to focus signals efficiently and, as a result, also lose their ability to distinguish signals from background noise and those coming from neighboring angular directions. There are a number of factors giving rise to defects on reflector antennas. These include thermal deformations, defects in the materials the antennas are made from, operational damage (damage incurred during the course of use), and so on.

Radio Tomography of Hidden Objects ■ 159

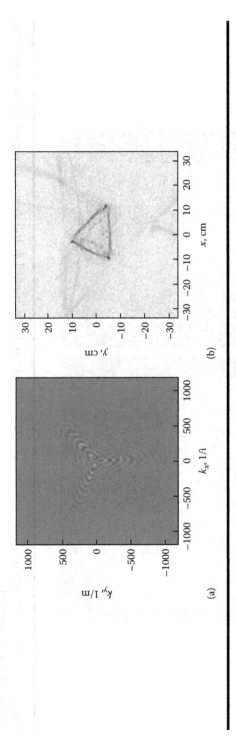

Figure 6.48 (a) Spatial spectrum and (b) reconstructed tomogram of test object.

160 ■ *Electromagnetic and Acoustic Wave Tomography*

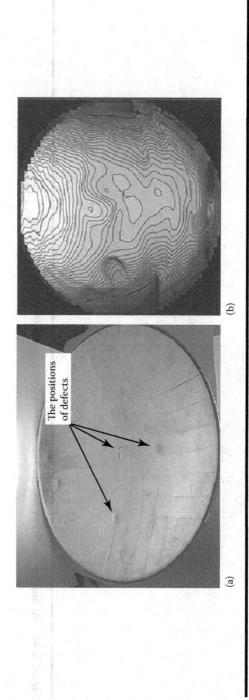

Figure 6.49 Surface of real parabolic antenna: (a) external view; (b) reconstructed distribution of deviations of its shape from that of a perfect reflector.

All of these lead to degradation of system performance or even to complete system breakdown. Large antennas mounted on communication and remote sensing satellites are especially sensitive to distortions. Actual replacement of such antennas is hardly possible, and the devising of curvature-correction systems is quite a difficult engineering task. Distortion of the optimal shape of a parabolic reflector ultimately results in partial or complete failure of the satellite system, which in turn leads to significant time and money lost on satellite repair.

The solution for tomographic measurement of the shape of a reflector antenna proposed here can be used to achieve additional focusing of signals in the antenna focal plane without resorting to mechanical correction of such distortions (deformations). Thus, the power gain of the antenna can be restored simply by suitable signal processing. Correcting for the effect of such deformations is possible on the mathematical level, provided the phase-amplitude spatial distribution of the field is completely recorded by an antenna array located near the reflector focal plane (see Figure 3.8). Thus, the problem that needs to be solved is split into two interrelated tasks: first, to define the geometrical surface flaws of the large satellite reflector antenna and, second, to take these flaws into account when restoring the antenna power gain without resorting to mechanical correction of the reflector.

To solve the first problem, it is proposed to use measurements of the field from a known reference source and radio tomography to reconstruct the surface curvature of the reflector antenna. This problem is solved by applying the fast Fourier transform, followed by reconstruction of the reflector antenna flaws using the phase distribution of the spatial spectrum of the recorded field (see Chapter 3). At the basis of this solution lies the assumption that all distortions of the spatial distribution are caused by geometrical distortions of the reflector antenna surface, and also by a high residual localization of the field near the focus, even in the presence of surface flaws.

The spatial frequency spectrum for a deformed reflector, calculated using the 2-D Fourier transform, is shown in Figure 6.50a. The asymmetry of the spectrum is caused by deformations. Using a perfect parabolic reflector here would give the spectrum shown in Figure 6.50b. To reconstruct the shape of the deformed dish, it is sufficient to recalculate the spatial frequency spectrum onto the aperture plane of the reflector and, then, using the inverse Fourier transform, to reconstruct the distribution of the complex amplitude. The shape of the reflector is uniquely determined by the phase distribution of the complex amplitude over the aperture.

A reference source located near the reflector is considered to be a convenient tool to eliminate the effect of deformations. The corresponding spatial frequency spectrum in the receiving array is found by convolving the spectrum of the reference source (known) with the transfer function of the focusing system, which includes within itself the effect of geometrical distortions. In the case of any other source, the spatial frequency spectrum in the receiving array also results from convolution of its (unknown) spectrum with the same transfer function of the focusing system as mentioned. Eliminating the imperfect transfer function of the focusing

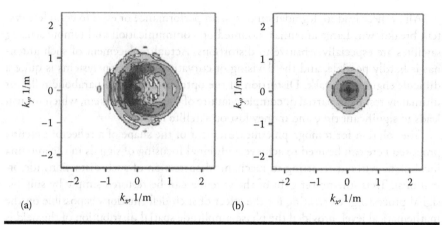

Figure 6.50 Spatial frequency spectrum in receiving array of combined reflector antenna (a) before and (b) after correction for antenna surface deformation.

system from these relations makes it possible to calculate the actual spatial frequency spectrum of this source.

Convolution of this spectrum with the transfer function of a perfect paraboloid yields maximum gain at the focal point. It is significant that all of this can be performed at the microprocessor level without resorting to any mechanical manipulations of the reflector surface. As a result, the antenna power gain can be restored to its maximum possible value. The equivalent directivity pattern (DP) of such an antenna is symmetrical, in contrast to the distorted initial DP (Figure 6.51). As illustrated in the figure, the maximum value of the amplitude has been increased, and the width of the DP has been reduced.

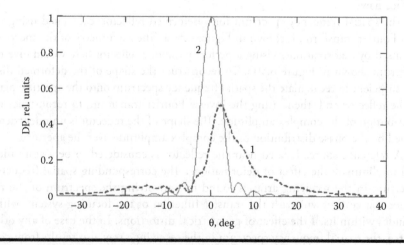

Figure 6.51 Directivity pattern of reflector antenna before (curve *1*) and after (curve *2*) correction of surface deformation.

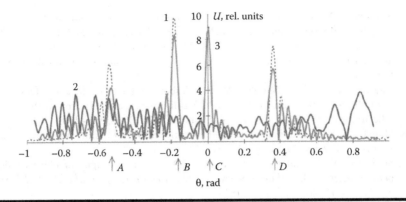

Figure 6.52 Example of simultaneous reception of signals from different directions (a–d): *1*: perfect antenna, *2*: defocused antenna, *3*: restored (compensated) antenna.

Thus, it becomes possible to reconstruct the shape and restore the power gain of a satellite-based antenna without resorting to mechanical correction of the curvature of the reflector. Both problems are interrelated and solved using computer technologies and, what is most significant, they permit the use of fast algorithms; that is, their solutions can be realized in real time. The effect of focusing–defocusing on signal detection from different directions is illustrated in Figure 6.52.

To conclude, we should emphasize that the radio tomography of opaque objects reduces to the reconstruction of the object shapes. Multi-angle wave sounding enables one to reconstruct their shapes and the corresponding spatial frequency spectra. At the same time, transmission tomography of the shapes of opaque objects is limited by the requirement that no area on the surface of the object overshadows other areas. There are fewer limitations due to shadowing requirements in unilateral location tomography. The approximation of geometrical diffraction theory suffices to describe the diffraction effects that arise. The results obtained using these methods have a wide range of application. The reconstruction and correction of distortions of the curvature of large parabolic reflectors by means of wave sounding is one example.

References

1. Balzovsky, E. V. and Yu. I. Buyanov, Ultra-wideband antenna element for the synthesize aperture, *Russian Physics Journal*, vol. 55, no. 8/2, 2012, pp. 60–61.
2. Satarov, R. N., I. Yu. Kuz'menko, T. R. Muksunov, et. al., Commutative ultrawideband -antenna array for radio vision, *Proceedings of III Scientific Conference Information-Measuring Equipment and Technologies*, Tomsk, Russia, 2012, pp. 208–211.

3. Satarov, R. N., I. Yu. Kuz'menko, T. R. Muksunov, et al., Commutative antenna array for radio tomography, *Russian Physics Journal*, vol. 55, no. 8, 2012, pp. 26–30.
4. Satarov, R. N., I. Yu. Kuzmenko, T. R. Muksunov, et al., Switched ultrawideband antenna array for radio tomography, *Russian Physics Journal*, vol. 56. no. 8, 2013, pp. 884–889.
5. Satarov, R. N., S. E. Shipilov, V. P. Yakubov, and E. O. Stepanov, Device for 2D-radiotomography on the basis of UWB-linear tact antenna array with focusing reflector, *Russian Physics Journal*, vol. 56, no. 8/2, 2013, pp. 198–200.
6. Shipilov, S. E., Method of synthesized aperture in the 3D-radiotomography, *Russian Physics Journal*, vol. 57. no. 9, 2013, pp. 70–71.
7. Sukhanov, D. Y. and K. V. Zav'yalova, 3D radio tomography of objects hidden behind dielectrically inhomogeneous shields, *The Russian Journal of Applied Physics*, vol. 60, no. 10, 2015, pp. 1529–1534.
8. Yakubov, V. P., S. E. Shipilov, D. Ya. Sukhanov, and A. K. Razinkevich, Ultra-wideband tomography of remote objects, *Russian Journal of Nondestructive Testing*, vol. 48, no. 3, 2012. pp. 191–196.
9. Yakubov, V. P., A. S. Mirontchev, A. G. Andretsov, and I. O. Ponomaryova, Artificial materials and radio tomography, *Russian Physics Journal*, vol. 53, no. 9, 2011, pp. 895–899.
10. Yakubov, V. P., S. E. Shipilov, and R. N. Satarov, Ultra-wideband sensing behind dielectric barriers, *Russian Physics Journal*, vol. 53, no. 9, 2011, pp. 887–894.
11. Yakubov, V. P., S. E. Shipilov, D. Ya. Sukhanov, and A. V. Klokov, *Radiowave Tomography: Achievements and Perspectives*, Tomsk, Russia: Scientific-Technical Literature (STL) Edition Group, 2016 (in Russian).
12. Balzovsky, E. V., V. I. Koshelev, and S. E. Shipilov, Detection of metallic objects behind wall during sounding ultra-wideband pulses, *Russian Physics Journal*, vol. 55, no. 8/2, 2012, pp. 24–28.
13. Sukhanov, D. Ya. and M. D. Sovpel', Ultra-wideband radiolocation on measurements of amplitude-phase field distribution at cylindrical surface, *Russian Physics Journal*, vol. 56, no. 8/2, 2013, pp. 203–206.
14. Sukhanov, D. Ya. and E. G. Permyakov, Estimation of one-dimensional distribution of dielectric permittivity at the flat surface based on bi-static measurements of the reflected field, *Russian Physics Journal*, vol. 56, no. 8/2, 2013, pp. 92–96.
15. Yakubov, V. P. and A. S. Mirontchev, Focusing by use of metamaterials, *Russian Physics Journal*, vol. 55, no. 9/2, 2012, pp. 84–85.
16. Yakubov, V. P., S. E. Shipilov, and R. N. Satarov, Ultra-wideband tomography of moving objects behind dielectric barriers, *Journal of Control and Diagnostics*, Special Issue, 2011, pp. 89–91.
17. Yakubov, V. P. and S. A. Slavgorodsky, Microwave tomography – experimental model, *VIII Joint International Symposium "Atmospheric and Ocean Optics. Atmospheric Physics"*. Symposium Proceedings, Eds V.A. Banakh and O.V. Tikhomirova, Tomsk: Institute of Atmospheric Optics SB RAS, vol. 8, 2001, pp. 128–129.
18. Yakubov, V. P. and S. A. Slavgorodsky, Model radiowave tomography, *Journal of Radioelectroniks*, no. 10, 2001, http://jre.cplire.ru/win/oct01/6/text.html.
19. Yakubov, V. P. and M. L. Masharuev, Method of double focusing for microwave tomography, *Microwave and Optical Technology Letters*, vol. 13, no. 4, 1996, pp. 187–189.

20. Yakubov, V. P., Yu. K. Tarabrin, and M. L. Masharuev, Regularization of inverse problems of microwave tomography in medicine, *Proceedings of SPIE*, vol. 2570, 1995, pp. 332–335.
21. Sukhanov, D. Y. and A. A. Murav'eva, Monochromatic ultrasonic transmission tomography, *Optoelectronics, Instrumentation and Data Processing*, vol. 51, no. 3, 2015, pp. 247–253.
22. Yakubov, V. P., E. D. Telpukhovskiy, G. M. Tsepelev, et al., Radio-wave tomography of inhomogeneous media, *Russian Physics Journal*, vol. 49, no. 9, 2006, pp. 923–928.
23. Koshelev, V. I., S. E. Shipilov, and V. P. Yakubov, Reconstructing the shape of an object in the narrow range-of-view super wideband radio location, *Journal of Communication Technology and Electronics*, vol. 44, no. 3, 1999, pp. 281–284.
24. Yakubov, V. P., D. V. Losev, and A. I. Maltsev, Wave tomography of absorbing media, *Journal of Communication Technology and Electronics*, vol. 49, no. 1, 2004, pp. 54–57.
25. Yakubov, V. P., S. E. Shipilov, and S. V. Ponomarev, Dual software and hardware refocusing of satellite reflectors, *Proceedings of CriMiCo – 2014 – 24th International Crimean Conference Microwave and Telecommunication Technology*, Art. No. 6959522, pp. 549–550.
26. Yakubov, V. P., S. E. Shipilov, and D. Ya. Sukhanov, Microwave tomography of radiopaque objects, *Russian Journal of Nondestructive Testing*, vol. 47, no. 11, 2011, pp. 765–770.
27. Yakubov, V. P., S. E. Shipilov, D. Ya. Sukhanov, and A. K. Razinkevitch, Ultrabroadband tomography of remote objects, *Russian Journal of Nondestructive Testing*, vol. 48, no. 3, 2012, pp. 191–196.
28. Koshelev, V. I., S. E. Sipilov, and V. P. Yakubov, Shape reconstruction in small-angle super-wideband radar using genetic functions, *Journal of Communication Technology and Electronics*, vol. 45, no. 12, 2000, pp. 1333–1338.
29. Klokov, A. V., V. P. Yakubov, S. E. Shipilov, and V. I. Yurchenko, Development of the vision systems based on radiowave using a Luneburg lens focusing for robots, *Proceedings of SPIIRAS*, vol. 2, no. 45, 2016, pp. 130–140.

Chapter 7
Proof of Specific Radio Tomography Methods

Vladimir Yakubov, Sergey Shipilov,
Dmitry Sukhanov, and Andrey Klokov

Contents

7.1	Ultra-Wide Band Incoherent Tomography	168
	7.1.1 Overview of Problem	168
	7.1.2 Spatial Testing with Non-Filled Aperture	169
	7.1.3 Time-Tact of Objects in the Frequency Domain	172
7.2	UWB Tomography of Non-Linear Inclusions	177
	7.2.1 Current Statement of Problem	177
	7.2.2 UWB Tomography of Non-Linearities	180
	7.2.3 Signal Processing and Algorithm of UWB Images of Non-Linear Inclusions	182
7.3	Doppler Tomography Experimental Proof	185
	7.3.1 Microwave Doppler Sensor and Location Sounding	185
	7.3.2 Reconstruction of Objects: Imitational Modeling	189
	7.3.3 Positioning System and Manual Doppler Scanning Device	191
	7.3.4 Doppler Sub-Surface Tomography	195
References		197

This chapter is based on results obtained and described in papers published by three different groups of researchers separately and mutually [1–36].

7.1 Ultra-Wide Band Incoherent Tomography
7.1.1 Overview of Problem

Incoherent tomography is taken to mean tomography in the case when the radio-wave transmitter and receiver are not synchronized, so that the relative phase difference between the transmitted and received signals cannot be measured. A number of widely used methods are available to reconstruct the radar images from incoherent radiation.

The method based on using a narrow-beam antenna or large-aperture antenna array is one of the most popular of such methods. A directional antenna allows one to measure the amplitude of the waves coming in from a certain direction, and also to reconstruct a 2-D radio image with high angular resolution. Multi-position measurements of the angular distribution of the amplitude of the field are required to obtain 3-D images. The 3-D image reconstruction techniques, in this case, are very similar to algorithms used in X-ray tomography. It should be noted that active, as well as passive, tomography is used for this purpose; the waves can be emitted by an outside source or the test objects themselves can act as radio-wave sources.

For example, stars, particle clouds, and relic background radiation are considered as wave sources in radio astronomy [1]. From 2001 to 2009, the NASA Wilkinson Microwave Anisotropy Probe (WMAP) space vehicle sent back data from orbit and enabled the construction of a celestial map with a resolution of 12′ (angle minutes) in various frequency ranges between 23 and 93 GHz [2].

Another radio tomography technique applied to incoherent radiation is based on the use of the radio-holographic principle for reception [3–6]. This method measures the amplitude of the interference field of the direct wave (reference wave) from the transmitter $E_0 = A_0 \exp\{i\varphi_0\}$ and the wave scattered by the test object (object wave) $E_1 = A_0 \exp\{i\varphi_1\}$ with the help of a non-directive antenna. Generally, the amplitude of the reference wave is far larger than the amplitude of the object wave: $A_0 \gg A_1$. It turns out that the amplitude of the interference pattern, in this case, is directly related to one of the quadratures of the object wave field:

$$A = |E_0 + E_1| = \sqrt{A_0^2 + 2A_0 A_1 \cos(\varphi_1 - \varphi_0) + A_1^2} = A_0 + A_1 \cos(\varphi_1 - \varphi_0) \quad (7.1)$$

The cosine field quadrature can be estimated as

$$C_1 \equiv A_1 \cos(\varphi_1 - \varphi_0) = A - A_0 \quad (7.2)$$

The phase is determined accurately to within a constant, namely, the phase of the reference wave. This constant phase shift does not affect focusing. This is all that is needed to apply the synthetic aperture technique and perform controlled focusing and, finally, to implement radio-tomographic synthesis. In fact, the presence of a reference wave, in a certain sense, enables the solution of the so-called *phase problem*. This technique is important for the terahertz waveband, for example, because standard high-frequency methods are not applicable to this band, as mixers and waveguide transmission lines require high accuracy in production, but this is very difficult to achieve. This approach is important for incoherent X-radiation as well, where the coherence is provided by stimulated radiation with high pump power. At the same time, phase reconstruction promises to deliver an increase in spatial resolution. The proposed approach is analyzed in more detail next.

For the sake of even-handedness, we note that in the case when only the amplitude of the object wave is measured and use of the two above-mentioned techniques (with narrow-beam radiation and analysis of the interference field of the reference and object signals) is impossible, iterative algorithms have been proposed in the literature to reconstruct the amplitude-phase field distribution [6, 7]. The Gerchberg–Saxton iterative algorithm is a good example of this [6]. The problem of reconstructing the image of the object from measurements of the amplitude of the scattered field in some region at some distance from the object is considered. The approximate shape of the scattering object is used in the initial step of the algorithm. Then, the wave field propagated from this object is calculated, whose intensity (or amplitude) is assigned on the basis of the shape of the scattering object and whose phase is determined by the phase of the incident radiation. Thus, the field in the measurement region is calculated, but its amplitude differs from the measured amplitude because of inaccuracies in the initial approximation; however, a phase distribution is obtained. Next, the amplitude of the calculated field is replaced by the measured amplitude, after which the inverse problem of reconstruction of the image of the scattering object is solved. The next iteration is performed on the basis of the obtained solution. The iterative process continues until the calculated amplitude coincides with the measured one. This algorithm converges under certain conditions. In practice, this algorithm is hardly ever used. The chapter is based on summarized results presented in references [7–13].

7.1.2 Spatial Testing with Non-Filled Aperture

Radio-holographic methods are considered to be a promising development of radio-tomographic systems in the terahertz and sub-terahertz bands, because they require measurements only of the field intensity and do not require any measurements of the phase. Phase measurements in these frequency bands are a complex technical problem. Nevertheless, phase information is needed to reconstruct images with resolution near the diffraction limit. However, in radio-holographic systems, partial preservation of the phase information in the intensity of the interference image is possible, due to interference of the reference and object signals. This makes it possible to reconstruct the radio image with the highest possible resolution, as shown in references [8, 9].

One of the problems related to the use of the radio-holography method according to the Nyquist–Shannon–Kotelnikov theorem is the need to measure the field amplitude with intervals of half a wave length, which requires a significant amount of measuring time under conditions of mechanical scanning or the use of expensive filled sensor arrays. A method that enables the use of sparse measurements of the field amplitude and does not result in artifacts and secondary maxima [10] will be covered in this section. This radio-holographic method for reconstructing radio images is based on sparse measurements of the amplitude of the interference field of the object signal and different reference signals (from different sources). Minimization of the level of artifacts and secondary maxima is assumed to obtain as a result of optimal placement of the radiating elements with respect to the measurement range.

The problem of narrowband radio holography based on sparse measurements of the field intensity using several reference sources is considered (see Figure 7.1). Intensity measurement points are marked with circles in Figure 7.1 and sources are marked with stars. It is assumed that the measurement points are located in a plane at a distance h_2 from the plane where the transmitters (sources) have been placed. The test object, in the shape of a stepped triangle, is placed at a distance h_1 from the plane with the transmitters. The transmitters are considered to be point sources and isotropic. The number of transmitters can be varied. The more transmitters that are used, the more accurate information about the scattering object can be obtained. It is assumed that the transmitters are not working simultaneously, but are clocked; that is, only one transmitter is working in a given time tic. As a result, different interference patterns will be obtained for different transmitters in the measurement plane.

To validate the feasibility of the proposed approach, we performed a numerical simulation of this problem for a predefined distribution of transmitting $\mathbf{r}_0$ and receiving $\mathbf{r}$ elements (Figures 7.2 and 7.3). A stepped triangle with a step size of 10 cm was used as the test object. The distance from the receiving array to the test object was 70 cm, while the distance between the array of transmitters and the test object was 50 cm. The radiation frequency was 20 GHz.

The scattered field $E_1(\mathbf{r})$ was calculated using the Kirchhoff approximation. The reference wave was assigned as the field of a point source $E_0(\mathbf{r}) = G_0(\mathbf{r} - \mathbf{r}_0)$. Thus, the

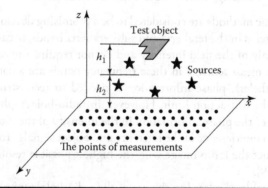

Figure 7.1 Layout of measurements.

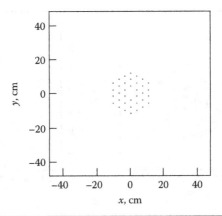

Figure 7.2 Matrix composed of 36 radiating elements.

information component of the resultant field according to Equations 7.1 and 7.2 is equal to

$$C_1 = |E_0 + E_1| - |E_0|$$

The tomography inverse problem can be solved using the inverse focusing technique, according to the formula

$$U(\mathbf{r}_F) = \sum_m \sum_n M(\mathbf{r}_F, \mathbf{r}_m, \mathbf{r}_{0,n}) E(\mathbf{r}_m, \mathbf{r}_{0,n}) \qquad (7.3)$$

where $C_1(\mathbf{r}_m, \mathbf{r}_{0,n})$ is substituted for $E_1(\mathbf{r}_m, \mathbf{r}_{0,n})$, and the focusing function is defined as in Section 3.1.4, as (we present it again for the reader's convenience):

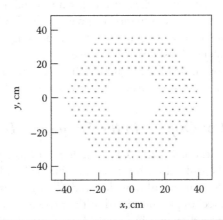

Figure 7.3 Matrix composed of 270 receiving elements.

172 ■ *Electromagnetic and Acoustic Wave Tomography*

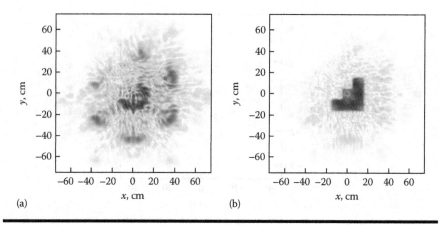

Figure 7.4 Reconstruction of image of test object: (a) using one source; (b) using 36 sources.

$$M(\mathbf{r}_F, \mathbf{r}_m, \mathbf{r}_{0,n}) = \exp\left\{-ik\left(|\mathbf{r}_F - \mathbf{r}_m| + |\mathbf{r}_F - \mathbf{r}_{0,n}|\right)\right\} \quad (7.4)$$

where:

$\mathbf{r}_m$ and $\mathbf{r}_{0n}$ are the transmitting and receiving points
$\mathbf{r}_F$ is the target focusing point

Figure 7.4a displays the test object for only one transmitter connected. Secondary maxima are observed due to the sparsity of the measuring matrix. Applying Equation 3.18 (see Section 3.1.4), the image was obtained for the case in which all 36 radiation sources are active (Figure 7.4b). No secondary maxima are evident, and the object image can be identified unambiguously. Increasing the image quality depends on overlaying interference images from different sources. This ensures in-phase addition of the central maximum and incoherent addition of the secondary maxima in the summation of images provided by different sources, according to Equation 3.18 (see Section 3.1.4). The focusing plane was taken to be coincident with the plane of test object placement.

Notice that the resolution of the image obtained is close to the diffraction limit for the system with the aperture considered here, and is equal to 1.5 cm.

7.1.3 Time-Tact of Objects in the Frequency Domain

As was described in Sections 6.2.1–6.2.3, the highest-quality 3-D image with a resolution close to the diffraction limit can be obtained with synthesis aperture radar (SAR) technology using ultra-wide band (UWB) signals. We consider a generalization of the technique considered in the present chapter for obtaining an image from incoherent radiation using a set of frequencies distributed in the UWB. The idea of this approach is based on the use of the amplitude technique at each

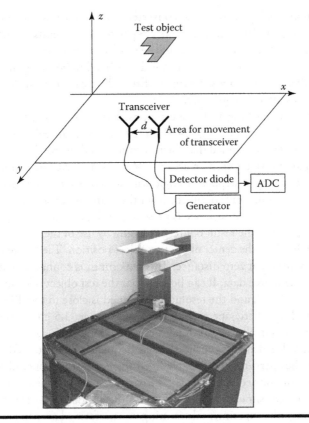

Figure 7.5 Diagram and experimental setup for measurements with frequency clocking.

frequency, with subsequent addition of the obtained interference patterns, according to Equation 3.19a (see Section 3.1.4). It is assumed that sounding is carried out in the clocked regime at different frequencies. This can be done by using a set of bandpass filters to isolate the required frequency bands, or by using dispersion prisms, which convert the time spectrum into a spatial wave spectrum. In our case, it can be realized by step (clocked) frequency switching of the backlight generator (the reference source).

Let us consider the measurement diagram shown in Figure 7.5 (top). The transmitting and receiving antennas are placed at a fixed distance d from each other and together form a transceiver. It is moved with a certain step in the xOy plane and enables measurement of the amplitude at different frequencies. The transmitting antenna is connected to the monochromatic signal generator. It emits radio waves in the direction of the test object and in the direction of the receiving antenna. The wave reflected from the object (object wave) interferes with the direct wave from the source (reference wave). A detector diode is connected to the receiving antenna and, together with an analogue-to-digital converter (ADC), measures the amplitude of the combined signal. The measurements are performed on emission of the monochromatic signal. Then, the frequency of the monochromatic signal is changed and the measurements are repeated. In this way, broadband radio-holographic measurements are realized. The proposed variant of sounding can be realized with a dual-axis scanner, as shown in Figure 7.5.

The frequency clocking technique proposed in this section was numerically simulated for the frequency range 5–10 GHz in the spatial region 1×1×1 m. In the calculations, it was assumed that the transmitter and the receiver were placed 10 cm apart and are moved within an area of 1×1 m with a step of 4 mm. The investigated inhomogeneity, a stepped triangle with a step size of 5 cm (see Figure 7.5, top), was placed at a distance of 30 cm from the scanning area. The result of radio tomogram synthesis at that distance is presented in Figure 7.6.

Figure 7.7 presents the result of pre-processing by use of the object's scanning at the sounding depth in the center of the test object position. The image was normalized to the maximum at zero distance. This maximum is connected with the reference signal in the source data. It can be seen that the test object is localized near the assigned value (30 cm), and the resolution obtained is close to the diffraction limit for systems with synthetic aperture in the UWB range: 1.5 cm in the transverse direction, and 6 cm in the longitudinal direction (along the z-axis).

This latter value depends on the clocked frequency band at 5 GHz. It should be noted that the aperture is completely filled there, and the proposed solution was realized using fast Fourier transform (FFT)-based fast algorithms at each frequency.

Thus, numerical simulation demonstrated the fundamental feasibility of frequency clocking. The experimental scanning was performed over an area of 80×80 cm with a step of 1 cm. A dual-axis mechanical scanner was used that consisted of a square frame with dimensions of 110×110 cm and included a moving carriage

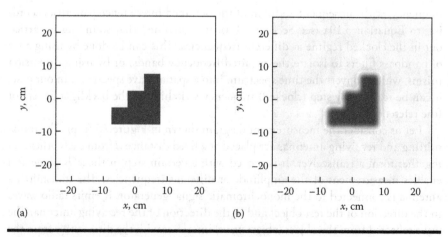

Figure 7.6 Test object and its radio tomogram.

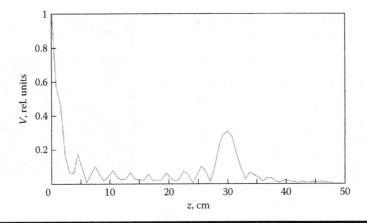

Figure 7.7 Image scanning of test object at scanning depth.

guided in two mutually perpendicular directions by steel ropes connected to the stepper motors. Each stepper motor controlled one of the two guide ropes. The transceiver module, consisting of two UWB antennas separated by a distance of 5 cm from each other, was fixed to the carriage at the cross-point of the guide ropes. One antenna was used for signal transmission, the other one for receiving the signal. The antennas were separated by an aluminum plate with a hole to reduce the direct signal from the transmitter to the receiver. However, the direct signal remained strong enough to be used as the reference signal.

Measurements were performed at 512 frequencies within the range from 8 to 18 GHz with a uniform step. Two stepped objects were used as the test object. The first one was a plaster object with a step size of 5 cm at a distance of 36 cm from the plane of transceiver movement (scanner plane). The second one was a metallic object with a step size of 10 cm at a distance of 56 cm from the scanner plane. Measurements at frequencies of 10 and 15 GHz are presented in Figure 7.8. A complex interference image is displayed there, suggesting that the measurement data contains information about the phase of the object wave. Reconstruction of images based on the experimental data was performed in two stages. Firstly, the reference signal amplitude was subtracted from the measured amplitude of the interference field. The next stage was to provide inverse focusing and receive the 3-D radio image.

The results of reconstructed images from a plaster and a metallic object are presented in Figure 7.9. The image of the plaster object was recovered with lower quality than the image of the metallic object; this depends on the relatively low reflection factor of plaster, in contrast to metal. The shadow from the plaster object is visible on the image of the metallic object placed at the longer range. As the plaster object is semitransparent, for waves of the chosen frequency band the metallic object placed behind it can be visualized in any case.

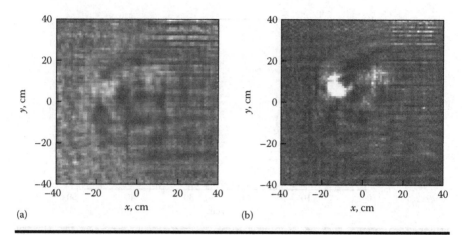

Figure 7.8 Field amplitude measurements at frequency of (a) 10 GHz and (b) 15 GHz.

The resolution of the images obtained is close to the diffraction limit for this frequency band and is about 1 cm. Moreover, there is range resolution in the obtained 3-D image, as evidenced by the distinct vision of both objects that were placed at different ranges. The evaluation of the range resolution in theory is 3 cm for the frequency band of 8–18 GHz.

An experiment with a stepped metallic object with step size of 5 cm was also conducted, when the object was placed 46 cm away from the scanner plane. Figure 7.10 shows the recovered image of the single test object.

Results obtained both experimentally and numerically, based on the methodology described in Section 3.1.4, revealed the potential of radio-wave tomographic synthesis

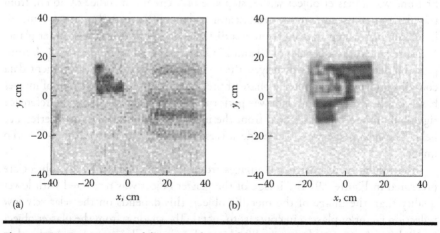

Figure 7.9 Reconstructed images of test objects: (a) plaster object, 15×15 cm in size at distance of 36 cm; (b) metallic object, 30×30 cm in size, 56 cm away from scanner plane.

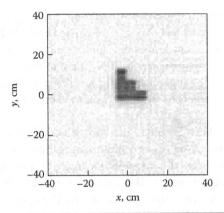

Figure 7.10 Tomogram of metallic test object at distance of 46 cm.

with incoherent radiation. The efficiency of usage of the spatial and frequency clocking of the measurements was demonstrated even by a few graphical illustrations.

This is important, since this method makes it possible to exceed the spatial resolution of conventional methods of incoherent tomography by several orders of magnitude. To realize the proposed approach, it is sufficient to provide interference of the reference and object fields and record the amplitude of the interference field. In the final analysis, it provides information about the relative phase of the object wave and ensures the possibility of focusing, which is needed for the radio tomography of test objects using the method of tomographic synthesis.

7.2 UWB Tomography of Non-Linear Inclusions

The problem analyzed in this section pertains to a process that is referred to as non-linear radiolocation (detection) (NRD), when non-linear inclusions or their absence in the monitored field are to be detected by distortions of the scattered field during sounding. In the first approximation, all sounded regions are linear, which underlies the linearity of the constitutive equations. However, this linearity breaks down if the transmitted power is increased. As a rule, this happens in regions filled with semiconductor electronic parts (diodes, transistors, microchips, etc.) or at bad electrical contacts between metal structures. The first-mentioned situations usually result in the appearance of even-order harmonics, while the latter typically lead to the appearance of odd-order harmonics. This is the basis for technologies of non-linear element detection [14–22].

7.2.1 Current Statement of Problem

Let us look at the current state of the solution to the problem of NRD. At the present time, the main task of NRD is to answer the question of how to provide countermeasures against industrial and economic espionage. Modern transmitters and recorders are so small that they can be carried and hidden nearly anywhere, being placed inside different household appliances, interior items, and building structures. Extended long-life batteries allow them to function for months or even years [24–32]. The wide range of modern non-linear radio detectors available in world and domestic markets today allows you to choose a detector that best meets your needs.

Back in the early 1970s, a NRD method was developed based on an analysis of harmonic emissions caused by an illuminating signal when it is reflected off a target. A few years later, this method was used by the Superscout non-linear junction detector (NLJD) system. The instrument typically includes an antenna, a transceiver, an indicator unit, and a power supply. The signal emitted by the NLJD antenna causes any currently known electronic device to generate a response signal. The principle of NLJD operation is based on flooding the suspect area or object with a monochromatic microwave signal. The test object re-emits the harmonics of the radar signal. This phenomenon is tied up with the fact that the induced potential difference on the non-linear object gives rise to a non-linear current that contains harmonics. The induced alternating current generates a re-emitted electromagnetic field that contains these harmonics.

High-quality NLJDs have the capability to compare the received signal intensity of both the second and third harmonics, so they assist the operator in discriminating between true semiconductor junctions and false junctions. This feature typically results in a much more expensive NLJD, because it means that the unit has two receivers with well-isolated channels (frequencies). Non-linear radio detection of objects is similar to the conventional detection of objects with active response in the identification mode. Thus, the power of the nth harmonic response (and the detection efficiency also) depends directly on the power of the radiation emitted by the detector and inversely on the square of the frequency and the order of the received harmonic. The lower the radiation frequency of the NLJD, the lower the signal attenuation and the response from the object. Thus, the frequency of the radiation emitted by the detector is one of the fundamental parameters of NRD. It is around 915 MHz for most NLJDs (the second harmonic is 1830 MHz and the third harmonic is 2745 MHz). The frequency of 888 MHz is rarely used (for which the second harmonic is 1776 MHz and the third harmonic is 2664 MHz). Generally speaking, detectors operate at one fixed frequency.

However, some detectors operate at multiple frequencies. As detectors with a limited frequency band often find themselves in conflict with other electronic devices, NLJDs with a wide frequency band and automatic free channel (frequency) selection are preferred. For example, in the United States, the NLJDs could interfere with cellphone signals in the frequency band at 888 MHz. Although the radiation power significantly increases detection capability, receiver sensitivity is equally important. NLJDs with low power output, but with a sensitive receiver, can have higher performance than high-power NLJDs with an insensitive receiver.

Moreover, high-power NLJDs can damage electronics and have an adverse effect on people's health. The modality of radiation is directly related to the power. Most NLJDs operate in the continuous mode, and their output power can hardly exceed 3 W. Some domestic non-linear detectors use pulsed mode and correspondingly have a number of advantages. Such NLJDs consume far less electric power, which reduces their power supply requirements. Moreover, when operating in pulsed mode with a peak power of 300 W, the mean power of irradiation of the operator (the specific absorption rate) is very low—far lower than in continuous mode with a power level of 3–5 W.

The three main functions of NLJDs are: detection, position location, and identification. Detection is successful if the response amplitude exceeds a threshold level. A visual or auditory signal keeps the user informed that the suspect object is located within the antenna illumination range. Location of the position of the object is performed through a comparison of the amplitude with the response bearing. The amplitude of the response signal increases as the antenna gets closer to the signal source during the search. The anisotropy of the directivity pattern allows the user to determine the direction to the source of the response signal from its maximum level.

Object identification is performed according to an analysis of the response signal from the object located within the antenna radiation range. In those NLJD models that simultaneously receive a response from the object in the second and third harmonics of the radiated signal, identification is realized by comparing the signals of both receiving paths by monitoring the linear indicator lights. Usually, objects with semiconductor junctions return a second harmonic signal that is 20–40 dB stronger than that of the third harmonic. Noisy metal contacts, in contrast, generate a third harmonic signal exceeding the signal of the second harmonic by 20–40 dB. NLJDs receiving only the second harmonic response require additional actions from the user to identify the object.

The problem here is that there are a number of objects containing non-linear metal contacts. These include reinforced concrete structures, different types of supports, furniture springs in contact with screw heads, nails contacting metallic objects inside a wall, electrical switches, contacts of fluorescent tubes, paper clips, and so on. Such objects give false alarms on the odd-numbered harmonics, but sometimes they produce even-numbered harmonic responses as well. Since these contacts are mechanically unstable, their current–voltage characteristic is unstable as well and is highly dependent on the mechanical state and history of the contacts. If low physical vibration is applied to a false junction, the crystal texture of corroded and bimetallic junctions may be destroyed, which leads to modulation of the response signal with the frequency of vibration. Vibration does not affect the p-n semiconductor junctions. The operator should test the search field, for example, by tapping the ground with a small rubber hammer, using both the indicator and the headphones. In this case, the response from electronic devices will not change, and no sound will be generated in the headphones. On tapping, metallic sources will cause disordered unstable display readings and noise (crackling) in the headphones.

Exploration with an NLJD usually starts with slow scanning of every surface in the search area with the radiation switched off. The operator should scan all surfaces thoroughly, including the walls, floor, ceiling, equipment, and so on. This is quite a time-consuming procedure (its rate is usually 0.2–0.4 m²/min); its aim is to detect devices that generate electromagnetic fields.

Then, with the irradiating signal switched on, walls and other surfaces should be scanned, holding the antenna at least 2–3 m away from the surface. This allows the operator to detect and isolate those objects that cause noise in the headphones. After removing these objects from the search area, the distance to the NLJD is reduced to 1–1.5 m, and the scanning procedure should be repeated. Finally, the distance should be reduced to 0.5 m or to immediate contact with the object. Then, a number of scanning operations are performed while gradually increasing the NLJD power from its lowest possible level up to its maximum value.

Scanning of flat surfaces is performed with a rate of 0.03 m²/s; complex surfaces are scanned at a slower rate. So, scanning of a small office (no more than 20 m²) takes about 2–3 h; a medium-sized office will take 3–4 h, but scanning of a large building demands 6–8 h or even several days.

Non-linear detection technology is applicable to the remote detection of small-sized objects such as radio-controlled devices, equipment for industrial espionage, small arms and light weapons, aircraft wreckage, portable transceivers (including ones that are switched off), and so on. The search objects can include special non-linear markers, which are used for secret marking of different objects and terrains (for precision-guided weapons) and also for people (e.g., lifeguards in inaccessible areas). Experiments have shown that the optimal frequency for remote detection of non-linear objects should not exceed 1 GHz, and it is reasonable to use high-power pulses with a low duty cycle ratio to extend the detector coverage range. If the pulse peak power is 50 kW and the duty cycle ratio is 1/1000, the airborne detection range for small-sized objects is 300–350 m. If an object is buried 0.1–0.2 m deep in the ground, the detection range is about 80–100 m.

7.2.2 UWB Tomography of Non-Linearities

The present section is aimed the possibility of usage of UWB signals for non-linear detection. As an example, we discuss here the well-known Luxembourg–Gorky effect (LGE) [23, 24], which, in our opinion, lies at the basis of an approach under consideration. The main idea underlying this effect consists in the occurrence of cross-modulation when radio waves propagate in non-linear media. Cross-modulation is the phenomenon that arises when a receiver tuned to a station on a broadcast frequency f_1 simultaneously receives signals from a high-power broadcast station operating at the carrier frequency f_2 (far enough removed from f_1). This effect was observed for the first time in 1933 in Eindhoven (the Netherlands), where the high-power Radio Luxembourg station, located in line with a Swiss station, could be heard during the Swiss broadcast. A similar phenomenon was observed in Gorky (former USSR), where high-power Moscow stations could be heard during the reception of radio stations located to the west of Moscow [23, 24]. The depth of such cross-modulation of the radio waves of both stations can reach 10% or even more, but it usually does not exceed 1%–2%. The GLE is a source of interference in radio reception.

The theory of the GLE was developed by the Australian physicists V. Bailey and D. Martyn (1934–1937), the Soviet physicists V. L. Ginsburg and A. V. Gurevich (1948), and then, experimentally proved by a group of researchers from NIRFI (Gorky, former USSR). On the other hand, the possible effect of cross-modulation in the ionosphere was mentioned first by the Soviet scientist M. A. Bonch-Bruyevich in 1932. The cause of the GLE is as follows: a high-power (pump) wave heats up, that is, perturbs ionospheric plasma at the region where another bandpass wave propagates, having a carrier signal with the transmitting information (see references [23, 24]). Both waves experience cross-modulation; that is, they can see each other [23]. This is based on the non-linear modulation effect. At the physical level, it is similar to the mixing of two signals in a solid-state mixer.

Let us consider how this effect can be applied to the UWB detection of non-linear inclusions. Let $E_0(t)$ be a pulsed UWB signal that irradiates a non-linear inclusion. Then, to within a constant factor, the scattered field can be written as

$$E_{scat}(t) = E_0(t) + g\big[E_0(t)\big] \qquad (7.5)$$

where $g(x)$ is the non-linearity. Generally, it can be assumed that

$$E_{scat}(t) \approx E_0(t)$$

If we add a high-power monochromatic illuminating wave

$$E_1(t) = A\sin(\omega t + \varphi_1)$$

to the UWB field, then the total scattered field can be represented as

$$E_{scat}(t) = E_0(t) + A\sin(\omega t + \varphi_1) + g\big[E_0(t) + A\sin(\omega t + \varphi_1)\big] \qquad (7.6)$$

The phase of the monochromatic illuminating signal is not correlated with the UWB signal, so after averaging we can write

$$\big\langle \tilde{E}_{scat}(t, A) \big\rangle = E_0(t) + \frac{1}{2\pi}\int_{-\pi}^{\pi} g\big[E_0(t) + A\sin(\varphi)\big]d\varphi \qquad (7.7)$$

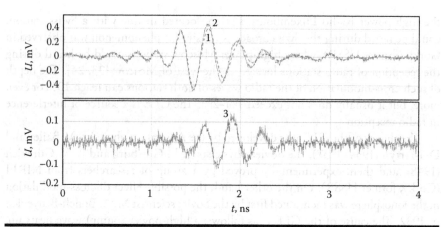

Figure 7.11 Signal shapes in UWB sounding of a D20 diode with illumination (curve *1*) and without (curve *2*), and the differential signal (curve *3*).

This means that a non-linear inclusion appears as an averaged scattered pulsed signal! This effect does not occur at a low level of illumination:

$$\langle \tilde{E}_{scat}(t, A \ll |E_0|) \rangle \approx E_0(t)$$

The difference between the obtained signals is the information value to identify the type of non-linearity. This conclusion was confirmed by numerical simulation. The proposed method is based on this effect.

It should be emphasized that although the harmonics are not recorded in contrast to the conventional method of non-linear detection, the non-linearity is manifested in the distortion of the shape of the UWB pulse. Conceptually, this corresponds to the GLE. Figure 7.11 shows typical shapes of signals received during UWB sounding of a D20 diode with illumination (curve *1*) and without illumination (curve *2*), as well as the differential signal (curve *3*). The diode was not connected to a circuit; it was just a freestanding factory-made unit. The frequency of the illumination signal was $f = 850$ MHz, and a focused UWB signal with a duration of 0.2 ns was used for sounding.

The illumination signal power at the output of the transmitting antenna was 30 W. As the figure shows, the differential signal stands at 20%–25%, which is quite a significant value. There is a certain inertia (delay) of the non-linear diode response. Similar characteristics are observed for other non-linear inclusions as well.

7.2.3 Signal Processing and Algorithm of UWB Images of Non-Linear Inclusions

A program chart of the experimental setup is presented in Figure 7.12. This program was applied within the framework of the *Volnorez* (i.e., Breakwater) project.

Proof of Specific Radio Tomography Methods ■ 183

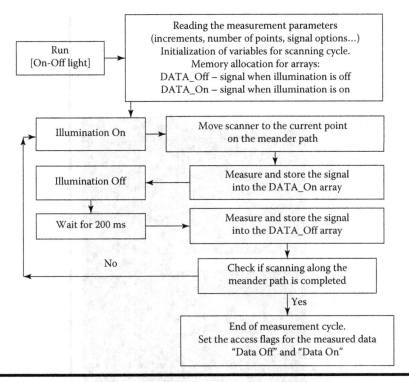

Figure 7.12   Program flowchart.

The program workflow is sequential UWB scanning of the test area with an assigned step. Sounding is performed in two sequentially activated modes: with illumination and without illumination by high-power monochromatic radiation.

Scanning is realized according to the program by automatic movement of the antenna unit. The collected arrays are sent to the program for tomographic processing and detection of non-linearities. The program's main window is illustrated in Figure 7.13. Layer-by-layer images of the sounded area are displayed in the pop-up windows. The source sounding data are shown in the first window. The waveform of the signal at a selected point is displayed at the bottom. The middle frame displays the tomogram of all the detected inhomogeneities at different depths. The upper right window displays a tomogram of the detected non-linearity or non-linearities. Images are displayed layer by layer in gray-scale.

Layer-to-layer transfer is performed interactively with the scrollbar: slide up for near layers, slide down for far layers. The depth of each layer (in meters) is displayed underneath. There was just one non-linear inhomogeneity in the given example (a microwave diode). Other inhomogeneities were 2×2 cm square-shaped pieces made of aluminum foil. Each object was taped to the foam plastic surface with adhesive tape. The experiments confirmed the feasibility of the proposed solution.

184 ■ *Electromagnetic and Acoustic Wave Tomography*

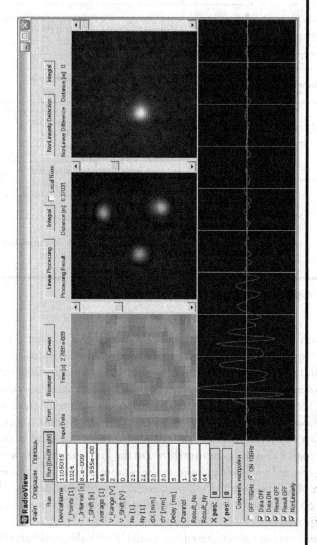

Figure 7.13 Interface of the program *RadioView*.

Finishing this section, we should emphasize that using UWB radiation for radio tomography of non-linearities of artificial origin creates numerous opportunities for their 3-D tomography. In contrast to the conventional non-linear detection method, it does not use the detection of harmonics to achieve this end. However, UWB radiation requires an external high-power microwave source to detect non-linearities.

7.3 Doppler Tomography Experimental Proof

As has already been pointed out, there are a number of advantages of using microwaves in remote detection of prohibited items concealed under clothing or in hand luggage over other forms of radiation. Thus, using high-intensity X-rays is potentially hazardous for individuals, and every time it is implemented it should be strictly regulated. This also pertains to particle radiation, that is, neutrons, protons, electrons, and so on. The use of ultrasound runs up against the fact that it is strongly attenuated as it penetrates air and clothing. The use of optical and infrared radiation is limited by their low ability to penetrate through clothes. The same disadvantages are encountered in the terahertz band and, moreover, the currently available element base for this band is not sufficiently developed and is not yet ready for extensive use. It follows that preference should be given to radio-wave systems in the microwave band.

In the first place, radio waves are almost completely safe and do not lead to negative health outcomes. This is their key distinction from X-rays. Secondly, the range of applications of radio waves is potentially quite wide: they have been successfully put to use in crowded spaces, during special forces raids for detecting and tracking the movement of people hiding behind building walls, detecting people who have been injured in natural disasters, and so on.

However, the microwave systems in current use are elaborate stationary systems of high cost—more than $250,000. Looking for an alternative approach based on Doppler measurements is the aim of this section. According to preliminary estimates, the alternate solution that we propose will cut the costs of such an equipment package more than a hundredfold. Moreover, the working instrument can be rather compact and operate under any conditions, including field conditions. Section 7.3 integrates the results of references [25–36].

7.3.1 Microwave Doppler Sensor and Location Sounding

Let us consider the operation of a microcircuit that goes under the name of a microwave radar-based motion detector module (RMM) of the type CON-RSM1700 [1], which operates at a frequency of 24 GHz (see Figure 7.14). Sensor size is 25.0×25.0×6.2 mm, and its market price is about €10. The radiating power does not exceed 16 dBm (40 mW), so an RMM is absolutely safe from the perspective

186 ■ *Electromagnetic and Acoustic Wave Tomography*

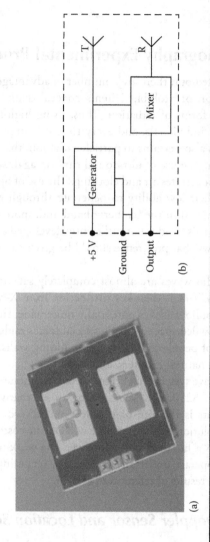

Figure 7.14 (a) Doppler motion sensor CON-RSM1700 and (b) its block diagram.

of human health. Such sensors can be easily combined into sensor arrays, thanks to their small size.

Basically, the sensor under discussion is a compact short-range radar station [2]. Key sensor elements are a reference-frequency generator, transmitting (T) and receiving (R) antennas, and a mixer (see Figure 7.14). The output signal is typically passed through a differentiating circuit. The reference-frequency generator produces a signal with frequency ω_0 of the form $u_0 = A_0\cos(\omega_0 t)$, which is fed to the radiating antenna (T) and then, somewhat attenuated, to the mixer. The incoming signal $u_1 = A\cos(\omega_0 t - 2kVt - \varphi)$ from the receiving antenna (R) is sent to the second input of the mixer. Here, A and φ denote the amplitude and phase of the signal reflected from the test object, and V is the velocity of the object relative to the sensor. Signal multiplexing and averaging are performed in the mixer:

$$U(t) = \overline{u_0 u_1} = \frac{1}{2} A_0 A \overline{\left[\cos(2kVt + \varphi) + \cos(2\omega_0 t - 2kVt - \varphi)\right]}$$

$$= \frac{1}{2} A_0 A \cos(2kVt + \varphi) \tag{7.8}$$

After the output signal has passed through the differentiating circuit, it takes the following form:

$$\frac{dU}{dt} = -VkA_0 A \sin(2kVt + \varphi) \tag{7.9}$$

or, in the general case,

$$\frac{dU}{dt} = -\frac{dr}{dt} kA_0 A \sin(2kr + \varphi) \tag{7.10}$$

where $r = Vt$.

Depending on whether the RMM moves in the forward or the reverse direction relative to the test object, the factor $V = dr/dt$ will be either positive or negative, and the Doppler shift frequency will either increase or decrease Equation 7.10. The case was considered here where the test object has a sharp boundary. Reflections from long parts of the object do not result in a Doppler effect and are not detected by the sensor.

For a constant speed, Equation 7.10, which describes the output signal of the RMM, coincides to within a constant factor with the quadrature component of the received signal

$$S(t) = A\sin(2kr + \varphi) \tag{7.11}$$

which contains information about the amplitude and phase of the signal reflected from the test object. This is clearly demonstrated in Figure 7.15, where the signal records are antisymmetric in relation to reversing the direction of motion of the RMM. Thus, the microwave motion sensor provides an objective measurement of one of the quadratures of the field and can be used for sounding objects of different shapes. On this basis, it is proposed to apply the Doppler effect in the tomographic examination of hidden objects, and in the reconstruction of object shapes using the SAR technique.

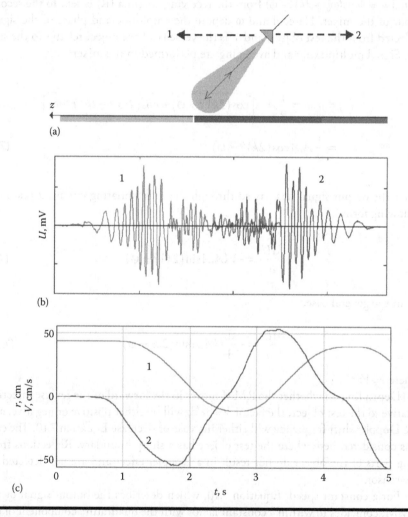

Figure 7.15 Explanation of operating principle of an RMM: (a) diagram of motions; (b) actual signal records when the sensor moves forward (curve 1) and backward (curve 2); (c) synchronous values of coordinates (curve 1) and sensor velocity (curve 2).

7.3.2 Reconstruction of Objects: Imitational Modeling

Thus, the basic idea of the method involves applying the Doppler effect as a means of continuous recording of information about the amplitude and phase of a received radio signal along the trajectory of motion. For a location system with a wide directivity diagram, the received signal is the result of interference of waves reflected from spatially distributed objects. Only local inhomogeneities will make a contribution to the signal here, but long homogeneous segments, even if they reflect strongly, will not affect the Doppler signal, as their contribution falls into the nonregistered zero Doppler frequencies. The recorded Doppler radio signal provides a basis for processing and reconstructing the inhomogeneity distribution using the SAR technique. The possibility exists to adjust the focusing depth and thereby to obtain the tomographic image. For processing of this type, the system function (SF) must be defined, which is the spatial and temporal location response to a point scatterer. The SF analogue in optics is the point spread function. Reconstruction of the SF can be achieved either theoretically by simulation with trajectory parameters taken into account, or experimentally, when a local segment of the image is selected corresponding to a point scatterer.

Let us elucidate the model of formation of the SF. If the location (detection) system moves at a velocity V along the Oy axis at a height H (Figure 7.16a), and the point scatterer is located at a point with coordinates $\mathbf{r}_0 = (x_0, y_0, 0)$, then the signal received at the observing point at some specified time t is written as follows:

$$E_0(x,t) = A_0 G^2 \frac{\exp\left\{2ik\left[(x-x_0)^2 + (Vt-y_0)^2 + H^2\right]^{1/2}\right\}}{\left[(x-x_0)^2 + (Vt-y_0)^2 + H^2\right]} \quad (7.12)$$

where:
- G is the antenna directivity diagram
- A_0 is the reflection coefficient
- $k = 2\pi f/c$ is the wave number for radiation with central frequency f

The parameter x here is the coordinate of the antenna system transverse to the direction of motion, that is, it is the coordinate of the position of the microwave sensor in the antenna array. The scanning direction is indicated by the arrow in Figure 7.16. The function $E_0(x,t)$ calculated in this way can be interpreted as the SF.

Figure 7.16b presents the form of the function $\text{Re}\{E_0(x,t)\}$ for the point scatterer $\mathbf{r}_0 = (0,0,0)$ in location scanning with microwave sensors at a velocity $V = 30$ cm/s at a height of $H = 30$ cm. The working frequency was $f = 24$ GHz. The calculation was performed for the beam angle of the transmitting-receiving antenna equal to 36° with the beam maximum oriented at an angle of 45° to the horizontal. The scanning step for the transverse coordinate was chosen to be $\Delta x = 5$ mm. As expected for such a wide beam, the SF was found to be weakly localized.

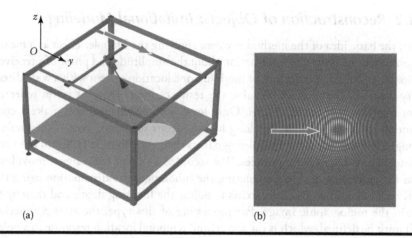

Figure 7.16 (a) Simulation modeling scheme and (b) calculated form of SF.

If there are several reflectors, or the tested (sounded) inhomogeneities have a distributed character, the radar image (wave projection) is more complicated. The example of simulation modeling, where four inhomogeneities (objects) are distributed diversely, is presented in Figure 7.17. Three of the inhomogeneities were assigned as point inhomogeneities, and the remaining one as a long rod (Figure 7.17a). This scene with inhomogeneities was configured intentionally to be similar to the full-scale experiment described in Section 7.3.3. The simulation was calculated for the case of 32 sensors.

Figure 7.17b displays the solution of the direct problem—calculation of the radar image for the chosen scene with inhomogeneities. Interference of signals coming from all four objects is clearly observed: traces of reflections from the rod are visible in the upper part of the image, and in the lower part, traces of the point-like inhomogeneities are seen. It is important that the influence of the directivity diagram of the microwave sensors was taken into account in the calculation. This is manifested in the asymmetrical intensity distribution of the reflections in the upper and lower parts of the image for the point-like inhomogeneities.

Processing the location signal $E(x,t)$ in the case of a complex distributed target can be performed with the SAR technique. It is sufficient to perform the computer-aided inverse focusing of the received radiation, or, what is completely equivalent, to apply 2-D matched filtering using the SF. This operation is performed using double convolution with the complex conjugate of the SF:

$$W(x, y = Vt) = E(x,t) \otimes E_0^*(x,t) \qquad (7.13)$$

Figure 7.17c shows the form of the function obtained $|W(x,y)|$. All of the reflectors are located at the positions where they were assigned initially. The reason for the achieved effect is obvious: the SAR technique provides radiation focusing with a

Proof of Specific Radio Tomography Methods ■ 191

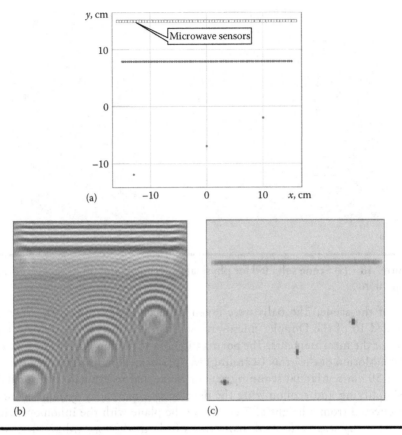

Figure 7.17 Results of simulation modeling: (a) source scene with four inhomogeneities; (b) solution of direct problem; (c) solution of inverse problem.

large aperture equal to the size of the scanning area or, more precisely, to the size of the spot that is illuminated by the directivity diagram on the sounded surface. A standard situation arises: the wider the directivity diagram, the more blurred the source location imaging, but the better the resolution for image processing using the SAR technique.

7.3.3 Positioning System and Manual Doppler Scanning Device

To experimentally check the working efficiency of the proposed approach, Doppler measurements were conducted at a frequency of $f = 24$ GHz.

A scene with inhomogeneities chosen for this purpose is shown in Figure 7.18a. A metallic rod 0.6 cm in diameter is located in the upper part of the scene, and three balls, which are 2.2 cm in diameter, are located in the lower

192 ■ *Electromagnetic and Acoustic Wave Tomography*

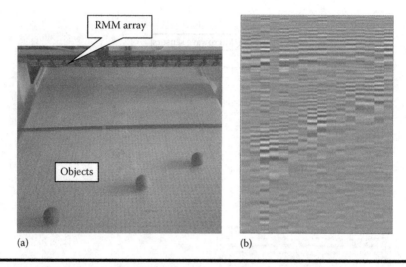

Figure 7.18 (a) Scene selected for physical experiment and (b) its measured wave projection.

part of the scene. The balls were taken from computer mice. An array that included 16 of the Doppler microwave sensors described in Section 7.3.1 was used for the measurements. The polarization of the microwave sensors was horizontal. Motion of the array (scanning) was performed manually with a speed of about 50 cm/s. Manual scanning was implemented to evaluate the possibility of simplifying and economizing the developed tomography system. Scanning was effected from a height of 7 cm above the plane with the inhomogeneities. The scanning was performed many times in both the forward and backward directions. A typical scanning result—a wave projection of the test scene—is shown in Figure 7.18b.

As manual scanning cannot be perfect—uniform and rectilinear—this must be taken into account in the processing of the experimental data. For this purpose, a precise positioning system was used that monitored all six parameters of the motion of the linear array of microwave sensors: the three Cartesian coordinates of the center of the array and the three rotation angles of the array in space (see Figure 7.19). For positioning, a Wintracker III magnetic tracking system for 3-D computer games was chosen. Judging from the experimental data, under conditions of manual scanning it is not possible using this system to reorient the motion of the array by an angle of less than 2°. However, the greatest impact on the quality of the location image was caused by the irregular motion of the operator's hand at the beginning and end of each scan.

After the coordinates of each sensor were recorded, the experimental data were recalculated (reduced) to a uniform Cartesian grid by spline interpolation. A sample cross-section of the obtained image (the central cross-section) is plotted by curve *1* in Figure 7.20a. The initial segment of this curve (for lesser values of *y*) plots the

Proof of Specific Radio Tomography Methods ■ 193

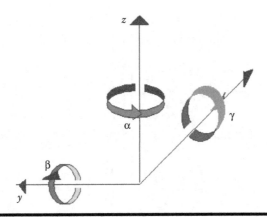

Figure 7.19 Angular rotations of linear array of microwave sensors.

response to the central ball in the test scene, and the following segment (for larger values of *y*) plots the response to the metallic rod.

The presented cross-section or, to be more exact, a segment corresponding to the central ball in the test scene, can be interpreted as the SF cross-section. Matching this segment with the function Equation 7.13, it is possible to reconstruct the ideal form of the SF. The ideal SF obtained in this way has the cross-section displayed by curve *2* in Figure 7.20a. The 2-D form of the SF is shown in Figure 7.20b. It should be noted that the reconstructed SF is similar to the SF shown in Figure 7.16b obtained by numerical simulation. The SF of the real system makes it possible using Equation 7.13 to reconstruct the distribution of inhomogeneities in the test scene. Here, $E(x,y)$ is understood as the reduced experimental

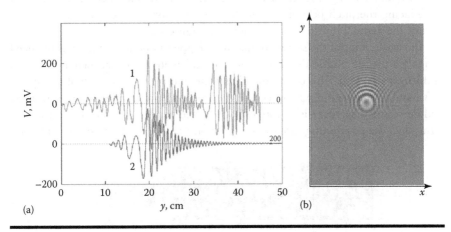

Figure 7.20 Reconstruction of SF: (a) distribution of central sensor signal along y-axis (curve 1: measurement result; curve 2: calculated SF); (b) reconstructed 2-D SF.

194 ■ *Electromagnetic and Acoustic Wave Tomography*

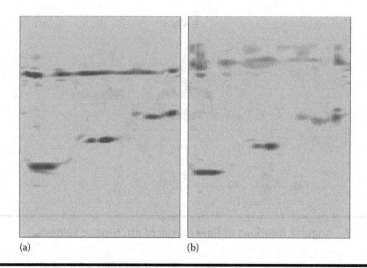

Figure 7.21 Reconstructed image of test scene: (a) with positioning system; (b) without positioning system.

data. The result of reconstructing the distribution of inhomogeneities in the test scene is presented in Figure 7.21a.

As can be seen from Figure 7.21a, the spatial distribution of inhomogeneities in the physical experiment has been faithfully reconstructed. Some of the errors in the image can be explained by the need to interpolate reduced data in the transverse cross-section; spline interpolation was used to increase the number of points in the transverse cross-section from 16 to 64, that is, by a factor of four.

Without resorting to interpolation, the transverse resolution could be increased by reducing the pitch between microwave sensors; for example, by a consecutive shift of the manual linear arrays, as shown in Figure 7.22.

The weak point of the system of manual Doppler radio tomography is the need to apply a precise positioning subsystem to recalculate the experimental data to a uniform spatial grid. This operation adjusts the non-uniform motion of the operator's hand at the beginning and end of the scanning process. However, if just the middle scanned segment is of interest, where the hand movement is almost uniform, then the positioning subsystem can be omitted. The average speed V of hand

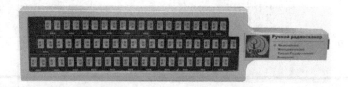

Figure 7.22 Manual microwave scanner.

motion in the center of the tested scene can be calculated by comparing the calculated SF with the experimental data.

In our case, that speed was 52 cm/s. This makes it possible to recalculate the timescale of scanning into the coordinates in the direction of motion of the microwave sensor array. Then, it is sufficient to apply the procedure in Equation 7.13, aided by the obtained SF. The result of this operation is presented in Figure 7.21b. Here, it is obvious that satisfactory quality is observed only in the central part of the scanned field, where the velocity of array motion is nearly uniform. But, often, this is entirely sufficient to come to a decision about the presence of hidden objects.

The technology of implementing Doppler microwave sensors proposed in this section for the development of a manual radio-detection scanner is feasible and quite promising from the point of view of practical applications. We suppose that this system will find wide use, primarily in security applications.

7.3.4 Doppler Sub-Surface Tomography

For the purposes of experimental testing of the applicability of the proposed approach for the tomography of inhomogeneities hidden under a planar interface of two media, a series of laboratory experiments was conducted at Fraunhofer Institute for Non-Destructive Testing (IZFP, Germany). Sounding was performed at the frequency $f = 24$ GHz using the moving horn-type antenna shown in Figure 7.23a. To eliminate clutter reflections as much as possible, the sounding height was lowered to $H = 30$ mm. Three different objects were sounded. Thus, Figure 7.23b presents the raw Doppler image of a wooden slat, a photograph of which is shown in Figure 7.23c.

Data processing was performed at National Research Tomsk State University, using the above-mentioned technique incorporating aperture synthesis. The result of SAR processing is shown in Figure 7.23d. To reconstruct the SF, a fragment was used containing the image of a small knot on the right-hand side of the slat. It can be seen that the image resolution has improved noticeably. The inhomogeneities hidden under the surface have become visible. The scanning direction is indicated by the arrow.

Similar results were obtained for metal plates with holes of different shapes (circular, triangular, and square holes, and slits) hidden under a plastic layer (Figure 7.24). To increase the resolution of the shapes of the holes, multi-angle sounding of each plate was performed. The results of data processing for six sounding angles are presented in Figure 7.24b. The obtained resolution is of the order of 1 cm.

These latter results are applicable, for example, in tomography and identification of hidden radio electronics or for crack detection. A significant point here is that the proposed SAR-based data processing reduces to performing a 2-D convolution, and

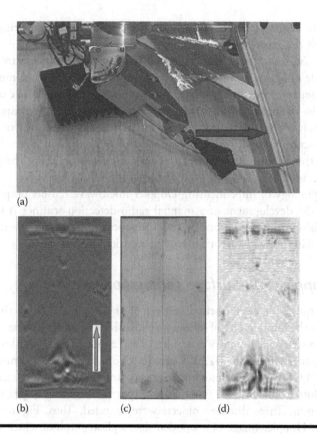

Figure 7.23 Doppler tomography of wooden slat: (a) measuring unit, (b) raw radar image, (c) photograph, and (d) result of SAR processing.

this is efficiently realized using the 2-D FFT. The time required for data processing does not exceed 30 s.

At the end of this section, we should like to outline that Doppler tomography is a special case of radio-wave tomography and mainly uses monochromatic radiation. The principal condition for signal detection in this type of tomography is relative motion of the transceiver and the object. The effect of the frequency shift that arises due to motion of the transceiver is used to select the information signal and disaggregate its quadrature components, which contain within themselves the amplitude phase of the reflected signal.

It has been demonstrated both theoretically and experimentally that radio-wave tomographic synthesis allows reconstruction of the image of small inhomogeneities. A design for a manual radio tomography based on Doppler motion sensors has been proposed.

At the same time, a disadvantage of Doppler tomography is the impossibility of radio vision of large, uniform, or stationary objects. Its relatively simple design,

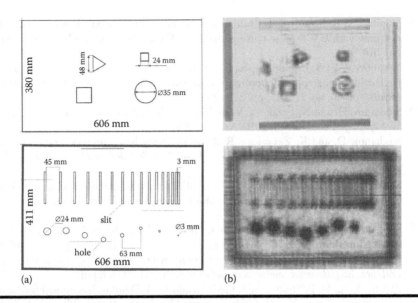

Figure 7.24 Doppler SAR tomography of metal plates: (a) holes in plates, showing their shapes; (b) their radio images.

the possibility of using standard motion sensors, and its overall low cost are considered among its obvious advantages. All of the obtained results were confirmed by numerical simulations and physical experiments.

References

1. Kraus, J. D., *Radio Astronomy*, London: Cygnus-Quasar Books, 1986.
2. Finkbeiner, D. P., Microwave interstellar medium emission observed by the Wilkinson microwave anisotropy probe, *The Astrophysics Journal*, vol. 614, no. 1, 2004, pp. 186–193.
3. Wang, S. and X.-C. Zhang, Pulsed terahertz tomography, *Journal of Physics D: Applied Physics*, vol. 4, 2004, pp. 1–36.
4. Bespalov, V. G. and A. Gorodetsky, THz pulse holography, *Journal of Holograph. Speckle*, vol. 5, no. 1, 2009, pp. 62–66.
5. Kitayoshi, H., Applied radio holography for EMC and instrumentation, *Proceedings of IEEE International Symposium on Electromagnetic Compatibility*, 16–20 May, 1994, Sendai, Japan, pp. 803–804.
6. Gerchberg, R. W. and W. O. Saxton, A practical algorithm for the determination of the phase from image and diffraction plane pictures, *International Journal for Light and Electron Optics*, vol. 35, 1972, pp. 237–246.
7. Sukhanov, D. Y. and K. V. Zav'yalova, Reconstruction of 3D radio images from multifrequency holographic measurements, *Technical Physics*, vol. 57, no. 6, 2012, pp. 819–823.

8. Shipilov, S., R. Satarov, I. Fedyanin, et al., Positioning for people behind barriers in real time with UWB pulse sensing, *MATEC Web of Conferences*, vol. 79, 2016, Art. No. 01079.
9. Yakubov, V. P., S. E. Shipilov, R. N. Satarov, and A. V. Yurchenko, Remote ultra-wideband tomography of nonlinear electronic components, *Journal of Technical Physics*, vol. 60, no. 2, 2015, pp. 279–282.
10. Sukhanov, D. and N. Erzakova, Visualization of broadband sound sources, *MATEC Web of Conferences*, vol. 79, 2016, Art. No. 01001.
11. Sukhanov, D. and K. Zavyalova, Radio tomography based on monostatic interference measurements with controlled oscillator, *MATEC Web of Conferences*, vol. 79, 2016, Art. No. 01040.
12. Sukhanov, D. Y. and K. V. Zav'yalova, 3D radio tomography of objects hidden behind dielectrically inhomogeneous shields, *Journal of Technical Physics*, vol. 60, no. 10, 2015, pp. 1529–1534.
13. Sukhanov, D. Y. and K. V. Zav'yalova, Ultrabroadband 3D radio holography in a stratified medium, *Journal of Technical Physics*, vol. 59, no. 12, 2014, pp. 1854–1858.
14. Johns, T., An overview of non-linear junction detection technology for counter surveillance, http://www.ccmeinc.com/pdfs/ORION_NLJDTech.pdf
15. Zharov, A. A., M. B. Stolbov, S. A. Gudkov, and V. M. Danilov, Instrument for defining of signals, Patent No. 2140656. Registered by Russian Patent Office, 27.10.1999 (in Russian).
16. Moving Detector of Nonlinear Transfer "NR-900EK", User Manual: UTDN 468 165 003 RE, "Groups – UTTA" Ltd. (in Russian).
17. STT GROUP's home page, http://stt-group.com/
18. Dyugovanets A. P., A. Yu. Dashchenko, A. V, Chervinko, Non-linear radar with target indicator, http://www.findpatent.ru/patent/247/2474840.html (in Russian).
19. Equipment for special forces: Non-linear radar, http://www.t-ss.ru/lokator.htm (in Russian).
20. NR-900S, Nonlinear locator, detector of nonlinear junction, http://www.arms-expo.ru/armament/samples/1422/60595/ (in Russian).
21. Yakubov, V. P., D. V. Losev, and A. I. Maltsev, Diagnostics of nonlinearities using scattered field disturbances, *Journal of Radiophysics and Quantum Electronics*, vol. 43, no. 7, 2000, pp. 645–651.
22. Yakubov, V. P., S. E. Shipilov, and D. Ya. Sukhanov, Method of detection of hidden nonlinear radioelectronic elements, Patent R.F. Application: No. 2012131727, 24.07.2012 (in Russian).
23. Filipp, N. D., N. Sh. Blaunstein, L. M. Erukhimov, V. A. Ivanov, and V. P. Uryadov, *Modern Methods of Investigation Dynamic Processes in the Ionosphere*, Kishinev, Moldova: Shtiintsa, 1991.
24. Blaunstein, N. and E. Plohotniuc, *Ionosphere and Applied Aspects of Radio Communication and Radars*, Boca Raton, FL: CRC Press, Taylor & Francis, 2009.
25. Radar-based motion detector module, Data sheet, http://www.produktinfo.conrad.com/datenblaetter/500000-524999/502371-da-01-ml-RADAR_BEWEGUNGSMELDER_RSM_1700_de_en.pdf
26. Petrou, M. and C. Petrou, *Image Processing: The Fundamentals*, New York: Wiley, 2nd edn, 2010.
27. Skolnik, M., *Radar Handbook*, New York: McGraw-Hill, 3rd edn, 2008.

28. Kuzmenko, I. Yu., T. R. Muksunov, and V. P. Yakubov, LFM locator for hidden objects tomography, *Proceedings of CriMiCo – 2014 24th International Crimean Conference Microwave and Telecommunication Technology*, 2014, vol. 2, Art. No. 6959738, pp. 1013–1014.
29. Satarov, R. N., I. Y. Kuz'menko, T. R. Muksunov et al., Switched ultra-wideband antenna array for radio tomography, *Russian Physics Journal*, vol. 55, no. 8, 2013, pp. 884–889.
30. Sukhanov, D. Y. and A. A. Murav'eva, Monochromatic ultrasonic transmission tomography, *Journal of Optoelectronics, Instrumentation and Data Processing*, vol. 51, no. 3, 2015, pp. 247–253.
31. Sukhanov, D. Y., Bi-static multipositional wave tomography, *Russian Physics Journal*, vol. 58, no. 1, 2015, pp. 26–34.
32. Sukhanov, D. Y. and K. V. Zav'yalova, 3D radio tomography of objects hidden behind dielectrically inhomogeneous shields, *Technical Physics*, vol. 60, no. 10, 2015, pp. 1529–1534.
33. Sukhanov, D. and K. Zav'yalova, Radio tomography based on monostatic interference measurements with controlled oscillator, *MATEC Web of Conferences*, vol. 79, 2016, Art. No. 01040.
34. Kuzmenko, I., V. Yakubov, and T. Muksunov, Relief restoration of complicated form objects by monochromatic microwave radiation, *MATEC Web of Conferences*, vol. 79, 2016, Art. No. 01047.
35. Fedyanin, I. S., V. P. Yakubov, V. P. Antipov, I. Yu. Kuz'Menko, and T. R. Muksunov, Using of movement microwave detector for radiovision tasks, *Proceedings of CriMiCo – 2014 24th International Crimean Conference Microwave and Telecommunication Technology*, 2014, vol. 1, Art. No. 6959394, pp. 282–283.
36. Yakubov, V., S. Shipilov, G. Parvatov, V. Antipov, and T. Muksunov, Complex security system for premises under conditions of large volume of passenger traffic, *MATEC Web of Conferences*, vol. 79, 2016, Art. No. 01007.

RADIO TOMOGRAPHY PRACTICAL APPLICATIONS

III

RADIO
TOMOGRAPHY
PRACTICAL
APPLICATIONS

Chapter 8

Ground-Penetrating and Geo-Radars

Vladimir Yakubov, Sergey Shipilov,
Andrey Klokov, and Nathan Blaunstein

Contents

8.1 Non-Contacting Sounding of Mines...205
    8.1.1 Random and Incomplete Ground-Penetrating Radar Technology..... 206
    8.1.2 Physical and Numerical Aspects of Mine Detection 210
    8.1.3 Signal Processing: Pre-Processing and Post-Processing.................. 214
8.2 Mine Detection and Imaging Using Diffraction Tomography Technique 218
References 221

The implementation of radio-tomographic techniques in the solution of various practical problems was described in Chapters 2–7. The theoretical frameworks of radio-wave tomography described in Chapters 2–5, and the possible techniques of testing of such phenomena described in Chapters 6 and 7, have been parallel validated by the results of numerical simulations. However, their experimental verification, which would provide the most objective estimation of current possibilities, is of far greater interest. The present chapter addresses the testing of methods of location tomography in the context of problems of geolocation of foreign objects (mines) buried in subsoil and sub-surface structures (tunnels, minerals, etc.), inside walls' structures, and so forth following the basic frameworks described in references [1–30].

From the beginning, it should be emphasized that the problem of detecting dielectric inhomogeneities hidden underground, behind barriers, or inside buildings' walls has become an ever more urgent task at the present time [1–9, 12–30].

First of all, it relates to the widespread use of plastic antipersonnel mines, whose detection is impossible by means of standard magnetic induction mine detectors [1–10, 23–28]. Minefields remaining after various local conflicts on large terrains create a number of risks for peaceful human activity. A problem that was initially a military and technical task has become a humanitarian one. Using radio-wave ground-penetrating radars is considered as the most promising direction in sub-surface detection of dielectric objects. These systems assist in non-destructive underground searches for industrial purposes such as detecting lost communication lines, monitoring the condition of gas and water supply pipelines, groundwater detection, ice, sand, salt, and prospecting for mineral resources [28–30].

Ground-penetrating radars, called also sub-surface or geo-radars, are widely used for archaeological purposes: for studying archaeological ground layers, searching for artifacts, and so forth. However, a number of problems arise in connection with difficulties in describing the interaction of waves with the medium and scattering inhomogeneities in contactless radio-wave sounding. These include interference created by strong wave reflections at the interface of the medium and their attenuation in the ground, and also frequency dispersion and multiple scattering numbers.

This raises the question of fast and reliable processing of the data obtained from ground-penetrating radars. A convenient tool for investigating inhomogeneities hidden underground is the synthetic aperture radar (SAR) technique in the wide frequency band. Scattered radiation in this sounding technique contains more definite information about scattering inhomogeneities in the medium. To obtain useful information from scattered radiation data, the inverse problem of sub-surface location must be solved; that is, the distribution of inhomogeneities in the medium must be reconstructed on the basis of measurements of the scattered field and information about the scattering inhomogeneities. Focusing the reflected signals is the basic technique for underground data processing.

The use of ground-penetrating radars dates back to 1971, when the first measurements of sea ice thickness were performed, based on the video pulse method introduced by M. I. Finkelstein in 1969. Unflagging interest in sub-surface sounding in dielectric media was manifested in the 1990s, owing to the widespread use of personal computers. Ground-penetrating radars are most widely used today in rail and road inspection, locating and monitoring communication lines and archaeological objects, estimating snow and ice cover, monitoring concrete structures, horizontal directional drilling and geological surveys, and searches for hidden objects [1–10, 21–30]. The following problems in the agricultural and forest sectors have been solved thanks to ground-penetrating radar: location of drain pipes, occurrence of ground water and hydric soils, determination of the productivity of land and the salinity level of soils, spatially variable fertilizer application control, watering control, location of sandy pockets, trunk and root analysis, and visual representation of tree cross-sections.

Despite more than 60 years of experience, there are still bottlenecks and unexplored facts. As experts have noted, the accuracy of determination of depth depends directly on the accuracy of determination of the permittivity of layers. Location-based methods are not widely accepted in agricultural communities, although these methods may be conducive to more efficient fertilizing, monitoring of soil consistency, early detection of the formation of plow sole, and so forth. Section 8.1 summarizes the results of references [1, 2, 12–30].

8.1 Non-Contacting Sounding of Mines

Radio-location sub-surface tomography of dielectric objects, including antipersonnel mines, is a challenging problem, whose solution is complicated by a number of factors. First of all, only location sounding is possible; in other words, unilateral (one-sided) sounding. There is no possibility in this case of introducing receiving or transmitting devices under the mines. Moreover, sounding should be performed without contact, since any ground contact is connected with the danger of triggering the explosive device. Wave reflections and irregularities at the interface of the medium and wave attenuation in the ground have a strong impact for this type of sounding, as do also frequency dispersion and multiple scattering. It is obvious that the problem can be solved only by using ultra-wide band (UWB) radiation and the SAR technique to process the data.

To test the location method proposed in Chapter 2 for the reconstruction of the distribution of inhomogeneities, an experiment with four dielectric test objects buried at different depths was carried out. The experimental setup presented in Figure 8.1 was developed at Magdeburg University (Germany) in cooperation with the authors of this monograph.

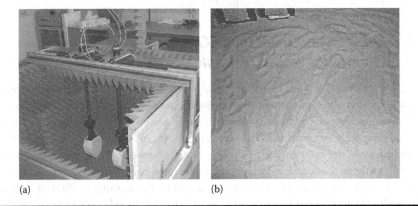

Figure 8.1 Experimental setup: (a) scanning system; (b) uneven relief of sand surface after burying test objects.

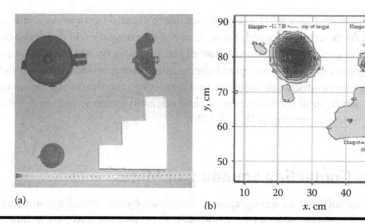

Figure 8.2 Example of UWB radio tomography of dielectric mines in wet sand, in frequency range 0.5–17 GHz: (a) photograph of test objects; (b) radio tomogram of object.

Three cases of plastic antipersonnel mines and a foam polystyrene object of stepped configuration, where each step was 5 cm in size, were placed in wet sand at depths varying from 1 to 25 cm. Photographs of these objects are presented in Figure 8.2a. In the course of the measurements, frequency scanning in the range from 0.5 to 17 GHz was implemented. The system of receiving and transmitting antennas mounted 14 cm apart was moved over a 50×50 cm square test area in the horizontal plane with a step of 1 cm at a height of 30 cm above the sand surface. The results of reconstructing the shapes of the test objects and their location depths in centimeters are presented in Figure 8.2b. It can be seen that the algorithm for reconstruction of the distribution of sub-surface inhomogeneities (distribution of mines in wet sand) works as it should work, and the obtained resolution was about 2 cm, which is sufficient for such types of objects.

8.1.1 Random and Incomplete Ground-Penetrating Radar Technology

The purpose of this section is to work up algorithms to reconstruct the radio tomogram of hidden objects and media when the matrix of measured wave projections is not complete, that is, when the measurements do not cover the area needed to apply fast reconstruction algorithms. This can happen, for example, when measurements in certain areas are not possible or there is a need to speed up the scanning process. Partial failure of the sensor matrix or the intentional use of a sparse matrix are also related to this case. Experimental data obtained by the authors in 2011 at Tohoku University (Sendai, Japan) were used to verify the developed method [1, 2].

Statement of Problem and Experimental Data. The experiment was carried out in the laboratory test facility at the Center for Northeast Asian Studies, Tohoku

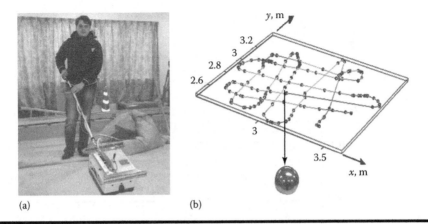

(a) (b)

Figure 8.3 Ground-penetrating radar and measurement scheme.

University (Sendai, Japan) with standard ground-penetrating equipment operating in real time and augmented by an GPS laser system for steady coordinate positioning (Figure 8.3a). The coordinate positioning was effected with an accuracy not worse than 1 mm, which is one order of magnitude better than for determination of the phase center of the antenna. The radar unit was repositioned manually along a relatively arbitrary path (Figure 8.3b).

A metal ball 15 cm in diameter was selected as the test object, and was buried in sand at a depth of about 20 cm. The approximate location of the test object is marked in Figure 8.4 by a cross. The shape of the reflected pulse signal was recorded at each point with a discretization of 512 points. The positions of the measurement points are marked with circles in Figures 8.3b and 8.4. Typical shapes of (a) the

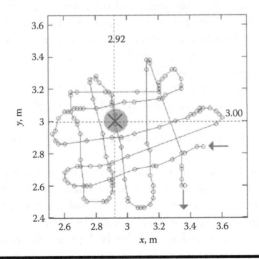

Figure 8.4 Actual measurement path.

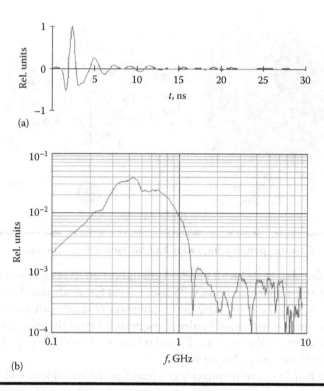

Figure 8.5 Shape of detecting signal and its spectrum.

observed signal and (b) its spectrum are shown in Figure 8.5. As is well known, the shape and spectrum of the signal are determined by the choice of the type of radiation antenna and the characteristics of the radar generator. An antenna unit with a UWB antenna of bow-tie type was employed in the experiment. The selected signals need to provide sufficient spatial resolution (3–5 cm) and the necessary penetration depth of the radiation into sand (up to 1 m).

Since the selected survey path is not a regular curve and does not completely cover the grid of observation points, as is required in the standard SAR technique, the measurements can be considered as typical for incomplete measurements. The measured data are represented as gray-scale levels in Figure 8.6, where they are plotted versus position along the measurement trajectory (horizontally) and time (vertically).

Figure 8.6a displays reflections from the air–sand interface (upper part of the figure) and traces from diffraction hyperbolas, which are caused by signal reflections from a test object hidden in the sand. The darker areas correspond to larger values, and the lighter ones to smaller values. The reflections from the air–sand interface are the strongest, and against this background the reflections from test objects have lower contrast. A simple subtraction of the average signal from the air–sand interface can be applied to enhance the contrast of the reflections from

Ground-Penetrating and Geo-Radars ■ 209

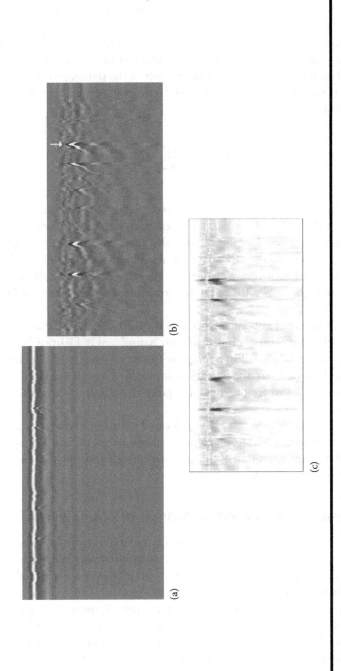

Figure 8.6 Ground-penetrating radar data.

the test object. The result of this operation is shown in Figure 8.6b. The time series of the signal records are halved by eliminating the less informative segments with a long time delay.

Next, the conversion to the amplitude image of the corresponding analytical responses was used at each measurement position to increase the contrast of the test image. Toward this end, all the time records of the signals $u(t)$ were subjected to the Hilbert transform, as a result of which the corresponding conjugate signals were calculated,

$$v(t) = \frac{1}{\pi} \int_{-\infty}^{\infty} \frac{u(\tau)}{\tau - t} d\tau \qquad (8.1)$$

and the analytical signals $w(t) = u(t) + iv(t)$ were then calculated, where the signal amplitudes are defined as $A(t) = |w(t)|$. The resulting amplitude distribution of the analytical signal is shown in Figure 8.6c. The areas where the ground-penetrating radar antenna moved directly above the test object stand out distinctly in the obtained image. One such area, which is the closest to the metallic ball, is indicated by the arrow in Figure 8.6b. The point marked by a cross in Figure 8.4 falls in this position.

It is of interest now to obtain a picture of the distribution of the analytical signal in the observation plane. This distribution is displayed in Figure 8.7a. It can be seen that the largest values fall in a region with central point $x = 2.9$ m and $y = 3.0$ m, that is, in the area where the test object is placed. This can be traced out more clearly in Figure 8.7b, where the measured points have been interpolated over the selected area.

Thus, it turns out that the test object is located somewhere below the point with the assigned coordinates, within a region with diameter 0.4–0.5 m. To improve the in-depth and in-plane resolution of the inhomogeneity, a focusing technique of any type should be implemented.

8.1.2 Physical and Numerical Aspects of Mine Detection

Three currently existing focusing techniques have proved themselves to be the most efficient (Chapter 2). These are: summation over diffraction hyperbolas (inverse focusing or diffraction summation method), multidimensional matched filtering implementing the system function, and the Stolt migration (group focusing). There are also other focusing techniques, for example, algebraic methods of reducing the problems to systems of linear equations, but their efficiency remains relatively low. Each of these three techniques has its own advantages and disadvantages. Let us briefly enumerate them based on results obtained in [1, 2, 10–22].

Theoretical Aspects. The summation over hyperbolas method is the simplest one. It performs a summation of the received signals introducing a signal delay to

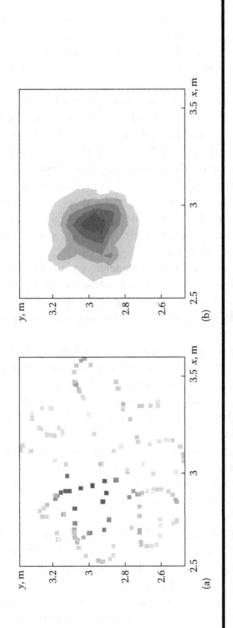

Figure 8.7 Distribution of analytic signal amplitude over observation plane.

compensate for the return path to some chosen focusing point. The tomographic distribution of inhomogeneities can be reconstructed by searching through a certain spatial region layer by layer with the focusing point. A disadvantage of this method is that the compensation is calculated individually for each point, so the total computation speed for testing large areas is low. Its advantages are complete versatility and adaptability to monochromatic signals and random spatial scanning.

The Stolt migration (see Chapter 2) employs the fundamental association between temporal frequencies and longitudinal spatial frequencies, which in the case of UWB signals enables fast transformation of signal time compression into longitudinal focusing using the fast Fourier transform. This is the main advantage of the method. The fact that this method is applicable just to the UWB signals and to plane-parallel scanning geometry is its obvious disadvantage.

Matched filtering occupies an intermediate place between the first and third methods listed. It bears noting that all three methods require *a priori* knowledge or step-by-step fitting of the background refractive index n, which is defined in terms of the wave propagation velocity $n = c/v$.

From the point of view of using incomplete or non-equidistant measurement data, only the first method—the sum over hyperbolas—is applicable as given, when the signal-to-noise ratio of the reconstructed image of the hidden objects is low. In the other two methods, interpolation of the source data over a uniform Cartesian grid is required.

Interpolation over skipping points can be performed, of course, in different ways, for example, by using spline interpolation or simple linear interpolation. But, it is a time-consuming process. We propose to use a variant of Fourier interpolation, where points in the measurement matrix where data are lacking are filled in with zeros. Such a radar response (such as that displayed in Figure 8.7a) is transformed into the spatial frequency domain. Ungrounded (false) values appear at high frequencies (Figure 8.8a) as a result of this zero filling interpolation. If those frequencies are filtered properly and the image is transformed back from the frequency domain to the spatial domain, this will provide the necessary interpolation. Only two fast operations are needed here, namely, the forward and the inverse Fourier transform.

Selecting the required filter is quite a difficult problem, but it can be avoided if the system function is transformed to the frequency domain as well. There are no *spurious* higher frequencies in the spectrum of the full system function. Now, it suffices to multiply the spectra of the image and the system function together, of course after taking the complex conjugate (Figure 8.8b). It then remains to take the inverse Fourier transform, so that both image interpolation and image focusing to the selected plane corresponding to the system function will be simultaneously realized. Thus, the described procedure of obtaining the tomogram with interpolation is accelerated by at least 20-fold in the scanning step, which is the slowest of all the required steps.

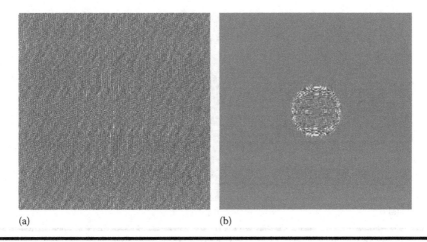

Figure 8.8 Example of spatial frequency distribution of a radar image (a) before and (b) after multiplying by the corresponding spectrum of the system function.

Concerning the final goal—a 3-D tomogram of the sub-surface inhomogeneities—there still remains the as yet unperformed operation of constructing or measuring the system function for each of the interesting layers. Here, it must be borne in mind that the refractive index of the lower background medium n practically always has significant contrast in comparison with the refractive index of air, which is equal to 1. Since the sounding antenna is located in the air but the inhomogeneity is located in the lower medium, strong refraction of the wave occurs on passage of the wave from one half-space to the other. This phenomenon obeys Snell's law:

$$n \sin \beta = \sin \alpha \qquad (8.2)$$

One further remark: if the possible depth of the inhomogeneities varies over a fairly narrow range, then in the construction of the system function, it is possible to use its value for some average depth, and not reconstruct for the entire range of depths.

Numerical Aspects. The problem of incomplete data processing was simulated to provide a convincing demonstration of the efficiency of proposed solutions and to practice the algorithm for 3-D tomogram reconstruction as well. The sounding path shown in Figure 8.9a and b was selected. Four point-like inhomogeneities were tested, which were located in the same plane at a depth of 15 cm. The sounding procedure for a 1×1 m square test area was simulated in the frequency band 6–12 GHz.

Figure 8.10a displays an image of the full radar data to the assigned group inhomogeneity. The measurements were simulated at a frequency of 10 GHz. A 256×256 image frame with a 4 mm step was recorded. If only 400 points are kept, that is, about 6% of all of the points marked in Figure 8.9b, then the radar data will be incomplete.

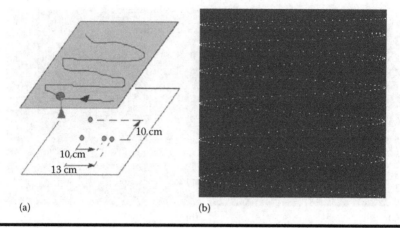

Figure 8.9 Sounding path of four point-like inhomogeneities.

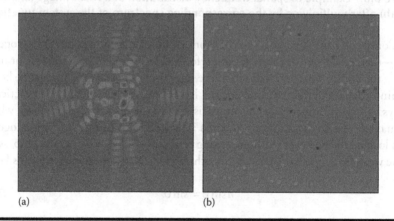

Figure 8.10 Radar signal of group uniformity with (a) complete and (b) incomplete data.

Figure 8.11 presents an example of reconstruction of the tomogram over complete and incomplete data for the four assigned inhomogeneities using the combined technique of matched filtering and Fourier interpolation. It is clear that a true reconstruction of the distribution of inhomogeneities is achieved, which confirms the adequacy of the proposed algorithm.

8.1.3 Signal Processing: Pre-Processing and Post-Processing

Having convinced ourselves of the potential feasibility of the method, let us turn to the processing of actual experimental data following results obtained in references [1, 2, 11–20]. The developed technique was applied to the experimental data described at the beginning of the section. To use this technique, the system

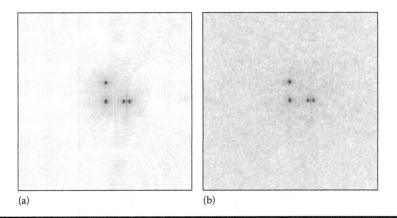

Figure 8.11 Test tomogram over (a) complete and (b) incomplete data.

function first had to be reconstructed. It would be better to perform special measurements over a sufficiently detailed grid of measurement points. However, this is not always possible in practice. In this situation, it is possible to use the most successful subset of the measurements. In our case, such a subset is indicated in Figure 8.6b. The vertical part of the measurement path near the cross marked in Figure 8.4 corresponds to this subset. This segment of the overall image is presented in Figure 8.12a. This segment can easily be interpreted as a diffraction hyperbola of the system function.

Using the technique presented enabled us to determine the average refractive index of the background medium (sand) from the diffraction hyperbola of the system function. That value turned out to be about $n \approx 2.0$. The true depth ($h \approx 20.0$ cm) of the test object, a metallic ball, was evaluated next. Those data made it possible to calculate the form of the system function. The experimental and calculated data are in good agreement. The full form of the calculated system function is displayed in Figure 8.13.

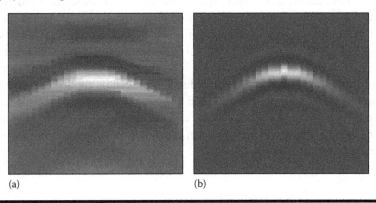

Figure 8.12 (a) Parts of experimental data and (b) calculated system function.

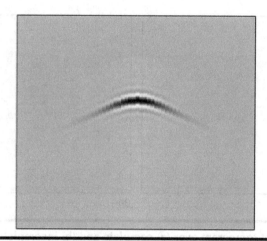

Figure 8.13 Reconstructed system function of ground-penetrating radar.

It now remains to perform the Fourier interpolation of incomplete data and matched filtering. Vertical and horizontal cross-sections of the reconstructed 3-D tomogram are shown in Figure 8.14. In addition, the conversion to the analytical signal was performed at each point of the xOy plane, and its modulus, the envelope of the signal, was taken. This increased the possibility of an unambiguous interpretation of the result. The data obtained are in good agreement with the results of direct measurements with a measuring rod.

The described algorithm is presented in the form of a control flow chart in Figure 8.15. The key element of the chart is the procedure for calculating and filtering the spatial frequency spectrum, which plays the role of interpolation of an incomplete matrix of source data followed by focusing of the image of the hidden object.

The studies conducted in this section confirm the efficiency of the developed method for 3-D tomogram reconstruction from incomplete experimental data using Fourier interpolation and multidimensional matched filtering, and also, the technique of using total internal reflection to determine the average refractive index of the background medium. This technique can be realized by using fast algorithms and can work in real time. The described technique can be applied to speed up the visualization of hidden objects in UWB sounding. Moreover, it can be directly employed in current geolocation systems and in advanced systems for radio tomography of hidden objects. Further gains in reduction of computation time are possible using the group focusing method.

Ground-Penetrating and Geo-Radars ■ 217

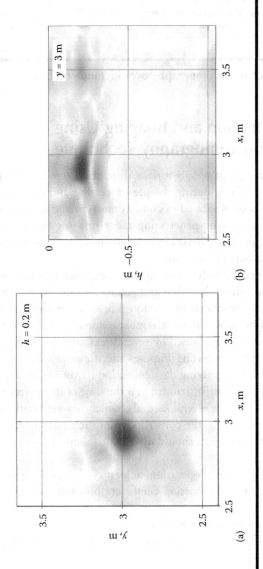

Figure 8.14 Two orthogonal cross-sections of tomogram reconstructed over incomplete experimental data.

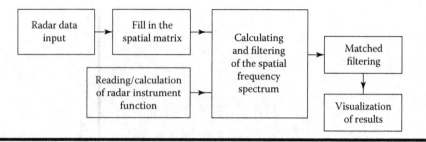

Figure 8.15 Flow chart of tomography over incomplete data.

8.2 Mine Detection and Imaging Using Diffraction Tomography Technique

At the same time, for proof of sub-surface diffraction tomography methods, discussed briefly in Chapter 3, during the pre-processing and post-processing stages of buried objects or objects located in clutter detection and identification, in references [21–28], special ground-penetrating radar imaging systems were designed and then the corresponding microwave measurements were carried out for optimization and improvements to each unit.

Thus, in references [23–30], the microwave ground-penetrating radar performed for mine and tunnel detection and reconstruction by the use of diffraction tomography methods, described in Section 3.3, was reported on. It was shown that the creation of portable microwave measurement units is an important part of the research and can be explained by several reasons. First of all, application of sub-surface tomography processing imposes specific requirements on the technical specifications of the underlying data collection hardware, which unfortunately cannot be satisfied by existing instruments, if the time of image reconstruction and convenience of work exceeds several seconds, which is a very critical parameter of mine detection. Therefore, for the construction of such sub-surface radar systems, the corresponding requirements should usually be followed:

- Wide frequency range of operation, at least twice initial frequency
- Ability to measure transmission coefficient on thousands of frequencies per second
- High microwave frequency stability
- Built-in hardware for compensation of large first reflection from air–medium interface
- Wide dynamic range and linearity to be able to recover weak useful signals from the target in the presence of strong background signals from the interface
- Minimum weight, size, and power consumption

Ground-Penetrating and Geo-Radars ■ 219

Figure 8.16 Moving laboratory setup of ground-penetrating radar for subsoil mine detection.

An experimental demonstration setup is presented in Figure 8.16.

We will start with the application of diffraction tomography in pre- and postprocessing of the models of plastic antitank and antipersonnel mines buried in the subsoil (mostly sand) medium, as shown in Figure 8.16. The corresponding radio tomograms are shown in Figure 8.17; the following results were obtained in references [25–28].

A principle of tomography image derivation allows the obtaining of images of a vertical cut in the case of 1-D scanning and both vertical and horizontal cuts in the case of 2-D scanning. Figure 8.17 shows print screen images obtained during real measurements for dielectric (antipersonnel) and metal (antitank) mines buried in subsoil media at depths varying from 11 to 18.5 cm. Here, in the left-hand panel of Figure 8.17, an image of buried mines (denoted by the yellow and red colors) is clearly seen in the soil with surrounding stones, as a specific clutter, for each slice (corresponding to a depth from 11 to 18.5 cm). The same shaped feature of buried mines (surrounded by stones, as clutter) is clearly seen from the right-hand panel, where each slice, which presents images in the vertical plane, corresponds to a specific depth from 11 cm to 18.5 cm.

In references [29, 30], windows of detection of mines buried into the sand and clay were found by the use of UWB with extremely short pulses. Thus, in Figure 8.18, a scheme of an experiment of detection of mines buried in the sand is presented.

The corresponding cross-section of the buried model of a mine in the frequency domain is shown in Figure 8.19, according to references [29, 30].

It is clearly seen that the response of the mine's profile was recorded by the radar along the trace from 1.2 to 1.5 m and its reaction in the frequency domain is much more visible at 3.5, 5, and 9 MHz.

220 ■ *Electromagnetic and Acoustic Wave Tomography*

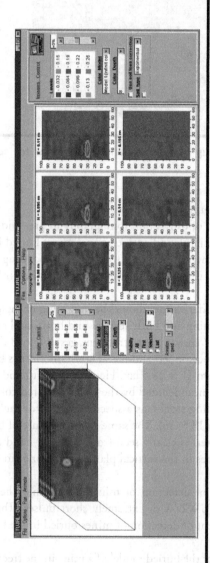

Figure 8.17 Images of mines in subsoil media (denoted by yellow and red colors) obtained by developed radar system: slices in the vertical plane along each depth (left), vertical photo of each slice of the mines buried at depths from 11 to 18.5 cm (right). (Adapted from Vertiy, A. A., S. P. Gavrilov, and G. Gençay, Microwave tomography systems for investigation of the ware structure, *4th International Conference on Millimeter and Submillimeter Waves and Applications*, San Diego, California, USA, 20–23 July 1998; Vertiy, A. A. and S. P. Gavrilov, Microwave imaging of cylindrical inhomogenities by using a plane wave spectrum of diffracted field, *Proceedings of International Conference on The Detection of Abandoned Land Mines*, Edinburgh, 12–14 October 1998; Vertiy, A. A., S. P. Gavrilov, I. Yoynovskyy, A. Kudelya, V. Stepanuk, and B. Levitas, GPR and microwave tomography imaging of buried objects using the sort-term (picosecond) video pulses, *Proceedings of 8th International Conference on Ground Penetrating Radar*, Gold Coast, Australia, 23–26 May 2000, pp. 530–534; Blaunstein, N., Ch. Christodoulou and M. Sergeev, *Introduction to Radio Engineering*, CRC Press, Taylor & Francis, Boca Raton, FL, 2017.)

Ground-Penetrating and Geo-Radars ■ 221

Figure 8.18 Scheme of experiment with miniature UWB radar detecting plastic mine buried in sand within clutter.

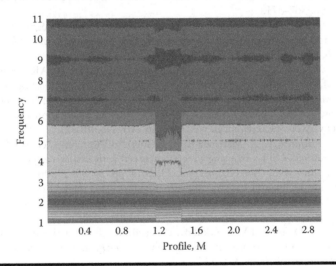

Figure 8.19 Reconstruction of shape and position of plastic mine versus radiated frequency (in MHz).

References

1. Yakubov, V. P., S. E. Shipilov, D. Ya. Sukhanov, and A. V. Klokov, *Radiowave Tomography: Achievements and Perspectives*, Tomsk, Russia: Scientific-Technical Literature (STL) Edition Group, 2016 (in Russian).
2. Yakubov, V. P., S. E. Shipilov, D. Ya. Sukhanov, and A. V. Klokov, *Wave Tomography*, Tomsk, Russia: Scientific Technology Publishing House, 2017.
3. Daniels, D. J., *Surface-Penetrating Radar*, London: Institute of Electrical Engineering, 1996.

4. Skolnik, M., *Radar Handbook*, New York: McGraw-Hill Education, 3rd edn, 2008.
5. Ulaby, F. T., R. K. Moore, and A. K. Fung, *Microwave Remote Sensing: Active and Passive*, Vol. 3, *Theory to Applications*, Norwood: Artech House, 1986.
6. Persico, R., *Introduction to Ground Penetrating Radar: Inverse Scattering and Data Processing*, IEEE Press Series on Electromagnetic Wave Theory, 1st edn, New York: Wiley-IEEE Press, 2014.
7. Jol, H. M., *Ground Penetrating Radar Theory and Applications*, Amsterdam: Elsevier, 2009.
8. Hamran, S., D. T. Gjessing, J. Hjelmstad, and E. Aarholt, Ground penetrating synthetic pulse radar: dynamic range and modes of operation, *J. Appl. Geophys.*, vol. 33, 1995.
9. IEE Proceedings on European International Conference on the Detection of Abandoned Land Mines: A Humanitarian Imperative Seeking Technical Solutions, Edinburgh, UK, 7–9 October, 1996, Conf. publication no. 431, Institute of Electrical Engineers, London, UK.
10. Wait, J. R., *Electromagnetic Probing in Geophysics*, London: Golem Press, 1971.
11. Groenenboom, J. and A. G. Yarovoy, Data processing for a landmine detection dedicated GPR, *Proceedings of 8th International Conference on Ground Penetrating Radar*, 2000, pp. 367–371.
12. Yakubov, V. P., D. Ya. Sukhanov, and A. V. Klokov, Radiotomography from ultra-wideband monostatic measurements on an uneven surface, *Russian Physics Journal*, vol. 56, no. 9, 2014, pp. 1053–1061.
13. Yakubov, V. P., A. S. Omar, D. Ya. Sukhanov, et al., New fast SAR method for 3-D subsurface radiotomography, *Proceedings of 10th International Conference on Ground Penetrating Radar*, 21-24 June, 2004, Delft, the Netherlands, pp. 103–106.
14. Klokov, A., A. Zapasnoy, and S. Shipilova, Jitter in geolocation data processing, MATEC Web of Conferences, vol. 79, 2016, Art. No. 01022.
15. Durán-Ramírez, V., A. Martínez-Ríos, J. Guerrero-Viramontes, et al., Measurement of the refractive index by using a rectangular cell with a fs-laser engraved diffraction grating inner wall, *Optics Express*, vol. 22, 2014, pp. 29899–29906.
16. Sukhanov, D. Ya. and V. P. Yakubov, Inclined focusing method in subsurface radar, *Technical Physics. The Russian Journal of Applied Physics*, vol. 51, no. 7, 2006, pp. 882–886.
17. Sukhanov, D. Ya. and V. P. Yakubov, Determining the refractive index of the background medium in subsurface tomography, *Radiophysics and Quantum Electronics*, vol. 50, no. 5, 2007, pp. 381–387.
18. Yakubov, V. P., S. E. Shipilov, and D. Ya. Sukhanov, Microwave tomography of radiopaque objects, *Russian Journal of Nondestructive Testing*, vol. 47, no. 11, 2011, pp. 765–770.
19. Epov, M. I., O. V. Yakubova, E. D. Telpukhovskii, et al., Method of ultra-wideband radio-pulse logging of horizontal wells, *Russian Physical Journal*, vol. 51, no. 9, 2008, pp. 949–957.
20. Kochetkova, T. D., A. S. Zapasnoy, A. V. Klokov, S. E. Shipilov, V. P. Yakubov, and A. V. Yurchenko, Ultra-wideband tomography of land cover, *Proceedings of SPIE – The International Society for Optical Engineering*, 2014, issue 9292, Art. No. 929250.
21. Yakubov, V. P. and D. Ya. Sukhanov, Solution of a subsurface radio-imaging inverse problem in the approximation of a strongly refractive medium, *Radiophysics and Quantum Electronics*, vol. 50, no. 4, 2007, pp. 299–307.

22. Klokov, A. V., A. S. Zapasnoy, A. S. Mironchev, V. P. Yakubov, and S. S. Shipilova, A comprehensive study of underground animals habitat, *Journal of Physics: Conference Series*, vol. 671, no. 1, 2016, Art. No. 012028.
23. Cloude, S. R., A. Milne, C. Thornhill, and G. Crips, UWB SAR detection of dielectric targets, in *IEE Proceedings on European International Conference on the Detection of Abandoned Land Mines*, Edinburgh, UK, 7–9 October, 1996, Conf. publication no. 431, pp. 114–118.
24. Jaureguy, M. and P. Borderies, Modeling and processing of ultra-wide band scattering of buried targets, in *IEE Proceedings on European International Conference on the Detection of Abandoned Land Mines*, Edinburgh, UK, 7–9 October, 1996, Conf. publication no. 431, pp. 119–123.
25. Vertiy, A. A., S. P. Gavrilov, and G. Gençay. Microwave tomography systems for investigation of the ware structure, *4th International Conference on Millimeter and Submillimeter Waves and Applications*, San Diego, California, USA, 20–23 July 1998, 5 pages.
26. Vertiy, A. A. and S. P. Gavrilov, Microwave imaging of cylindrical inhomogenities by using a plane wave spectrum of diffracted field, *Proceedings of International Conference on The Detection of Abandoned Land Mines*, Edinburgh, UK, 12–14 October 1998, 5 pages.
27. Vertiy, A. A., S. P. Gavrilov, I. Yoynovskyy, A. Kudelya, V. Stepanuk, and B. Levitas, GPR and microwave tomography imaging of buried objects using the sort-term (picosecond) video pulses, *Proceedings of 8th International Conference on Ground Penetrating Radar*, Gold Coast, Australia, 23–26 May 2000, pp. 530–534.
28. Blaunstein, N., Ch. Christodoulou and M. Sergeev, *Introduction to Radio Engineering*, Boca Raton, FL: CRC Press, Taylor & Francis, 2017.
29. Bezrodni', K. P., V. B. Boltintzev, and V. N. Ilyakhin, Geophysical testing of injection section of the behind-installed space by method of electromagnetic pulse ultra-wide band sounding, *Journal of Civilian Construction*, no. 5, 2010, pp. 39–44 (in Russian).
30. Boltintzev, V. B., V. N. Ilyakhin, and K. P. Bezrodni', Method of electromagnetic ultra-wide band pulse sounding of the surface medium, *Journal of Radio Electronics*, no. 1, 2012, 39 pages, Moscow (in Russian).

Chapter 9

Sub-Surface Tomography Applications

Vladimir Yakubov, Sergey Shipilov,
Andrey Klokov, and Nathan Blaunstein

Contents

9.1 Subsoil Structures Detection and Visualization 226
    9.1.1 Tunnels and Tubes (Pipelines) Detection and Identification 226
    9.1.2 Geolocation of Rough and Uneven Terrain Surfaces 226
    9.1.3 Subsoil Structures Detection and Reconstruction of Images 233
9.2 Experimental Reconstruction of Foreign Structures Buried in Subsoil
    Medium 240
9.3 Minerals Detection and Identification 244
9.4 Geolocation of Special Sub-Surface Man-Made Structures and Objects 247
9.5 Millimeter-Wave Sub-Surface Tomography Applications 257
    9.5.1 Theoretical Background of Radiometer Operation 258
    9.5.2 Typical Application of Millimeter-Wave Radiometers 260
References 263

This chapter is based on results obtained experimentally in references [1–7, 26–31].

9.1 Subsoil Structures Detection and Visualization

9.1.1 Tunnels and Tubes (Pipelines) Detection and Identification

This series of experiments was carried out for detection and identification of tunnels and guiding structures. We present only one example from many similar published in references [1, 2, 4]. Thus, in Figure 9.1a, the first experimental setup of a ground-penetrating radar (GPR) antenna (top panel) is shown with the photograph of the cross-view of the real tunnel (middle panel). Figure 9.1b presents the obtained image of a real tunnel at a depth of 35–40 cm after processing using diffraction tomography, the elements of which are briefly described in Chapter 3.

Another set of experiments, reported in reference [3], is related to detection and imaging of the subsoil tubes and pipelines. We present here only few of them as the result of usage of the diffraction tomography technique described in Chapter 3. Thus, in Figure 9.2, the 2-D imaging of a plastic tube with diameter of 10 mm buried into the subsoil medium at a depth of 15 cm is shown.

It is clearly seen that at the previous ($H = 11$ cm) and subsequent depths ($H = 19$ cm), imaging is very pure, whereas at the actual depth of the buried object, located at 15 cm, the shape and the location of the pipe are seen quite vividly.

These examples, even only a few, show that the created experimental prototypes of the ultra-wide band (UWB) and extremely-short-pulse (ESP) radar systems with special pre-processing and post-processing, based on elements of diffraction tomography, allow us to obtain images of buried objects with sufficiently high resolution and image quality, as well as processing the experimental data at a rate sufficient for image reconstruction within the process of scanner movement. It was shown in [1–4] that the maximum speed of the surface analysis does not exceed 10 cm/s, and it is limited by the maximum velocity of scanner movement in a transverse direction. Increase of velocity of the analysis in a longitudinal direction is possible only by refusal of mechanical scanning and a transition to a parallel multi-channel principle of receiving data about spatial distribution of microwave fields over the surface. The examples presented show the efficiency of the pre-processing, based on novel approaches in diffraction tomography for the detection and identification of different kinds of buried mines, small (antipersonnel) to large (antitank), as well as tunnels at depths up to 10–15 m.

9.1.2 Geolocation of Rough and Uneven Terrain Surfaces

This section describes three techniques for sub-surface image reconstruction based on geolocation scanning over an uneven surface following the results discussed in references [1, 2].

Inverse Focusing. In this section, we consider the profile of sub-surface sounding shown in Figure 9.3. It is assumed that the radiating and receiving antennas are

Sub-Surface Tomography Applications ■ 227

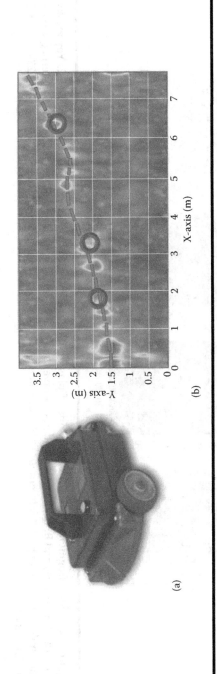

Figure 9.1 (a) Antenna unit of OKO-2 ground-penetrating radar with central frequency 1700 MHz. (b) Image of real tunnel after processing using elements of diffraction tomography. (Adapted from Yakubov, V. P., S. E. Shipilov, D. Ya. Sukhanov, and A. V. Klokov, *Radiowave Tomography: Achievements and Perspectives*, Scientific-Technical Literature (STL) Edition Group, Tomsk, Russia, 2016; Yakubov, V. P., S. E. Shipilov, D. Ya. Sukhanov, and A. V. Klokov, *Wave Tomography*, Scientific Technology Publishing House, Tomsk, Russia, 2017.)

228 ■ *Electromagnetic and Acoustic Wave Tomography*

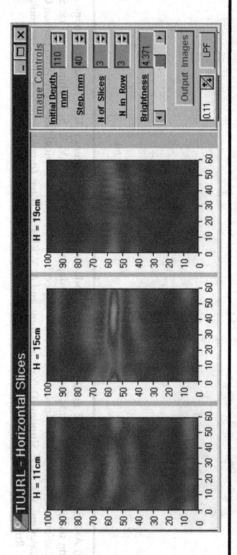

Figure 9.2 Three fragmental (sliced) tomograms of plastic tube. (Adapted from Vertiy, A. A., S. P. Gavrilov, I. Yoynovskyy, A. Kudelya, V. Stepanuk, and B. Levitas, GPR and microwave tomography imaging of buried objects using the sort-term (picosecond) video pulses, *Proceedings of 8th International Conference on Ground Penetrating Radar*, Gold Coast, Australia, 23–26 May 2000, pp. 530–534.)

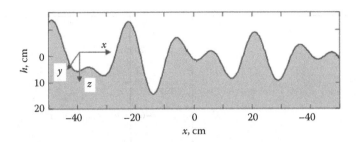

Figure 9.3 Uneven surface. (Adapted from Yakubov, V. P., S. E. Shipilov, D. Ya. Sukhanov, and A. V. Klokov, *Radiowave Tomography: Achievements and Perspectives*, Scientific-Technical Literature (STL) Edition Group, Tomsk, Russia, 2016; Yakubov, V. P., S. E. Shipilov, D. Ya. Sukhanov, and A. V. Klokov, *Wave Tomography*, Scientific Technology Publishing House, Tomsk, Russia, 2017.)

integrated and move along the uneven surface $h(x,y)$. The synthesized frequency bandwidth was taken to extend from 5 to 10 GHz.

Two cases of inhomogeneities were considered: (1) a single inhomogeneity and (2) a group inhomogeneity. Corresponding radar responses are presented in Figure 9.4. We note that the solution of the problem for a flat surface would be take the form of a diffraction hyperbola, but the radar response for an uneven surface resembles the shape of that surface.

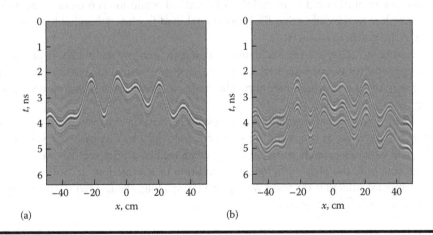

Figure 9.4 Solution of direct problem: (a) for a single inhomogeneity; (b) for four inhomogeneities. (From Yakubov, V. P., S. E. Shipilov, D. Ya. Sukhanov, and A. V. Klokov, *Radiowave Tomography: Achievements and Perspectives*, Scientific-Technical Literature (STL) Edition Group, Tomsk, Russia, 2016; Yakubov, V. P., S. E. Shipilov, D. Ya. Sukhanov, and A. V. Klokov, *Wave Tomography*, Scientific Technology Publishing House, Tomsk, Russia, 2017.)

230 ■ *Electromagnetic and Acoustic Wave Tomography*

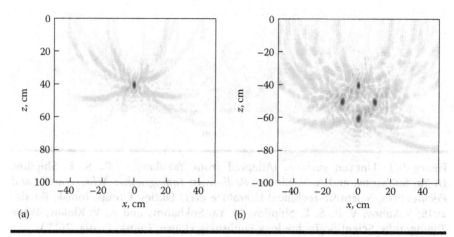

Figure 9.5 Images of point-like inhomogeneities reconstructed by inverse focusing: (a) for a single inhomogeneity; (b) for four inhomogeneities. (From Yakubov, V. P., S. E. Shipilov, D. Ya. Sukhanov, and A. V. Klokov, *Radiowave Tomography: Achievements and Perspectives*, Scientific-Technical Literature (STL) Edition Group, Tomsk, Russia, 2016; Yakubov, V. P., S. E. Shipilov, D. Ya. Sukhanov, and A. V. Klokov, *Wave Tomography*, Scientific Technology Publishing House, Tomsk, Russia, 2017.)

Figure 9.4 presents the results of processing of the data displayed in Figure 9.5 using the inverse focusing method described in Chapter 3. It can be seen that the transverse resolution is 3.5 cm and the longitudinal resolution is 6.8 cm. These values are close to the diffraction limit per the selected frequency bandwidth and the size of the synthesized aperture.

Inverse focusing is a relatively universal and easy-to-use technique. It provides the opportunity to consider surface irregularities and different velocities of wave propagation in a stratified medium. The main drawback of this technique is the large volume of calculations.

Phase Screen Approximation. It is supposed in the phase screen approximation that the curvature of the scanning surface is not high. In this case, the values of the field from an uneven surface can be transferred to a relatively flat surface by correcting only the phase difference (by addition or subtraction) according to the following formula as if the wave were propagating along the normal to the average surface at each frequency [1, 2]:

$$E(\mathbf{r}) \rightarrow E(\mathbf{r})\exp\{-2iknh(x,y)\} \quad (9.1)$$

where $k = \omega/c$, and n is the refractive index of the lower medium.

Next, using the group focusing algorithm (analogue of the Stolt migration) the tomogram of the distribution of scattering inhomogeneities in the lower medium can be calculated as if the medium interface were planar. The images so obtained

Sub-Surface Tomography Applications ■ 231

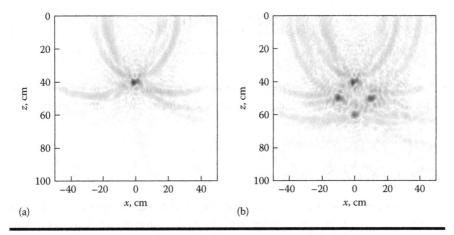

Figure 9.6 Reconstructed images of point-like scatterers in phase screen approximation: (a) for a single inhomogeneity; (b) for four inhomogeneities.

are displayed in Figure 9.6, rearranged from references [1, 2]. The assigned distributions of inhomogeneities are reconstructed, but the level of artifacts has increased. Here, the inaccuracy of the approximation becomes apparent. For lower curvatures, the errors will be smaller.

Huygens–Fresnel Interpolation. Interpolation based on the Huygens–Fresnel principle, as mentioned in Chapter 2, can be considered as the more accurate approximation. Let us consider it. The scattering inhomogeneities can be replaced by in-phase field sources at the doubled frequency if the measurements are performed with a transceiver, in line with Equation 2.11 (see Chapter 2). This approximation does not take into account the falloff of the field with distance after scattering; however, since this measurement is less critical than the phase differences, it can be neglected. Thus, the direct problem for the scattered field (Equation 2.11) is transformed into the problem of searching for the field sources [1, 2, 4]:

$$E(\mathbf{r}) = \iiint_{V_1} j(\mathbf{r}_1) G_2(\mathbf{r}_1 - \mathbf{r}) d^3 \mathbf{r}_1 \qquad (9.2)$$

where, as in Chapter 2, $j(\mathbf{r}_1) \approx k^2 \Delta\varepsilon(\mathbf{r}_1)/(4\pi|\mathbf{r}_1 - \mathbf{r}|)$ is the distribution of sources, whereas in fact it is the distribution of inhomogeneities; $G_2(\mathbf{r}) = \exp(ik^2|\mathbf{r}|)/(4\pi|\mathbf{r}|)$ is the Green function for the lower homogeneous space of the background medium at the doubled frequency, where $\mathbf{r}$ is the point of radiation and reception, which lies on the curvilinear surface.

Let us calculate the field $E(\mathbf{r}_0)$ on the average plane $z = 0$ from the known field $E(\mathbf{r})$ on the curvilinear surface S using the Huygens–Fresnel principle as follows [1, 2, 4]:

$$E(\mathbf{r}_0) = 2 \iint_S E(\mathbf{r}) \frac{\partial G_2(\mathbf{r}-\mathbf{r}_0)}{\partial n} dS \qquad (9.3)$$

where S is the curvilinear surface, where the measurements are to be performed. Here, without incurring great error, we can make the following substitution:

$$\frac{\partial G_2(\mathbf{r})}{\partial n} \approx \frac{\partial G_2(\mathbf{r})}{\partial z} \qquad (9.4)$$

It is important to note here that the field $E(\mathbf{r}_0)$ is defined in a plane, and the fast algorithm of group focusing (analogue of the Stolt migration: see Chapter 2) can be applied to this field. The images shown in Figure 9.7 were obtained as a result of applying the interpolation in Equation 9.2 and the group focusing algorithm. The quality of the reconstructed images is substantially improved and the level of artifacts has been reduced.

The quality of the reconstructed image is comparable to the quality of the image obtained by applying inverse focusing. In other words, the Huygens–Fresnel interpolation is actually equivalent in image quality to inverse focusing, but it allows the use of fast algorithms.

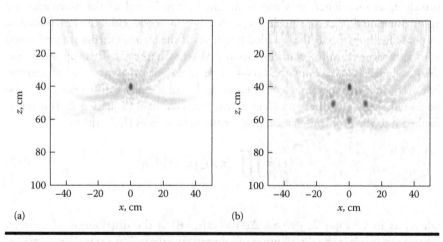

Figure 9.7 Images of point-like inhomogeneities reconstructed by inverse focusing: (a) for a single inhomogeneity; (b) for four inhomogeneities. (From Yakubov, V. P., S. E. Shipilov, D. Ya. Sukhanov, and A. V. Klokov, *Radiowave Tomography: Achievements and Perspectives*, Scientific-Technical Literature (STL) Edition Group, Tomsk, Russia, 2016; Yakubov, V. P., S. E. Shipilov, D. Ya. Sukhanov, and A. V. Klokov, *Wave Tomography*, Scientific Technology Publishing House, Tomsk, Russia, 2017.)

9.1.3 Subsoil Structures Detection and Reconstruction of Images

This section describes the effectiveness of radar tomography for soil cover research. A comprehensive approach is proposed that combines the application of the OKO-2 GPR (Figure 9.8), the conventional method of cross-sectioning applied in soil science, and relevant soil-testing parameters with mobile and laboratory research on the permittivity of hidden soil layers. The combination of several techniques enabled us to relate the electrical properties of soil to soil moisture and density, and thereby detect the location of deep aquifers and reconstruct the actual local topography. This research was performed within a specific forest district in Russia.

Contact Measurements of Soil Characteristics. Several traces were cut under different soil conditions. Figure 9.9 (bottom) presents one of the selected traces within the geomorphological slope with a difference in levels of 3 m in the specific forest district near the tested experimental area. The area is covered by green moss pine forest that is 80–90 years old, and the shrub layer consists of mountain ash undergrowth. Crown density is 40%. Soil cover includes bilberry, green moss, sedge, and several rare plants. Projective cover is 10%–15%. The macrorelief is a terrace above the Tom flood plain. The mesorelief comprises the mid-section and base of the slope with southern exposure, and the steepness is 3°–5°. The parent material is pine forest sands (Figure 9.9, top).

Sod-podzolic soil with loamy sand prevails in the sounding field. This type of soil is characterized by a thin humus horizon. The organic layer here does not exceed 3%–4%. A small amount of humus, lack of structural aggregates, and a predominance of a sandy fraction in the soil texture predetermine the high density

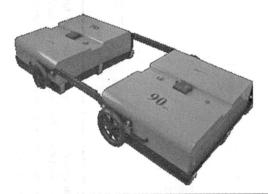

Figure 9.8 Antenna unit of OKO-2 ground-penetrating radar with central frequency 90 MHz. (From Yakubov, V. P., S. E. Shipilov, D. Ya. Sukhanov, and A. V. Klokov, *Radiowave Tomography: Achievements and Perspectives*, Scientific-Technical Literature (STL) Edition Group, Tomsk, Russia, 2016; Yakubov, V. P., S. E. Shipilov, D. Ya. Sukhanov, and A. V. Klokov, *Wave Tomography*, Scientific Technology Publishing House, Tomsk, Russia, 2017.)

234 ■ *Electromagnetic and Acoustic Wave Tomography*

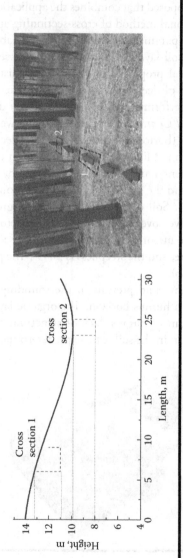

Figure 9.9 Geomorphological slope profile and land view. (From Yakubov, V. P., S. E. Shipilov, D. Ya. Sukhanov, and A. V. Klokov, *Radiowave Tomography: Achievements and Perspectives*, Scientific-Technical Literature (STL) Edition Group, Tomsk, Russia, 2016; Yakubov, V. P., S. E. Shipilov, D. Ya. Sukhanov, and A. V. Klokov, *Wave Tomography*, Scientific Technology Publishing House, Tomsk, Russia, 2017.)

of the structure of soil horizons over the entire soil profile. The firmest soils are alluvial horizons (B-horizons) that contain pseudo-fiber layers. Pseudo-fibers consolidated by organic mineral colloids slow down the process of filtration of atmospheric moisture and prevent water evaporation from deep horizons. There are short-term regenerative conditions over the pseudo-fiber layers and under them. Morphologically, this is manifested by smoke-blue tones and ferrous iron staining, and also by increased humidity. Two cross-sections of soil were cut on this trace marked with a dashed line in Figure 9.9.

Figure 9.10 illustrates the morphology of sod-podzolic soil in the middle part of the slope (cross-section 1). Soil samples were selected from genetic horizons of the cross-section.

The complex permittivity of the samples was determined in a laboratory environment. The permittivity was measured by the waveguide–coaxial method with an R2M scalar network analyzer. The match between the calculated and measured values of the transmission coefficient and the standing wave ratio was monitored on a frequency of 90 MHz. The temperature was 25°C. The density of samples in the measuring cell was taken into account. The value of the permittivity was renormalized for each sample with allowance for the density: first, it was divided into the density of the soil sample in the measuring cell; then, it was multiplied by

Figure 9.10 Sod-podzolic soil profile of cross-section 1. (Adapted from Yakubov, V. P., S. E. Shipilov, D. Ya. Sukhanov, and A. V. Klokov, *Radiowave Tomography: Achievements and Perspectives*, Scientific-Technical Literature (STL) Edition Group, Tomsk, Russia, 2016; Yakubov, V. P., S. E. Shipilov, D. Ya. Sukhanov, and A. V. Klokov, *Wave Tomography*, Scientific Technology Publishing House, Tomsk, Russia, 2017.)

the dry density in the cross-section. The dry density is the mass of a unit volume of absolutely dry soil in the undisturbed structure.

The soil moisture in the samples was measured by the dry-and-wet weight method, in which the water content is calculated as the ratio of the weight of water in the sample to the dry weight of the sample. In this case, samples weighing from 10 to 25 g were dried in an oven at 105°C for 3 h and then weighed. Then, they were oven dried once again for 3 h and weighed once more. The weight difference between weighings was 0.004 g on average, resulting in 0.1% error in the moisture measurement.

Figure 9.11 presents summary results of contact laboratory experiments for cross-section 1. This cross-section provides a clear view of three rust-colored pseudo-fiber layers 0.5–6 cm thick. They are distinguished by an increased value of permittivity and density (see numbers presented in Figure 9.11). An increased moisture content between the layers can also be seen.

Similar results were obtained for sod-podzolic soil from the dell (cross-section 2). The results are shown in Figures 9.12 and 9.13.

Results of Geological Survey. A geo-radar survey was conducted with an OKO-2 commercial GPR company (Moscow region, Russia). A subsystem with bipolar sound pulses and a central frequency of 90 MHz was decided on for soil sounding, based on its penetrating power. Sounding was conducted in a straight line, so that real cross-sections were cut along this line (see Figure 9.10).

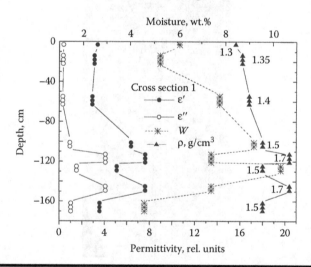

Figure 9.11 Depth-wise distribution of moisture, dry density, and permittivity for cross-section 1. (Adapted from Yakubov, V. P., S. E. Shipilov, D. Ya. Sukhanov, and A. V. Klokov, *Radiowave Tomography: Achievements and Perspectives*, Scientific-Technical Literature (STL) Edition Group, Tomsk, Russia, 2016; Yakubov, V. P., S. E. Shipilov, D. Ya. Sukhanov, and A. V. Klokov, *Wave Tomography*, Scientific Technology Publishing House, Tomsk, Russia, 2017.)

Sub-Surface Tomography Applications ■ 237

Figure 9.12 Sod-podzolic soil profile (cross-section 2).

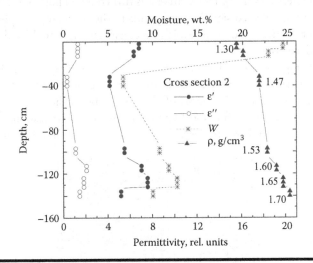

Figure 9.13 Depth distribution of moisture, dry density, and permittivity for cross-section 2. (Adapted from Yakubov, V. P., S. E. Shipilov, D. Ya. Sukhanov, and A. V. Klokov, *Radiowave Tomography: Achievements and Perspectives*, Scientific-Technical Literature (STL) Edition Group, Tomsk, Russia, 2016; Yakubov, V. P., S. E. Shipilov, D. Ya. Sukhanov, and A. V. Klokov, *Wave Tomography*, Scientific Technology Publishing House, Tomsk, Russia, 2017.)

238 ■ *Electromagnetic and Acoustic Wave Tomography*

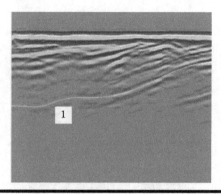

Figure 9.14 **Geo-radar reference profile.**

Since the profile was traced along the slope in a top-down way, and sounding was conducted by geo-radar set up directly on the ground, geometrical distortions showed up in the radar image (see Figure 9.14). Horizontal movement of the sounding antenna is plotted along the abscissa and the time delays of the reflected signals are plotted along the ordinate. The reflected signal intensity is plotted in gray-scale. The level of the underground aquifer is shown by curve *1*. This curve was plotted using the least squares method. The aquifer level should actually be horizontal. Correction for such distortions results in the image presented in Figure 9.15, according to references [1, 2]. Here, curve *1* is turned into a horizontal line, and the upper edge of the reflected signals corresponds to the actual topography. Thus, the presence of an aquifer made it possible to reconstruct the actual local topography along the radar path.

The next stage in geo-radar data processing was to compress the images. The external manifestation of this effect is the presence of the so-called *diffraction hyperbolas* in the images. The traces of these hyperbolas show up in Figure 9.16 as curves *2* and *3*. The diffraction hyperbolas are the radar response to a point-like inhomogeneity and the analogue of the point spread function in optics. The

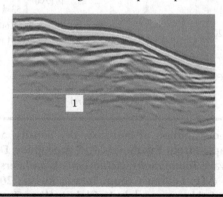

Figure 9.15 **Geo-radar profile accounting for topographic relief.**

Sub-Surface Tomography Applications ■ 239

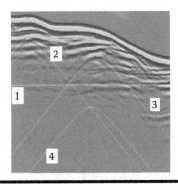

Figure 9.16 Diffraction hyperbola traces.

aperture angle of the hyperbolas depends on the average refractive index n of the medium in which the inhomogeneities occur. By varying the aperture angle of the approximating curves, it is possible to determine the average refractive index of the soil. In this case, n is seen to be approximately equal to 2.5 [1, 2]. Attenuation of the signal along the trace is shown in Figure 9.17.

This value corresponds to a value of the permittivity equal to $\varepsilon' \approx 6.5$, which is close to the values presented in Figures 9.11 and 9.13. Elimination of the point spread function is performed by large aperture synthesis or, equivalently, by 2-D matched filtering.

However, before matched filtering can be applied, another operation needs to be performed, namely, correction for exponential attenuation of the radiation with penetration depth into the medium, that is, into the soil. Of course, this can be done only on average for the background medium.

The sequence of operations is as follows: (1) for all positions of the antenna, the time series (vertical records) are transformed into the amplitudes of the corresponding analytical signal; (2) the obtained amplitude profiles are averaged over

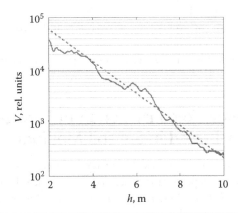

Figure 9.17 Exponential attenuation of radiation.

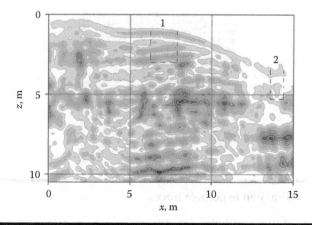

Figure 9.18 Radar cross-section of sounding path.

all antenna positions. The average attenuation of the relative amplitude of the analytical signal, obtained by this method, is presented in Figure 9.17. Its exponential approximation is represented by the dashed line. Correction for attenuation makes it possible to increase the relative contribution of the response signals from the deep layers [1, 2].

Combined execution of the above-mentioned procedures (matched filtering and amplitude correction) delivers the final result—a vertical cross-section along the radar path (see Figure 9.18). The results show that a number of features in the obtained cross-section match up well with the results of contact measurement.

This work demonstrates a high correlation between the structural features revealed by non-destructive radar sounding and by contact measurement. The procedures required for correct processing of the radar data are:

- Data correction for relief (local topography)
- Reconstruction of the average refractive index and vertical positioning of the data
- Amplitude correction for attenuation of radiation with depth

9.2 Experimental Reconstruction of Foreign Structures Buried in Subsoil Medium

In this section, to prove the geometrical multilayer optical model described in reference [29], a set of actual experiments described in references [5, 6] was carried out. Schematically, this UWB/ESP radar system is shown in Figure 9.19.

The main goal of such radar system operation was an increase in the depth of sounding, and the resolution that was finally achieved was due to:

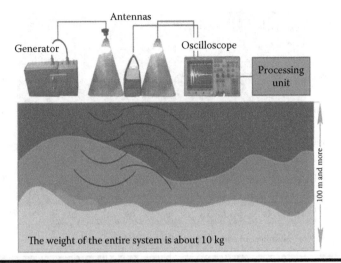

Figure 9.19 Scheme of UWB/ESP radar system used for sub-surface structure detection and identification. (From Bezrodni', K. P., V. B. Boltintzev, and V. N. Ilyakhin, *Journal of Civilian Construction*, 5, 39–44, 2010; Boltintzev, V. B., V. N. Ilyakhin, and K. P. Bezrodni', *Journal of Radio Electronics*, 39, 2012.)

- Increase of generated pulse power
- Good conditions of matching transmitter–receiver antennas with the environment and the subsoil media
- Wide band of operating frequencies of the receiver, creating an effective algorithm of signal processing of the obtained data

As was mentioned in reference [4], performance of an accurate radar experiment is a complex process that requires knowledge of the subsoil media at depths of up to 10 m at least, by use of an object large enough to retain the signals very clearly. In both studies, a UWB/ESP GPR with a frequency range from 1 to 200 MHz was used. The radar was completed by different antennas with various frequency ranges available for a center (carrier) frequency that varied from 15 MHz to 150 MHz and with sizes in the unfolded form of 1–10 m, accordingly.

The transmitter has the following parameters: signal voltage of 10 kV, pulse power of 10 mW, and pulse duration of 1–20 ns. The transmitter power is regulated with attenuators between 0 and 120 dB. The data acquisition and processing program allow real-time combination of GPR scanning and GPS data, as well as building 3-D images, taking into account the relief with the accuracy of the GPS receiver. We have used existing GPR software with our matches to obtain automatic results, including tables.

One of the results of imaging multilayered structures was reported in references [5, 6] where a UWB/ESP wave profile was performed consisting of 30 measurements ranged at a distance from each other of 2 m. The profile crossed two metro

242 ■ *Electromagnetic and Acoustic Wave Tomography*

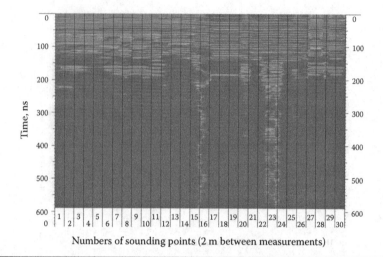

Figure 9.20 Imaging of subsoil medium with two tunnels of Metropolitan. (Adapted from Bezrodni', K. P., V. B. Boltintzev, and V. N. Ilyakhin, *Journal of Civilian Construction***, 5, 39–44, 2010; Boltintzev, V. B., V. N. Ilyakhin, and K. P. Bezrodni',** *Journal of Radio Electronics***, 39, 2012.)**

tunnels. To define the positions of tunnels and their depths, mutual correlation functions (MCF) of reflected signals (of the meter and decimeter bands) were used. This result is shown in Figure 9.20, as the signal response in the time domain for each of the 30 points of measurement lining the measured trace above the terrain, with a multilayered subsoil medium as a strong clutter for reconstruction of the buried tunnel structures.

It is clearly seen that the two tunnels were 18–20 m apart from each other. Recording signals of extremely short periods reflected from the tubes' surfaces allowed estimation of the depth of such tunnels. Unfortunately, in references [5, 6], these estimations were not presented. At the same time, such estimations were carried out in reference [4] based on the geometrical optics approach described there. Next, we will briefly describe the radar experiments and the corresponding estimations not only of the depth of the tunnel, but also of the permittivity of its walls.

Thus, according to reference [4], we present the characteristic experimental data that were based on the ESP returning from the train tunnel built at the depth of the multilayer soil medium, which at the beginning was not known accurately. This experiment, as well as the other typical experiments, was based on a program of UWB/ESP radar and allows separation of the transmitter and the receiver operation. Unlike standard GPR operational work, the transmitter remained fixed in place. In this case, the radar output imaginary picture presents the lines going up and down at a certain angle along the experiment. The possible work area was not limited to 8 m, as was usually the case in other similar experiments. At the same time, the depth of the tunnel was not accurately known; the depth in relation to the

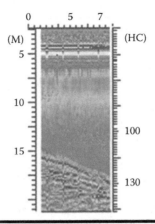

Figure 9.21 Imaging of the subsoil structure regarding special subsoil experiment. (From Blaunstein, N., Ch. Christodoulou, and M. Sergeev, *Introduction to Radio Engineering*, CRC Press, Boca Raton, FL, 2017.)

tunnel entrance was measured with the radar system. Ultimately, it was measured to be close to 16.5 m.

In Figure 9.21, the tomogram obtained after performing the corresponding processing technique, described in reference [4], and received during the experiment, is presented.

Then, estimations of the real depth of the tunnel structure were measured and reconstructed by the corresponding GPR and obtained after the use of the geometrical-optic model described for the multilayer soil structure in reference [4]. The corresponding graphs, depicted in Figure 9.22, show the results after the calculation of the depth to the tunnel, L_1, for 15 different points. Here, L_1 is the depth of the tunnel roof, and the line denoted "L_1 calculated by GPR" is the calculated depth $\underline{L_1}$ using the developed geometrical-optic multilayer subsoil model [4].

In this case, the relative error cannot be calculated, because the accurate depth of the tunnel from the beginning was known and roughly equals ~16.6 m. We can say that average error can be changed over the range 0.2 to 0.5 m.

From the data obtained here, the average permittivity was estimated in reference [4], and the results of these computations are depicted in Figure 9.23, as a function of the distance between the transmitting and the receiving antennas, L_0 (in meters). It is seen that because, during this experiment, the depth of the buried object was measured *a priori*, we observe a narrow range of changes of the permittivity of the tested soil, which is not so far from the real measured parameter equaling $\overline{\varepsilon}$ ~4.2 by use of geological methods.

The accuracy of calculation can be improved with a decrease of distance between the transmitting and the receiving antennas. It should be outlined that other similar experiments giving the accuracy of the tunnel depth detection varied

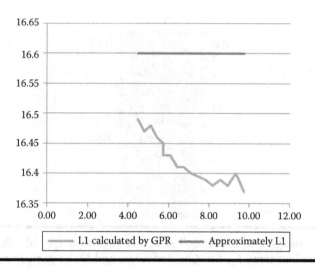

Figure 9.22 The depth L_1 calculated by GPR and L_1 obtained from simulation versus L_0 (in meters). (From Blaunstein, N., Ch. Christodoulou, and M. Sergeev, *Introduction to Radio Engineering*, CRC Press, Boca Raton, FL, 2017.)

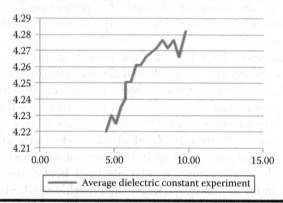

Figure 9.23 The average permittivity $\bar{\varepsilon}_r$ as a function of L_0 (in meters). (From Blaunstein, N., Ch. Christodoulou, and M. Sergeev, *Introduction to Radio Engineering*, CRC Press, Boca Raton, FL, 2017.)

over a range from 0.2 to 0.7 m, and the relative permittivity of the subsoil guiding structure ranged from $\bar{\varepsilon}_r \sim 4.1$ to 4.5.

9.3 Minerals Detection and Identification

Generally speaking, each ground-penetrating or sub-surface radar using different kinds of antennas and operating at any height above ground surface and ground

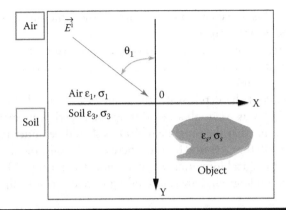

Figure 9.24 Simple sketch of problem where object is buried in subsoil medium with the parameters ε_3 and σ_3 differing from those in the air (ε_1 and σ_1). (From Blaunstein, N., Ch. Christodoulou, and M. Sergeev, *Introduction to Radio Engineering*, CRC Press, Boca Raton, FL, 2017.)

clutter should find the desired buried targets, as any anomalies of background permittivity and permeability of the subsoil or other dielectric or conductive medium. This can obviously be schematically sketched in a simple manner as in Figure 9.24, where, according to reference [4], we reflect a simple geometry of incidence of the GPR's electromagnetic wave of the electrical field amplitude $\mathbf{E}^i$ at the air–soil boundary, without entering into discussions of the multilayer structure of the subsoil medium; this aspect was discussed in reference [4].

The subsoil medium is a frequency dispersive environment [4] and, therefore, is characterized by the complex permittivity $\varepsilon = \varepsilon_0 \varepsilon_r = \varepsilon_0(\varepsilon'_r + j\varepsilon''_r)$ and the complex conductivity $\sigma = \sigma' + j\sigma''$. If so, we can introduce relations between the normalized parameters $\varepsilon_r = \varepsilon'_r + j\varepsilon''_r$ (to that in free space, ε_0), and $\sigma_r = \sigma'_r + j\sigma''_r$ (to $\sigma_0 \sim 1$ S/m related to a weak conducting dry subsoil media):

$$\sigma_r = j\omega\varepsilon_r \qquad (9.5)$$

Moreover, in terms of wave-propagation equations, described in reference [4], we can introduce the parameters ε_r and σ_r, which can be presented in combination with their real and imaginary parts as

$$\sigma_r + j\omega\varepsilon_r = \sigma'_r + \omega\varepsilon''_r + j\omega\left(\varepsilon'_r - \frac{\sigma''_r}{\omega}\right) \qquad (9.6)$$

A set of experimental tests carried out in various sand and ground media by use of a UWB/ESP remote sensing system was described in references [4–6] (see also bibliography therein), where 45 experimental measurements were carried out. During measurements, at each point of observation an electromagnetic pulse signal was radiated to the sub-surface environment, and, at the same time, its response was registered by the radar complex.

During measurements, the experimentally determined relative effective dielectric permittivity of the rocks on the surveyed area of $\varepsilon_r = 10$ was found (corresponding to 1 ns at 0.0475 m).

For calculation of the depth scale, the complex effective dielectric permittivity value $\varepsilon'_{ef} = 11.5$ was used, obtained during the experiment carried out in the test area. This means that the timeline of 400 ns should be understood as a depth interval, equal to 19.0 m. On the basis of these calculations, the maximum depth of the surveyed geological formation was determined as equal to 133 m. Results of one of the tested underground areas (according to references [4–6]) are shown in Figure 9.25.

As an analysis method for the signal spectral characteristics and presenting information, the following evaluation criteria of frequency spectra was chosen. A frequency interval from 70 to 200 MHz in the reflected electromagnetic signals indicated the presence of minerals (poly-metallic elements) in the area sections.

In the frequency interval 70–120 MHz, poly-metallic elements with high values of permittivity, permeability, and conductivity were found, accompanied by a

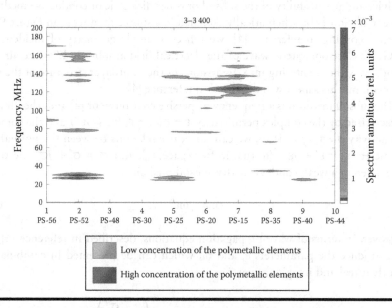

Figure 9.25 Distribution pattern of frequencies with characteristic values of spectral density of signal spectrum and its relationship with the concentration of poly-metallic elements in points along tested area. (Adapted from Blaunstein, N., Ch. Christodoulou, and M. Sergeev, *Introduction to Radio Engineering*, CRC Press, Boca Raton, FL, 2017; Bezrodni', K. P., V. B. Boltintzev, and V. N. Ilyakhin, *Journal of Civilian Construction*, 5, 39–44, 2010; Boltintzev, V. B., V. N. Ilyakhin, and K. P. Bezrodni', *Journal of Radio Electronics*, 39, 2012.)

high density of metal components. In the frequency interval 120–200 MHz, polymetallic elements with low values of permittivity, permeability, and conductivity were found, accompanied by a low density of metal components.

The value of the signal spectral power depends on the concentration of objects buried in the subsoil medium, which can therefore easily be differentiated and assigned with a high or low concentration of poly-metallic elements. In such manner, two types of poly-metallic element concentration were allocated in the test area, with high and low concentration.

Another interesting result of real-time testing of hot slants buried in the subsoil sand media is shown in Figure 9.26, according to references [5, 6]. The soil surface relief is shown from the top to the bottom layer, where sand, sand with stones, sand with clay, and clay layers are shown.

9.4 Geolocation of Special Sub-Surface Man-Made Structures and Objects

In current studies of the formation of radiation wave projections using different sounding techniques, we showed that to solve the inverse problem, namely, the problem of reconstructing the distribution of inhomogeneities, it is necessary, first of all, to localize the radiation in the propagation medium. Only under this condition does reconstruction of the structure of inhomogeneities become achievable, that is, in real time and with acceptable accuracy. Realization of the localization effect can be achieved in two ways that are conceptually equivalent. These are computer-based focusing and physical focusing. A combination of both techniques is also possible. In this context, the requirement arises of developing a working model that could enable radio tomography of different man-made objects and subsoil structures in the medium according to both schemes: the scheme with reflection (location) and the scheme with ray transmission. Of course, in this case, localization of radiation is mandatory. To prove the efficiency of the working model, we have used it in studies on simulated media that are typical for a petroleum reservoir.

A special bipolar generator was used as the transmission unit in the developed model radio tomograph with a pulse shape similar to that of one period of a sine wave with an amplitude of ±8 V and a duration of 100 ps at the 0.1 level of amplitude with a pulse repetition rate of 100 kHz. A stroboscopic oscilloscope was used to record the signals. A double reflector focusing system constructed in the Cassegrain configuration (Figure 9.27) was used to localize the region of interaction of radiation with matter. Such a design delivers lateral localization of radiation in the order of 3–5 cm. The test substance (simulated medium) was placed into a measuring cell with planar boundaries.

Siliceous sand was used as the carrier fraction, and oil from a special oil was used as the filling fraction. A salt solution was prepared from distilled water and

248 ■ *Electromagnetic and Acoustic Wave Tomography*

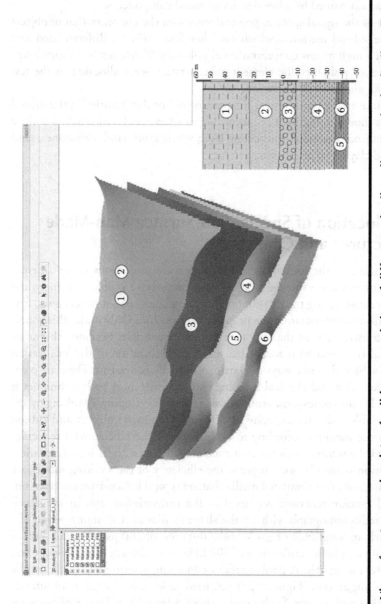

Figure 9.26 View of post-processing data of soil layers consisting of different subsoil media: (1) mixed subsoil media; (2) dry sand; (3) wet sand; (4) sand with stones; (5) sand with clay; (6) clay. (Adapted from Bezrodni', K. P., V. B. Boltintzev, and V. N. Ilyakhin, *Journal of Civilian Construction*, 5, 39–44, 2010; Boltintzev, V. B., V. N. Ilyakhin, and K. P. Bezrodni', *Journal of Radio Electronics*, 39, 2012.)

Sub-Surface Tomography Applications ■ 249

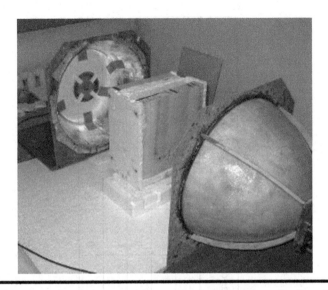

Figure 9.27 UWB radio-tomography setup for sounding simulated media of petroleum reservoir. (From Yakubov, V. P., S. E. Shipilov, D. Ya. Sukhanov, and A. V. Klokov, *Radiowave Tomography: Achievements and Perspectives*, Scientific-Technical Literature (STL) Edition Group, Tomsk, Russia, 2016; Yakubov, V. P., S. E. Shipilov, D. Ya. Sukhanov, and A. V. Klokov, *Wave Tomography*, Scientific Technology Publishing House, Tomsk, Russia, 2017.)

table salt. The percentage amounts of all components were taken according to Table 9.1 [1].

The fundamental problem here is to investigate the peculiarities of UWB pulse propagation in the specific media of a petroleum reservoir. The specific practical purpose is to investigate the propagation of UWB radiation in simulated media of a petroleum reservoir and evaluate the possible use of UWB radiation for well logging and navigation of drilling equipment in horizontal drilling.

Interest in such research has grown more intense all over the world, and in Russia also. It is assumed *a priori* that the use of UWB radiation provides relatively deep penetration into oil-bearing media while maintaining sufficient spatial resolution. This research is based on the results of UWB experiments and should give a direct answer to the question of the realistic use of UWB techniques for radio-wave well logging in the exploitation of hydrocarbon deposits.

According to general ideas about the structure of the media of a petroleum reservoir, it was believed that the oil-bearing stratum lies between the gas cap and the water-bearing stratum, with the clay cap in between. It is supposed that radio-wave sounding should be performed within the oil-bearing stratum, that is, directly from the horizontal water-injection borehole filled with water–clay sludge.

Using a collimated beam helps evaluate the actual penetrating ability of the UWB radiation by comparing the forms of the incident and transmitted pulses

Table 9.1 Medium Structure in a Petroleum Reservoir

No.	Layer	Water (%)	Salinity (g/L)	Clay (%)	Sand (%)	Oil (%)	Gas (Methane) (%)	Depth (m)
1	Water–clay layer (oil well)	88	3	12	0	0	0	0.2
2	Oil-bearing stratum	6	17	0	85	9	0	5
3	Gas cap	4.5	17	0	85	0	10.5	1
4	Clay cap	3	17	97	0	0	0	2
5	Water-bearing stratum	15	17	0	85	0	0	3

and also evaluating the electrical properties of the test media. Two records of such signals are presented in Figure 9.28.

The calculated spectra of the incident $S_x(f)$ and transmitted $S_y(f)$ signals enabled us to find the transfer function of a planar layer $W(f) = S_y(f)/S_x(f)$, and, in the case of a planar layer, to determine the complex permittivity of the filling substance $\varepsilon = n^2$ using the formula [1, 2]:

$$W(f) = \frac{4n}{(n+1)^2 - (n-1)^2 \exp(2iknd)} \exp(iknd) \qquad (9.7)$$

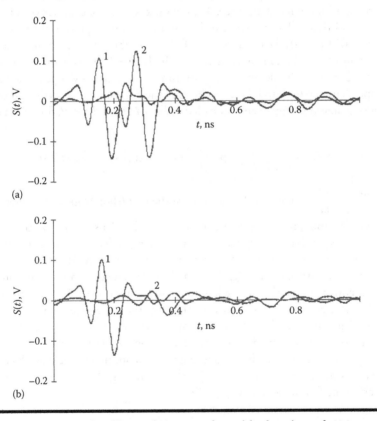

Figure 9.28 Shape of collimated UWB pulse with duration of 100 ps while empty cell (curve 1) and cells (curve 2) filled with (a) sand and with (b) oil-bearing substance were examined in the open space. The cell thickness was 50 mm. (From Yakubov, V. P., S. E. Shipilov, D. Ya. Sukhanov, and A. V. Klokov, *Radiowave Tomography: Achievements and Perspectives*, Scientific-Technical Literature (STL) Edition Group, Tomsk, Russia, 2016; Yakubov, V. P., S. E. Shipilov, D. Ya. Sukhanov, and A. V. Klokov, *Wave Tomography*, Scientific Technology Publishing House, Tomsk, Russia, 2017.)

where d is the layer depth and $k = 2\pi f/c$ is the wave number for free space. This model considers all possible wave re-reflections that arise within the layer when a plane wave penetrates it. The obtained parameters of the main fractions can be used to determine the blend parameters (parameters of mixtures) according to the so-called *refraction model*, according to which,

$$n = \sum_j n_j w_j \qquad (9.8)$$

where w_j is the volume fraction of the jth component with complex index n_j.

Figure 9.29 plots the permittivity values of the main media in a petroleum reservoir as a function of frequency. The obtained spectra of complex blend permittivity (permittivity of mixtures) can be used to estimate pulse transmission and reflection within pure layers and any of their combinations as well. Calculated attenuation of a radiation pulse is plotted in Figure 9.30 as a function of penetration depth in a simple oil-bearing stratum. The signal envelope M (amplitude envelope) is understood to be the modulus of the corresponding analytical signal.

The calculations employed a so-called *balanced optimal signal*, for which [1, 2]

$$S_x(t) = S_0 \left(\frac{t}{T}\right)^p \left\{\exp\left(-\frac{t}{T}\right) - m^{p+1} \exp\left(-m\frac{t}{T}\right)\right\}, \quad t \geq 0 \qquad (9.9)$$

Here, $T_0 = T \frac{p+1}{m-1} \ln(m)$ is the duration of the first (short) lobe of the pulse at the zero level. We took, in our current computations, $T_0 = 1$ ns and $p = m = 6$.

It can be seen that the attenuation per unit length decreases smoothly from 25 dB/m at the surface to a value of 5 dB/m at a depth of 10 m. This is caused by the rapid rate of absorption with depth of the high-frequency components compared with the slower attenuation of the low-frequency components. In the case of radio sounding of a petroleum reservoir directly from the water-injection borehole, the strongest influence on the reflection of waves comes from the boundary of the borehole itself. The weighted differential technique (WDT) for extraction of weak signals from deep layers involves extracting the contribution of the boundary reflection $|Y(t)|$ from the integral signal $|Y(t)|$ and normalizing. That technique reduces to calculating the following quantity [1, 2]:

$$A(t) = \frac{y(t)}{|Y(t)|} \approx \frac{|S(t)|}{|Y(t)|} - 1 \qquad (9.10)$$

This technique makes it possible to clearly delineate the difference between upward and downward sounding (see Figure 9.31), which is especially important for navigating the drilling equipment as well as tomographic reconstruction of the 3-D structure of the media in a petroleum–gas reservoir.

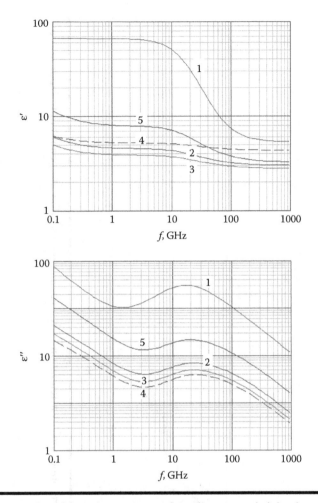

Figure 9.29 Real and imaginary parts of medium permittivity in a petroleum reservoir: curve *1*: water–clay layer (well); curve *2*: oil-bearing stratum; curve *3*: gas cap; curve *4*: clay cap; curve *5*: water-bearing stratum. (From Yakubov, V. P., S. E. Shipilov, D. Ya. Sukhanov, and A. V. Klokov, *Radiowave Tomography: Achievements and Perspectives*, Scientific-Technical Literature (STL) Edition Group, Tomsk, Russia, 2016; Yakubov, V. P., S. E. Shipilov, D. Ya. Sukhanov, and A. V. Klokov, *Wave Tomography*, Scientific Technology Publishing House, Tomsk, Russia, 2017.)

In the final analysis, the obtained result delivers a positive response to the question of the realistic applicability of UWB techniques to radio-wave well logging for exploitation of hydrocarbon fields.

Another set of experiments, described in references [5, 6] (see also bibliography therein), have shown that the use of UWB/ESP sounding for searching

254 ■ *Electromagnetic and Acoustic Wave Tomography*

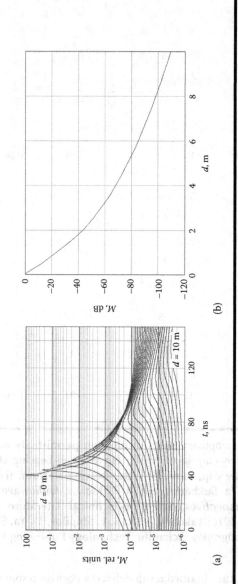

Figure 9.30 (a) Shape of UWB pulse envelope and (b) attenuation of its maximum depending on penetration depth into oil-bearing stratum.

Sub-Surface Tomography Applications ■ 255

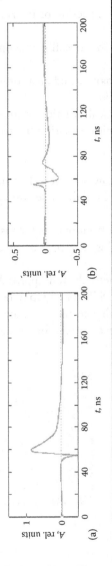

Figure 9.31 Shape of reflected pulse after weighted differential processing for sounding of petroleum reservoir in (a) downward and (b) upward direction. (From Yakubov, V. P., S. E. Shipilov, D. Ya. Sukhanov, and A. V. Klokov, *Radiowave Tomography: Achievements and Perspectives*, Scientific-Technical Literature (STL) Edition Group, Tomsk, Russia, 2016; Yakubov, V. P., S. E. Shipilov, D. Ya. Sukhanov, and A. V. Klokov, *Wave Tomography*, Scientific Technology Publishing House, Tomsk, Russia, 2017.)

underground areas consisting of man-made (i.e., artificial) and other buried objects and structures allows:

- Determination of the width and quality of the construction or man-made structure
- Determination of contours of the structure or freezing subsoil material and its conductivity
- Determination of empty areas in the subsoil media up to 60–70 m (tunnels, bunkers, holes)
- Determination of tectonic disturbances and areas of water inside mountains behind each man-made construction
- Estimation of engineering–geological situation inside mines and tunnels
- Estimation of engineering–geological situation in areas under man-made or natural accidents

As a characteristic example, we show schemes of sounding of underground man-made structures, usually constructed in the collectors and other constructions in the tunnels (see Figure 9.32a and b).

The corresponding recording signal from one such construction (namely, that presented in Figure 9.32a) is shown in Figure 9.33, obtained in the decimeter and meter wave length band.

Finally, we can outline that the methods and approaches considered in this chapter demonstrate the efficiency of UWB tomography techniques for geo-radar research. The suggested approaches enhance the accuracy of current techniques and provide new efficient methods, for example, for sounding of the media in a

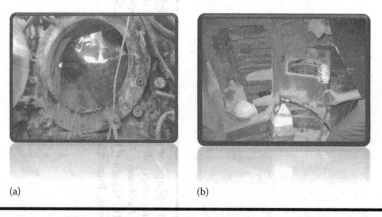

Figure 9.32 Various man-made metallic constructions inside tunnel. (From Bezrodni', K. P., V. B. Boltintzev, and V. N. Ilyakhin, *Journal of Civilian Construction*, 5, 39–44, 2010; Boltintzev, V. B., V. N. Ilyakhin, and K. P. Bezrodni', *Journal of Radio Electronics*, 39, 2012.)

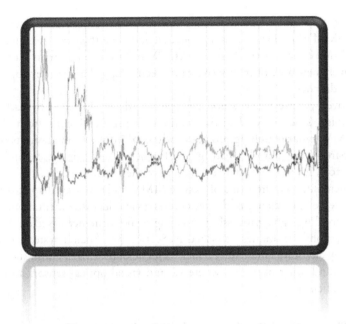

Figure 9.33 Recorded ESP signals of decimeter (dark curve) and meter (light curve) wave length band. (From Bezrodni', K. P., V. B. Boltintzev, and V. N. Ilyakhin, *Journal of Civilian Construction*, 5, 39–44, 2010; Boltintzev, V. B., V. N. Ilyakhin, and K. P. Bezrodni', *Journal of Radio Electronics*, 39, 2012.)

petroleum–gas reservoir and navigating the drilling equipment in horizontal and angular drilling.

Not without practical interest is random and incomplete scanning, as well as scanning techniques over a curved surface. All results are confirmed by the numerical simulation presented in Chapter 3 and partially in this chapter, and by good agreement with experimental data. The results of practical radio sounding of soils at depths down to 6 m in combination with contact measurements of the vertical profile parameters are important [1–6].

9.5 Millimeter-Wave Sub-Surface Tomography Applications

For the last few decades, microwave and millimeter-wave (MMW) radar, both active and passive (the latter is usually called *radiometers* in the literature: see references [7–19]), became important technical facilities in various remote sensing applications of the Earth's environment, such as ground subsoil medium, atmosphere, and ocean. Such kinds of radiometers were developed in the mid-1980s to the early 1990s simultaneously in many countries [18–26]. Recently, several kinds of passive MMW radiometers were operated and were then reported in references [27–31]. MMW radiometry, based on MMW passive radar systems (radiometers), is a relatively new area of microwave radar technology that was developed during the recent decades.

Moreover, passive radar is non-detectable and not degraded by path error or speckle noise. Essentially, detecting various thermic emissions from the target under MMW-radar searching provides additional information about its peculiarities compared with that obtained by low-frequency radar or optical and infrared (IR) detectors, for more accurate estimation of the target response.

The increasing requirements of passive MMW-radar practical applications dictate an increase in efficiency of the system design and its instrumentation, as well as the corresponding signal processing and imaging performance based on the diffraction tomography techniques described in Section 3.4. Such a system should pass through any environmental clutter, such as fog, clouds, drizzle, rain, snow, smoke, vegetation, and other obstacles, where IR and visual optical sensors and cameras become inefficient.

9.5.1 Theoretical Background of Radiometer Operation

The principle of usage of MMW radiometers is based on the fundamentals of Planck's radiation law, according to which a body in thermodynamic equilibrium at a temperature T exceeding the absolute temperature of 0K radiates energy (power) [19–23]:

$$P = k_B B_\omega T \tag{9.11}$$

where k_B is Boltzmann's constant and B_ω is the MMW-radar bandwidth (which can be easily converted in the wave length range). This law strictly applies only to a black body, and is therefore called the *black-body law of radiation*. Usually, in the literature, a black body is defined as an ideal body that absorbs all incident energy without any reflection, and also radiates energy at the same rate as it absorbs energy, thus maintaining thermal equilibrium. The density of the radiated energy per area in square meters and the wave length in microseconds is presented in Figure 9.34, according to Planck's law [19, 31]. It is seen that the maximum radiation of the ideal black body is located at the wave length surrounding $\lambda = 10$ μm. At the same time, at the range between 3 and 8 mm, there is also a possibility of detecting radiated energy from a black body arriving at the radiometric receiver.

A non-ideal black body will partially reflect incident energy, or, conversely, will not radiate the same power as would a black body at the same temperature. Usually, to verify this difference, a new characteristic, called *emissivity*, is introduced:

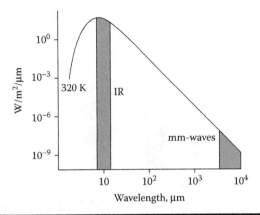

Figure 9.34 Visualization of Planck's law for a black body.

$$e = \frac{P_{nbb}}{P_{bb}} = \frac{P}{k_B B_\omega T} \qquad (9.12)$$

It is seen from Equation 9.12 that emissivity is a measure of the power radiated by a non-ideal black body relative to that radiated by an ideal black body at the same temperature, from which it follows that $0 \leq e \leq 1$, and $e = 1$ for a perfect body, the radiated properties of which are similar to the black body.

We should notice that, for radiometric purposes, a *brightness temperature* T_B is usually introduced to quantify a radiated energy (in the same manner as a noise power N_0 is presented in terms of effective temperature T_{eff} [7]), that is:

$$T_B = e \cdot T \qquad (9.13)$$

Here, T is the physical temperature of the real body. Equation 9.13 shows that for the radiometric receiver, a body is cooler with respect to a real body temperature, since $0 \leq e \leq 1$.

Now, we will consider the process of how the antenna of the passive radiometer differentiates the noise power from different sources located in the surroundings.

We can now formulate the main objective of any MMW radiometer: to infer information about the desired area from the measured T_B and then, via signal processing, to find relations between the brightness temperature and the physical conditions of the tested area (the same radiometric mechanism occurs for each body or scene under consideration). For example, the power reflected from a uniform layer of snow over soil can be treated as plane wave reflection from a multilayer dielectric region, leading to the development of an algorithm that gives the thickness of the snow layer in terms of measured brightness temperature at various frequencies.

260 ■ Electromagnetic and Acoustic Wave Tomography

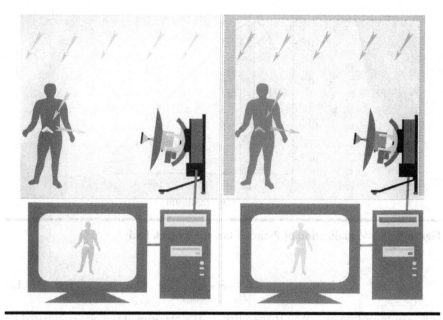

Figure 9.35 Simplified illustration of indoor (left) and outdoor (right) environment influence on contrast between temperatures of air, human body, and concealed foreign object (plastic pistol).

The same situation is observed in indoor/outdoor environments with the contrast between the air temperature, the temperature of the human body, and the temperature of a foreign object concealed under clothes, as illustrated in Figure 9.35.

9.5.2 Typical Application of Millimeter-Wave Radiometers

There are many practical applications of passive MMW radiometric systems. These systems have gained an increasing interest over the past years due to their imaging capability to see through fog, clouds, drizzle, dry snow, smoke, and other obstacles, where IR and optic systems become inefficient. The minimum atmospheric absorption of MMW radiation occurs near 35 GHz and 94–95 GHz. Therefore, these bands are used for MMW imaging. The choice of operating frequency band depends on the particular application. It was shown in references [18–27] that at 35 GHz, a better temperature contrast is obtained of a scene compared with 94–95 GHz, where higher spatial resolution is achieved. The main problem consists of obtaining an image in real time in such a way as with IR thermal imagers. This can be fulfilled by the use of simultaneous reception of radiation from the different parts of a scene. For simultaneous reception, the creation of an array with a large number of receiving channels and a multi-beam quasi-optical antenna are required. The passive MMW imaging systems were developed from a single-channel scanning imager to the fully beam-steering array, which contains more than thousands of receivers [12–21].

In references [27–30], a passive MMW imager, which occupies an intermediate position between the single-channel imager and the fully beam-steering array, was described. Thus, the 3 mm wave passive system was designed especially for demonstration of the possibilities of the single-channel MMW radiometer for remote observation and for experimental (evaluation) purposes. At the same time, the main further advantage to using the scanning imager is the possibility of obtaining a wide field of view [18]. The 32-channel 8 mm wave passive scanning system contains a linear array of radiometric sensors as the quasi-optical multi-beam antenna [18, 28–30].

We present here only one from a number of various applications that has an important practical meaning related to the usage of passive detection of concealed foreign objects under clothes on the human body. Here, designs of a single-channel 3 mm passive imaging system (radiometer) are based on the fact that all bodies radiate, absorb, reflect, and pass electromagnetic energy; that is, a human body reacts as a black body (see Figure 9.34 and discussion in Section 9.5.1). At the same time, if a foreign body is hidden in clothes, the quantity of the energy that is radiated, absorbed, reflected, and passed depends on the body's material. Since, as shown in Figure 9.35, millimeter waves pass well through clothes' material, the human body absorbs them, which is equivalent to its own thermal radiation. Passing now through the foreign body, this radiation partly is reflected and partly absorbed by that body's material.

As a result (if measurements are carried out in indoor conditions), the radiometric sensor will see the contrast between the radiation of a body with temperature around $T_b = 309K$ and the indoor temperature, which is usually around $T_i = 279K$. Thus, passive formation of the images in a 3 mm wave length band represents an effective non-invasive method for the detection and recognition of any foreign object hidden in clothes near the human body. Of course, there will be other problems of environmental influence (clutter), signal processing, and real system practical realization. In references [27–30], a single-channel demonstration radiometric imaging system has been reported with mechanical rotation of the antenna operating at the 90–94 GHz frequency band, based on the heterodyne receiver with fluctuation sensitivity in compensation mode at not less than $0.05K/(Hz)^{1/2}$.

As the result of imaging without the corresponding post-processing, we present in Figure 9.36 one from the numerous images of the foreign objects concealed under clothes near the human body (tests were done at ranges from 5 to 10 m from the radiometer).

As is clearly seen, with direct measurements of the contrast, using such an imaging system, we cannot differentiate and identify the embedded objects. Only using post-processing based on the framework of diffraction tomography [31] (see also Section 3.4) can the metal ring be identified, modeling a concealed mine as an

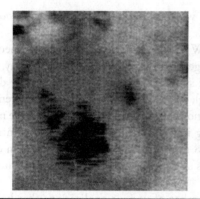

Figure 9.36 Image of foreign objects hidden in clothes before pre-processing by use of 3 mm single-channel demonstration radiometer.

explosive structure, and a plastic pistol, modeling a real weapon hidden in clothes near the human body (see Figure 9.37).

A vivid example of MMW radiometer applications allows us to summarize that the radiometric imaging system can be successfully used in indoor/outdoor applications for imaging and identification of foreign objects at short ranges (from a few meters up to 10–15 m).

Figure 9.37 Imaging after post-processing based on diffraction tomography theoretical framework. (From Blaunstein, N., Recognition of foreign objects hidden in clutter by novel method of diffraction tomography, *Proceedings of IEEE Conference on Radar Applications*, Kiev, Ukraine, 2010.)

References

1. Yakubov, V. P., S. E. Shipilov, D. Ya. Sukhanov, and A. V. Klokov, *Radiowave Tomography: Achievements and Perspectives*, Tomsk, Russia: Scientific-Technical Literature (STL) Edition Group, 2016 (in Russian).
2. Yakubov, V. P., S. E. Shipilov, D. Ya. Sukhanov, and A. V. Klokov, *Wave Tomography*, Tomsk, Russia: Scientific Technology Publishing House, 2017.
3. Vertiy, A. A., S. P. Gavrilov, I. Yoynovskyy, A. Kudelya, V. Stepanuk, and B. Levitas. GPR and microwave tomography imaging of buried objects using the short-term (picosecond) video pulses, *Proceedings of 8th International Conference on Ground Penetrating Radar*, Gold Coast, Australia, 23–26 May 2000, pp. 530–534.
4. Blaunstein, N., Ch. Christodoulou, and M. Sergeev, *Introduction to Radio Engineering*, Boca Raton, FL: CRC Press, Taylor & Francis, 2017.
5. Bezrodni', K. P., V. B. Boltintzev, and V. N. Ilyakhin, Geophysical testing of injection section of the behind-installed space by method of electromagnetic pulse ultra-wide band sounding, *Journal of Civilian Construction*, No. 5, pp. 39–44, 2010 (in Russian).
6. Boltintzev, V. B., V. N. Ilyakhin, and K. P. Bezrodni', Method of electromagnetic ultra-wide band pulse sounding of the surface medium, *Journal of Radio Electronics*, No. 1, 39 pages, 2012, Moscow (in Russian).
7. Johnston, S. I., (ed.), *Millimeter Wave Radar*, Dedham, MA: Artech House, 1980.
8. Button, K. J. and J. C. Wiltse (eds), *Infrared and Millimeter Waves, Volume 4*, New York: Academic Press, 1981.
9. Wiltse, J. C., Millimeter wave trends, *Journal of Millimeter Wave Technology*, vol. 337, pp. 7–9, 1982.
10. Hovanessian, S. A., *Radar System Design and Analysis*, Dedham, MA: Artech House, 1984.
11. Hovanessian, S. A., Detection of non-moving targets by airborne MMW radars, *Microwave Journal*, vol. 29, no. 2, pp. 159–166, 1986.
12. Currie, H. C. (ed.), *Principles and Applications of Millimeter Wave Radars*, Norwood, MA: Artech House, 1988.
13. Hovanessian, S. A., *Introduction to Sensor Systems*, Norwood, MA: Artech House, 1988.
14. Goldsmith, P. F., C. T. Hsieh, G. R. Huguenin, J. Kapitzky, and E. L. Moore, Focal plane imaging systems for millimeter wavelength, *IEEE Transaction on Microwave Theory and Techniques*, vol. 41, no. 10, pp. 1664–1675, 1993.
15. Yujiri, I., H. Agravante, M. Biedenbender, G. S. Dow, M. Flannery, et al., Passive millimeter wave camera, *Proceedings of SPIE*, vol. 3064, pp. 15–22, 1996.
16. Appleby, R., R. N. Anderton, S. Price, N. A. Salmon, G. N. Sinclair, et al., Compact real-time (video rate) passive millimeter-wave imager, *Proceedings of SPIE*, vol. 3703, pp. 13–19, 1998.
17. Yujiri, I., H. Agravante, S. Fornaca, B. Hauss, R. Johnson, et al., Passive millimeter wave video camera, *Proceedings of SPIE*, vol. 3378, pp. 14–19, 1998.
18. Denisov, A. G., Possibilities of designing matrix millimeter imaging systems at the base of superconducting elements, *Proceedings of SPIE*, vol. 3064, pp. 144–147, 1996.
19. Yujiri, I., H. Agravante, S. Fornaca, B. Hauss, R. Johnson, et al., Passive millimeter wave video camera, *Proceedings of SPIE*, vol. 3378, pp. 14–19, 1998.

20. Ferris, D. D. and N. C. Currie, Overview of current technology in MMW radiometric sensors for law enforcement applications, *Proceedings of the SPIE Conference on Passive Millimeter-Wave Imaging Technology*, April, 2000, pp. 61–71.
21. Lang, R. J., L. F. Ward, and J. W. Cunningham, Close range high-resolution W-band radiometric imaging system for security screening applications, *Proceedings of the SPIE Conference on Passive Millimeter-Wave Imaging Technology*, April, 2000, pp. 34–39.
22. Gaiser, P. W., K. M. St. Germain, E. M. Twarog, G. A. Poe, W. Purdy, et al., The WinSAT space borne polarimetric microwave radiometer sensor description and early orbit performance, *IEEE Transactions on Microwave Theory and Technology*, vol. 52, no. 11, pp. 2347–2361, 2004.
23. Suess, H. and M. Soellner, Fully polarimetric measurements of brightness temperature distribution with a quasi-optical radiometer system at 94 GHz, *IEEE Transactions on Geoscience and Remote Sensing*, vol. 43, no. 5, pp. 473–479, 2005.
24. Luthi, T. and C. Matzler, Stereoscopic passive millimeter-wave imaging and ranging, *IEEE Transactions on Microwave Theory and Technology*, vol. 53, no. 8, pp. 2594–2599, 2005.
25. Yujiri, I., M. Shoucri, and P. Moffa, Passive millimeter wave imaging, *IEEE Microwave*, vol. 4, no. 3, pp. 39–50, 2003.
26. Martin, C. A., W. Manning, V. G. Kolinko, and M. Hall, Flight test of a passive millimeter-wave imaging system, *Proceedings of SPIE*, vol. 5789, pp. 24–34, 2005.
27. Gorishnyak, V. P., A. G. Denisov, S. E. Kuzmin, V. N. Radzikhovsky, and B. M. Shevchuk, Radiometer imaging system for the concealed weapon detection, *Proceedings of CriMiCo'2002*. Sevastopol, Ukraine, pp. 187–188, September 9–13, 2002.
28. Radzikhovsky, V. N., V. P. Gorishnyak, S. E. Kuzmin, and B. M. Shevchuk, 16-channels millimeter-waves radiometric imaging system, *MSMW'2001 Symposium Proceedings*. Kharkov, Ukraine, pp. 466–468, June 4–9, 2001.
29. Radzikhovsky, V. N., V. P. Gorishnyak, S. E. Kuzmin, and B. M. Shevchuk, Passive millimeter-wave imaging system, *Proceeding of CriMiCo'2001*. Sevastopol, Ukraine, pp. 263–264, September 10–14, 2001.
30. Gorishnyak, V. P., A. G. Denisov, S. E. Kuzmin, V. N. Radzikhovsky, and B. M. Shevchuk, 8-mm passive imaging system with 32 channels, *Proceeding of EurRAD-2004*, Amsterdam, the Netherlands, October 2–14, 2004, 6 pages.
31. Blaunstein, N., Recognition of foreign objects hidden in clutter by novel method of diffraction tomography, *Proceedings of IEEE Conference on Radar Applications*, Kiev, Ukraine, 2010, 5 pages.

Chapter 10
UWB Tomography of Forested and Rural Environments

Vladimir Yakubov, Sergey Shipilov, and Andrey Klokov

Contents
10.1 Tomography of Forested Areas..265
10.2 Tomography of Wooden Structures and Constructions............................269
10.3 Applications of Focusing Method in Rural Environments 273
References 278

The discussions of this chapter generalize the results of references [1–19].

10.1 Tomography of Forested Areas

During the recent years, interest in boreal forests, which are the lungs of the Earth, accumulating carbon, has risen significantly. Hence, there is an increased interest in the development of methods for remote monitoring of the condition of forest covers. Radio-engineering space-based assets in the microwave range are especially efficient for forest sounding. These are all-weather high-performance systems. For an adequate assessment of forest conditions, an unambiguous correspondence between radio-physical measurements and the results of conventional methods for measuring forest parameters must be established. The potential for accuracy of the

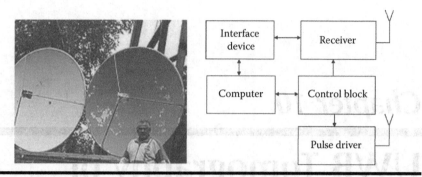

Figure 10.1 External view of antenna system and block diagram of UWB radar for forest tomography. (From Yakubov, V. P. et al., *Radiowave Tomography: Achievements and Perspectives*, Scientific-Technical Literature (STL) Edition Group, Tomsk, Russia, 2016; Yakubov, V. P. et al., *Wave Tomography*, Scientific Technology Publishing House, Tomsk, Russia, 2017.)

method identified by experimental research on an ultra-wide band (UWB) radar functional prototype has become a subject of interest.

The authors carried out tests of the pulsed radar prototype to acquire measurements of the parameters of a normal larch forest in the test polygon of the Forest Institute, Siberian Branch of the Russian Academy of Sciences. The exterior of the antenna system and a block diagram of the UWB radar prototype for forest tomography are presented in Figure 10.1.

For testing of forest areas, a microwave pulser (pulse driver) with a 50 ps edge was developed and tested (see Figure 10.2). The bandwidth of the radiated signal, extending from 500 to 17 GHz, overlapped the frequency band of forest semitransparency. The transverse resolution of the UWB radar is defined by its directivity

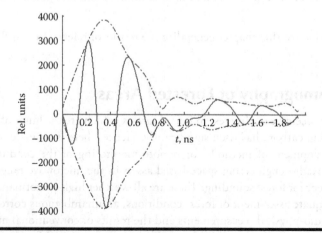

Figure 10.2 Waveform of radiated pulses.

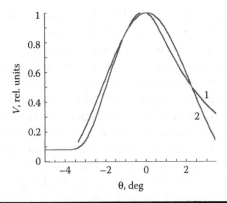

Figure 10.3 Evaluation of UWB radar pattern.

pattern (DP). Results of the actual DP evaluation, obtained from the signal reflected from a corner reflector (CR), are presented in Figure 10.3 (curve *1*).

The amplitude of the analytical signal was used. Curve *2* plots the approximation, which was used in subsequent tomographic processing. The pattern width on the half-power level of the radar was 4°–5°. The snail-type UWB antennas described in Chapter 7 were used as feed horns, as basic elements of the transmitting and receiving antennas to provide the required bandwidth.

In forest tomography, it should be taken into account that the forest is simultaneously the object of sounding and the medium for radiation propagation as well. It can be assumed in the first approximation that wave scattering arises on certain forest inhomogeneities (trunks, branches, leaves, and needles), but direct and reflected radiation penetrates the forest canopy with multiple attenuation. This means that the amplitude of the reflected signals will decay exponentially with distance from the scattering point. A graph of the experimental dependence of the amplitude of the analytical signal, which corresponds to the azimuth-averaged

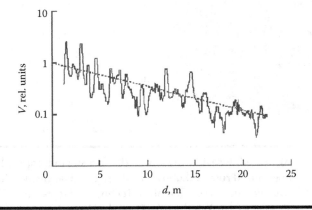

Figure 10.4 Attenuation of UWB radiation.

radar response, is presented in Figure 10.4. The slanting dashed line describes the exponential attenuation averaged over multiple angles.

It follows that the first operation of forest tomogram reconstruction is to align the multi-angle radar data by renormalizing them with respect to their average dependence. The second operation is compression of the radar response over time (distance). This is a quite routine procedure, which is realized using matched filtering. It results in a maximum increase in the signal-to-noise ratio: the contributions of individual inhomogeneities become more localized and the noise is averaged out. The third, and less conventional, operation is deconvolution; that is, the operation of removing image blurring over the azimuth (bearing angle) due to the finiteness of the antenna DP. Here, the approximating antenna DP is used, which was mentioned in the previous section. Deconvolution is performed using Wiener filtering with regularization.

Forest sounding was performed by scanning the azimuth (bearing angle) over the range from −10° to +10° and the elevation angle over the range from 0° to 15°. A view on the tested area is shown in Figure 10.5 (left) and the reconstructed image of the forest tested area is presented in Figure 10.5 (right) as a gray-scale plot. The circles indicate the locations of trees in the landscape plan of the survey area.

A similar radio tomogram for another elevation angle is shown in Figure 10.6.

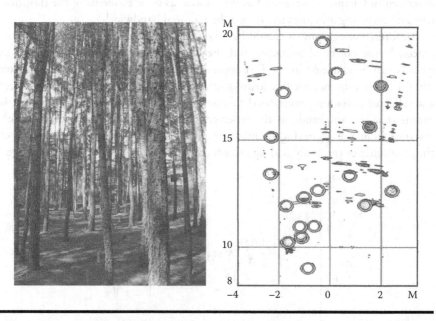

Figure 10.5 (a) Tested forested area and (b) forest landscape plan (circles) with forest tomogram. (From Yakubov, V. P. et al., *Radiowave Tomography: Achievements and Perspectives*, Scientific-Technical Literature (STL) Edition Group, Tomsk, Russia, 2016; Yakubov, V. P. et al., *Wave Tomography*, Scientific Technology Publishing House, Tomsk, Russia, 2017.)

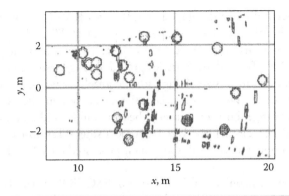

Figure 10.6 Forest landscape plan (circles) and forest tomogram. (Adapted from Yakubov, V. P. et al., *Radiowave Tomography: Achievements and Perspectives*, Scientific-Technical Literature (STL) Edition Group, Tomsk, Russia, 2016; Yakubov, V. P. et al., *Wave Tomography*, Scientific Technology Publishing House, Tomsk, Russia, 2017.)

Figures 10.5 and 10.6 demonstrate good agreement of the obtained results with the landscape plan, even up to the positions of individual trees. Such agreement was achieved in 70% of cases. Besides the marked trees, some additional inhomogeneities are visible in the tomogram, which seem to be large branches of the forest canopy.

A further increase in spatial resolution can be obtained, if we take into account the possibility of transforming from multi-angle measurements of the distribution of the forest radar response to the equivalent spatial frequency spectrum. With the help of the Fourier transform, this spectrum is brought into a one-to-one correspondence with the spatial field distribution in an equivalent transverse aperture. The spatial field distribution reconstructed in this way can then be used for computer-aided focusing of the recorded field at any required distance and angle using the synthetic large aperture technique. The larger the inhomogeneity that is located at the focusing point, the more intense the response will be. By varying the location of the focusing point, it is possible to achieve full scanning of the sounded area so that a tomogram of the inhomogeneous medium (the forest, in our situation) can be obtained. Of course, the proposed procedure works only within the limits of the Fresnel diffraction zone (see discussions presented in Chapter 2).

10.2 Tomography of Wooden Structures and Constructions

This section discusses results of field tests of the nanosecond radar that were obtained in collaboration with personnel of the Department of Physical Problems, Buryat Scientific Center, SB RAS (Ulan-Ude, Russia). The testing was conducted at an arbitrarily chosen site with a CR inserted into the scene as a test object.

The nanosecond radar developed at IHCE SB RAS (Tomsk) radiated short pulses with a duration of 10 ± 2 ns, peak power of 40 W, and a pulse repetition rate of 5 kHz on a carrier frequency of 10 GHz. The width of the pattern of this radar was $\alpha = 2.5°$. These characteristics provided range resolution of objects that were larger than 1.5 m in size. A logarithmic amplifier was used to increase the dynamic range of the radar. The radar orientation system allows scanning over a large angular range, both in azimuth (bearing angle) and elevation angle. The waveform of the received pulse was recorded using a Tektronix digital oscilloscope with a sample rate of 1 GS/s and then processed on a NoteBook computer in MathCad.

A nondescript field on one of the test sites of BSC SB RAS near Istomino village was selected as the test area. This field was situated at a distance of 60 m behind a wooden fence on uneven ground covered with scant vegetation. The only object located on this wasteland was an unfinished wooden house. A trihedral CR with an edge size of 30 cm situated at a height of 2 m was used for the purpose of calibration. During the measurements, one CR was placed inside the house in a window opening, the other one outside the house, on the green. Constant visual contact between the radar and the CR ensured accurate alignment of the optical axis of the CR with the direction to the radar.

Tomographic measurements were preceded by measurements of the attenuation of the response with distance to the CR and measurements of the radar DP. It was confirmed that the amplitude of the radar response A fell off inversely as the square of the distance d^2 up to the target against the background of which a relatively small interference contribution was observed due to the effect of the Earth's surface. It was shown that, although the width of the DP was not large ($\alpha = 2.5°$), the contribution of the side lobes was significant due to the extended dynamic range of the radar, even when the radar antenna axis was raised above the horizon.

Tomographic measurements of the test site involved recording the radar response as a function of the azimuth angle α with a step of 1° within a range of ±13° relative to some average direction. Per its elevation angle, the antenna axis of the radar was oriented precisely toward the CR. Each recorded response matched the position of a corresponding reflection point in the Cartesian system:

$$x = d \cos\alpha, \ y = d \sin\alpha \qquad (10.1)$$

where:
$d \ = \ c\tau/2$
$\tau \quad$ is the corresponding signal time delay

The radar was taken to be located at the origin of the coordinate system. The resulting two-dimensional radar image $A(x,y)$ for transformation to the tomographic image S was renormalized for geometrical attenuation with range:

$$S(x,y) = d^2 \cdot A(x,y) \qquad (10.2)$$

That procedure had already been used previously for forest tomography (Section 10.1). A tomographic image, reconstructed in that way, is presented in Figure 10.7 in gray-scale, using the Contour Plot option in MathCad. The actual positions of the objects are marked off here with dashed lines. Locations of two CRs (objects 2) are shown as triangles. The entire image is bounded by the scanning area in bearing angle and range.

Let us discuss the obtained result. In Figure 10.7, object *1*, corresponding to the wooden fence, has a straight-lined shape kinked at the point ($x = 0$, $y = 57$ m), as can be easily identified in the image. The position of the back post, clearly rising up above the fence, is marked by a circle. Two CRs (objects *2*) are imaged quite distinctly and are localized both in range d and in azimuth α. The front wall (object *3*) of the unfinished house, which faces the radar, is visible in the image. Object *4* corresponds to a reflection from some hill with quasi-sinusoidal channels that are parallel to the fence on the test site. This hill was covered by the radar side lobes. Objects *5* and *6* are manifestations of the effect of multipath propagation of radio waves. This is proven by the fact that they are located at exactly the same azimuth as object *4* and are nearly equidistant; consequently, they are the result of double-hop and triple-hop reflections of signal *4* from fence *1*.

Object *7* is of the greatest interest. First, there were no visible objects at this point at the test site. Second, the observed distance between the radar and object *7* is close to the distance to the adjacent CR. This suggests that the phantom object *7* belongs to the actual CR *2*. The low dependence of the radar response of the CR on its orientation within the range $\pm 45°$ (Figure 10.8) is key to understanding these phenomena. The situation arises in this regard when the signal transmitted by the radar to point *7* is scattered over the edge of fence *1* and propagates to CR *2*, which reflects the signal directly back (Figure 10.9a).

The scattering diagram of the CR is filled out in gray in Figure 10.9a. Next, the signal returns to the radar from the initial direction. This results in the phantom object *7* (shown in Figure 10.7), which is represented as object *2'* in Figure 10.9a.

One possible application of the effect discovered connected with the scattering characteristics of the CR is of interest. The idea here is the possible application of passive illumination for low-reflective target detection, for example, stealth aircraft (Figure 10.9b). The idea is to use the spatially distributed field from the CRs *2*, which is set up on the approach to a radar-protected zone. In this case, the absence of direct reflections from the radio-location (radar) target *1* can be compensated by secondary signals of the CR, which makes it possible, in fact, to realize bi-static radio-location using a single-position radar without large losses. The phantom objects *2'* will be observed to lie along the direction to the actual low-reflective target. The chain of phantoms *2'* allows one to detect the actual target and determine the direction to it.

272 ■ *Electromagnetic and Acoustic Wave Tomography*

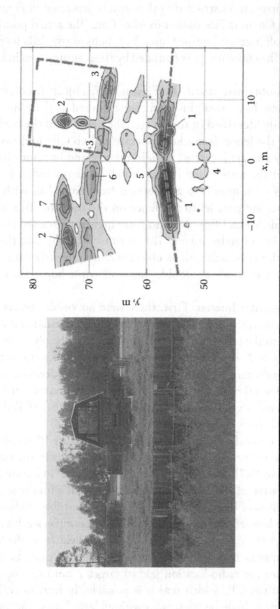

Figure 10.7 (a) Photograph and (b) radio-tomographic image of objects in the test area. *1:* fence; *2:* CRs; *3:* wooden house. (From Yakubov, V. P. et al., *Radiowave Tomography: Achievements and Perspectives*, Scientific-Technical Literature (STL) Edition Group, Tomsk, Russia, 2016; Yakubov, V. P. et al., *Wave Tomography*, Scientific Technology Publishing House, Tomsk, Russia, 2017.)

UWB Tomography of Forested and Rural Environments • 273

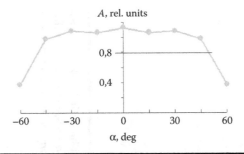

Figure 10.8 Dependence of the reflectivity of the CR on its azimuthal orientation.

The results of the flying target recognition presented here confirms the potential of using the nanosecond radar considered to sound the environment and verify the aerospace sounding data.

10.3 Applications of Focusing Method in Rural Environments

A further increase of spatial resolution can be achieved by the possibility of transforming from multi-angle measurements of the distribution of a forest radar response to the equivalent spatial frequency spectrum that can be taken into account by the use of the refocusing technique described in Section 3.3. We will not repeat all the formulas and corresponding illustrations mentioned there; we only will point out the technical aspects of this approach and discuss the theoretical framework obtained via the acquired experimental data.

Thus, with the help of the Fourier transform, the spatial frequency spectrum can be presented as being in one-to-one correspondence with the spatial field distribution in some equivalent transverse aperture, as shown in Figure 10.10. After the spatial field distribution is reconstructed, the synthetic aperture technique can be used and the recorded field can be focused at any predefined distance and angle within the Fresnel diffraction zone, where possible.

If there is an actual inhomogeneity at the focusing point (as presented by Figure 3.8, similarly to Figure 10.10, see Chapter 3), the intensity of the response will be directly proportional to the degree of non-uniformity. If there is no inhomogeneity, that is, scatterer, at the focusing point, the response will be close to zero. By varying the position of the focusing point, it is possible to scan the entire sounded area so that the tomogram of the inhomogeneous medium (the forest, in this situation) can be obtained. The procedure proposed here is computer-based, and is equivalent to the synthetic large aperture technique described briefly in Chapter 3 (see also references [1, 2] and bibliography there). The transverse spatial resolution of inhomogeneities can be markedly improved. In the first approximation,

274 ■ *Electromagnetic and Acoustic Wave Tomography*

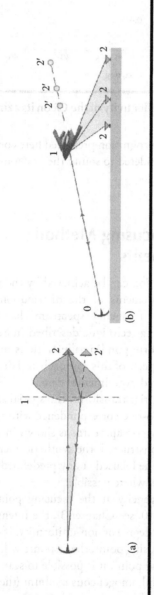

Figure 10.9 Generation of false images (phantoms) of CRs: (a) on the test site (top view), (b) in quasi-bi-static location of low-reflective targets (side view). (From Yakubov, V. P. et al., *Radiowave Tomography: Achievements and Perspectives*, Scientific-Technical Literature (STL) Edition Group, Tomsk, Russia, 2016; Yakubov, V. P. et al., *Wave Tomography*, Scientific Technology Publishing House, Tomsk, Russia, 2017.)

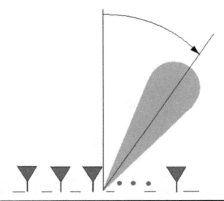

Figure 10.10 Directivity pattern and equivalent aperture of antenna.

the resolution in the focusing area achieved by using a synthetic aperture radar system is estimated as

$$\Delta x = 2\lambda z/2D \qquad (10.3)$$

where:
- λ is the wave length
- z is the distance from the aperture to the focusing point
- $2D$ is the size of the synthetic aperture

This estimate is the distance between the first zeros.

It is obvious that the proposed procedure works only within the Fresnel diffraction zone (the near field zone), which is defined by the condition

$$z \leq z_D \qquad (10.4)$$

where:
- $z_D = D^2/\lambda$ is the diffraction length of the light beam for a wave beam with radius D (the aperture radius in our situation)
- λ is the wavelength of the radiation [8]

In other words, z_D is the distance within which the phase difference from the center to the edge of the aperture is significant enough to be accurately measured.

Generally speaking, this zone is called the Fresnel diffraction zone (see Chapter 2). For example, for the wave length $\lambda = 0.3$ m and the aperture width $D = 3$ m, the diffraction length z_D is estimated as 30 m. For $\lambda = 0.1$ m, it increases to $z_D = 90$ m. For a frequency equal to 7.1 GHz ($\lambda = 0.04$ m) and an aperture width $D = 1.5$ m, the diffraction length $z_D = 53$ m. This latter distance covers the entire extent of exponential attenuation of the field in forests as measured in units of the attenuation length.

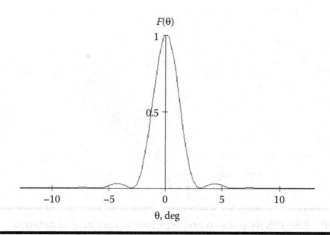

Figure 10.11 Directivity pattern.

A scanning system having a DP with a width of 3° in the azimuthal plane was numerically simulated to validate the efficiency of the proposed approach (Figure 10.11). Such a DP is realized for the aperture $2D = 6$ m and the wave length $\lambda = 0.5$ m. Eight points inside the aperture were used in the calculation, which implies a discretization step of 7.5 cm (0.15λ).

Angular scanning was performed in the range (−90°, +90°) under the assumption that one CR was placed at the point x = 5 m, y = 6 m. The angular spectrum so obtained was recalculated to the spatial frequency spectrum and then to the equivalent field distribution over the aperture. The focusing technique described was applied to the obtained equivalent field distribution. The focusing of the field in the vicinity of the focusing point achieved in this way is displayed in gray-scale in Figure 10.12. The boundary of the DP of the antenna for sounding from the point $x = 0$, $y = 0$ is indicated by dashed lines.

It can be seen that the best focusing of the field (indicated by the arrows) is realized at the position of the CR. The size of the focused field is estimated as 0.5 m, while the energy spread width calculated from the DP at this distance is estimated as 3.7 m. Thus, the resolution has been improved by roughly sevenfold thanks to the focusing effect. Note that the distance between the first zeros in this region is somewhat greater, by about 1 m; however, it is preferable to use the value of the pattern width as the estimate.

Using pulsed radiation and matched filtering, it is possible to obtain the 2-D tomogram, that is, the spatial distribution of inhomogeneities in the sounding plane. For this purpose, it is sufficient to focus at all frequencies of the pulsed signal spectrum:

$$S(\mathbf{r},\tau) = \int S_0(f) E(\mathbf{r},f) \exp\{i2\pi f \tau\} df \qquad (10.5)$$

UWB Tomography of Forested and Rural Environments ■ 277

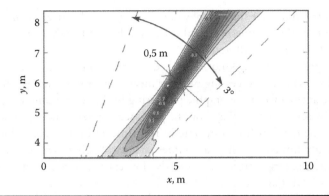

Figure 10.12 Field amplitude distribution around focusing point. (Adapted from Yakubov, V. P. et al., *Radiowave Tomography: Achievements and Perspectives*, Scientific-Technical Literature (STL) Edition Group, Tomsk, Russia, 2016; Yakubov, V. P. et al., *Wave Tomography*, Scientific Technology Publishing House, Tomsk, Russia, 2017.)

where:
$S_0(f)$ is the spectrum of the sounding pulse
$\tau = 2r/c$

In this way, additional scanning over the range will be achieved and the overall spatial resolution of inhomogeneities will be improved. Figure 10.13 displays an example of reconstruction of a tomogram of three scatterers using the procedure described for a pulse with a duration of 100 ps. The spatial distribution of the function $|S(\mathbf{r}, \tau = 2r/c)|$ is the tomogram.

Thus, employing the equivalence between angular radar scanning and measurements of the spatial frequency spectrum of the scattered field makes it possible to realize controlled spatiotemporal focusing using the synthetic large aperture

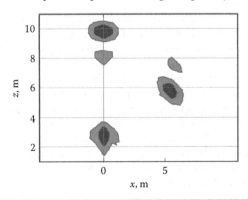

Figure 10.13 Tomogram of three inhomogeneities using pulsed radiation.

technique for radio-wave tomography of a distribution of scatterers and enables an improvement of the transverse spatial resolution by not less than sevenfold within the limits of the Fresnel diffraction zone.

The proposed method is not described in the literature, as it is a new technique for tomographic processing of radar measurements. Its implementation will increase the accuracy of tomographic processing in the realization of angular measurements by several times. Based on the proposed approach, new mobile scanning systems can be developed that operate in real time and under conditions of strict size and weight limitations, such as in forested areas. Such systems can also be used for monitoring of large traffic flows and also for security purposes in airports, subway stations, and stadiums.

Pulsed radio-wave scanning can be used to realize tomography of the natural environment, for example, forests or various structures. Radio-wave tomography enables the reconstruction of forest landscapes.

Tomographic processing of radar data is based on the focusing methods. These techniques are used to increase the transverse resolution. A shorter wavelength and/or a wider bandwidth must be used to increase the longitudinal resolution. The proposed refocusing method works only within the limits of the Fresnel diffraction zone and provides a sevenfold improvement of the transverse resolution.

References

1. Yakubov, V. P., S. E. Shipilov, D. Ya. Sukhanov, and A. V. Klokov, *Radiowave Tomography: Achievements and Perspectives*, Tomsk, Russia: Scientific-Technical Literature (STL) Edition Group, 2016 (in Russian).
2. Yakubov, V. P., S. E. Shipilov, D. Ya. Sukhanov, and A. V. Klokov, *Wave Tomography*, Tomsk, Russia: Scientific Technology Publishing House, 2017.
3. Daniels, D. J., *Surface-Penetrating Radar*, London: The Institute of Electrical Engineering, 1996.
4. Klokov, A., V. Yakubov, and M. Sato, Ultra wideband radiolocation of a forest, *Proceedings of the 2010 IEICE General Conference*, Sendai, Japan, The Institute of Electronics, Information and Communication Engineers, March 16–19, 2010, p. 298.
5. Mironov, V. L., V. P. Yakubov, E. D. Telpukhovsky, and S. N. Novik, Height dependence of electromagnetic field inside the forest canopy at meter and decimeter wavelengths, *Proceedings of International Geoscience and Remote Sensing Symposium (IGARSS)*, Denver, CO, vol. VI, 2006, pp. 2962–2964.
6. Mironov, V. L., S. N. Novik, A. V. Klokov, et al., Space – time and frequency – polarization variations in the electromagnetic wave interacting with the forest curtains, *Proceedings of International Geoscience and Remote Sensing Symposium (IGARSS)*, Barcelona, Spain, 2007, p. 200.

7. Mironov, V. L., V. P. Yakubov, E. D. Telpukhovsky, et al., Spectral study of microwave attenuation in a larch forest stand for oblique wave incidence, *Proceedings of International Geoscience and Remote Sensing Symposium (IGARSS'05)*, Seoul, Korea, 2005, pp. 3204–3207.
8. Telpukhovskiy, E. D., V. P. Yakubov, V. L. Mironov, et al., Wideband radar phenomenology of forest stands, *Proceedings of IEEE International Geoscience and Remote Sensing Symposium (IGARSS)*, Toulouse, France, July 21-25, 2003, pp. 4265–4267.
9. Yakubov, V. P., E. D. Telpukhovskiy, V. L. Mironov, et al., Measured spectrum and polarization of wideband radar signal from forest stand, *Proceedings of IEEE International Geoscience and Remote Sensing Symposium (IGARSS)*, Anchorage, AK, 20–24 September, 2004, pp. 1235–1238.
10. Yakubov, V. P., E. D. Telpukhovskiy, K. Sarabandi, et al., Attenuation and depolarization data measured for scattered field inside larch canopy, *Proceedings of IEEE International Geoscience and Remote Sensing Symposium (IGARSS)*, Toulouse, France, July 21-25, 2003, pp. 4195–4197.
11. Wood, R. W., *Physical Optics*, Charleston, SC: Na bu Press, 2010.
12. Mironov, V. L., V. P. Yakubov, E. D. Telpukhovskiy, et al., Spectral study of microwave attenuation in a larch forest stand for oblique wave incidence, *Proceeding of International Geoscience and Remote Sensing Symposium (IGARSS)*, vol. 5, 2005, Art. No. 1526522, pp. 3204–3207.
13. Yakubov, V. P., E. D. Telpukhovskiy, G. M. Tsepelev, et al., Measured spectrum and polarization of wideband radar signal from forest stand, *Proceedings of International Geoscience and Remote Sensing Symposium (IGARSS)*, vol. 5, 2004, pp. 3471–3473.
14. Magazinnikova, A. L. and V. P. Yakubov, Dual mechanism of radio wave propagation in forests, *Radiotekhnika i Elektronika*, vol. 44, no. 1, 1999, pp. 5–9 (in Russian).
15. Telpukhovskiy, E. D., V. P. Yakubov, V. L. Mironov, et al., Wideband radar phenomenology of forest stands, *Proceeding of International Geoscience and Remote Sensing Symposium (IGARSS)*, vol. 7, 2003, pp. 4265–4267.
16. Magazinnikova, A. L. and V. P. Yakubov, Attenuation of coherent radiation in forest regions, *Microwave and Optical Technology Letters*, vol. 19, no. 2, 1998, pp. 164–168.
17. Magazinnikova, A. L. and V. P. Yakubov, Dual mechanism of radio wave propagation in forests, *Bandaoti Guangdian/Semiconductor Optoelectronics*, vol. 19, no. 6, 1998, pp. 5–10.
18. Yakubov, V. P., E. D. Telpukhovskii, and G. M. Tsepelev, Pulse sensing of a forest canopy, *Russian Physics Journal*, vol. 46, no. 8, 2003, pp. 784–790.
19. Mironov, V. L., E. D. Telpukhovsky, V. P. Yakubov, et al., Space-temporal and frequency-polarization variations in the electromagnetic wave interacting with the forest canopy, *Proceeding of International Geoscience and Remote Sensing Symposium (IGARSS)*, 2007, Art. No. 4423370, pp. 2574–2576.

Chapter 11

Detection of Live People in Clutter Conditions

Nathan Blaunstein, Felix Yanovsky,
Vladimir Yakubov, and Sergey Shipilov

Contents

11.1 Finding Living People Buried in Sub-Soil Media 281
11.2 Detection and Identification of Living People behind Walls 287
References 289

11.1 Finding Living People Buried in Sub-Soil Media

As was mentioned in Chapters 2–5, one of the important tasks of sub-surface, ground-penetrating, or geo-radar systems operation is to detect living people buried under ruins in clutter conditions after disasters observed after seismic events, catastrophes, and explosions (see Figure 11.1a), detection and localization of living people under snow layers (see Figure 11.1b), and localization of living people in smoky conditions, on moving underground stairways (see Figure 11.1c), and so on [1–10].

Thus, a prime problem of dealing with the consequences of earthquakes is the exact allocation of available rescue resources. First of all, efforts of rescue teams should be applied to buildings, under which living people could be situated. The main aim of this chapter is the presentation of a new approach of technical means dedicated to the detection of living people under ruins, in rubble, and behind walls [1–10].

The currently available sub-surface video pulse radar and active radiometric systems are microwave devices used to search for people. The first type of device gives images of cavities and the other may fix the mobility of persons. As a rule, this

282 ■ *Electromagnetic and Acoustic Wave Tomography*

Figure 11.1 Different possibilities of successful usage of UWB, CW, and ESP radar systems.

Table 11.1 Estimated Parameters of ESP/UWB Radar

Frequency range	100	1000 MHz
Pulse duration	1	10 ns
Maximum depth of detection (thickness of wall)	—	up to 5 m
Average radiated power	—	<250 mW

information is not enough to search for living people, but may be, for example, for immovable or unconscious persons. In this case, the main physical signs of life are breathing and a heartbeat. These features are exactly those signs of the live people that can be much more easily identified by microwave band waves.

References [7–10] suggested a system in its final incarnation in the form of an extremely-short-pulse (ESP) ultra-wide band (UWB) radar, which can operate in the ultra-wide band of frequencies in the optimum chosen frequency range. Its technical data are presented in Table 11.1.

The realization of measurement on a particular set of frequencies has to minimize the fading effect inherent in single-frequency systems. The operating principle of the system to search for people under rubble or walls is shown in Figure 11.2.

The main aim of this radar is to search for a reflected signal from objects in a situation where the recorded reflected signal is from a human body trapped under ruins or behind walls. Signals from a human body arriving to the receiver, reflected after the corresponding pre-processing via a wireless channel, are transmitted by wireless channel on a laptop for visualization of the reaction of the human body in the space–time–frequency domain. The physical model of the future industrial prototype used in the concrete testing experiment is shown in Figure 11.3.

This physical model of the real prototype of UWB radar was used in measurements carried out behind a wall of 70 cm thickness (see Figure 11.4a). The movements of the person on the opposite side of the wall (Figure 11.4b) and the signal replicating these movements in the frequency domain is clearly seen in Figure 11.4c.

The main signs of human life are breathing and heartbeat. Both of these processes mainly cause the phase and amplitude modulation of the reflected signal. The proposed UWB radar has to be able to detect exactly this modulation in the frequency

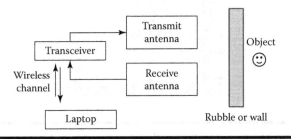

Figure 11.2 Block scheme of the UWB radar to detect living people.

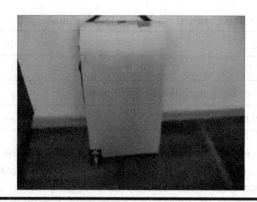

Figure 11.3 View while testing "Demo" prototype.

range 0.1–10 Hz and suppress all other spurious signals (Figure 11.5). A similar experiment was carried out with a stationary, breathing person (Figure 11.5a) and the corresponding post-processing data is clearly seen in Figure 11.5b.

The comparison of the obtained experimental data shows the same amplitude and phase random modulation, but a faster rate (see Figure 11.5b) was observed by searching for the person's motion, as shown in Figure 11.4c.

This task can be fulfilled using certain technical means and specific software. A synthesized microwave generator with a low level of phase noise, a highly sensitive

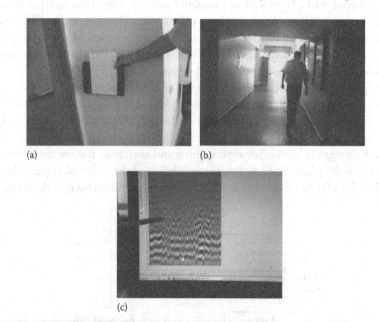

Figure 11.4 (a) Thickness of the wall of 70 cm, (b) real-time experiment, and (c) pre-processing measured data.

Detection of Live People in Clutter Conditions ■ 285

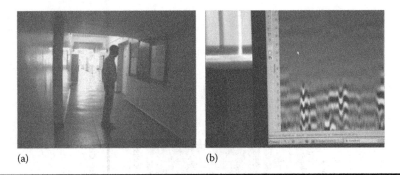

Figure 11.5 (a) Real-time experiment and (b) pre-processing measured data.

quadrature receiver, and a pair of narrow-beam antennas with a low level of interconnection and side lobes can be referred to as the technical basement of the ESP/UWB radar system. But, this is not enough, because the main responsibility for correct signal recognition lies with the software package, which realizes the methods of signal processing and recovery discussed in Chapter 2.

Another methodology, based on the ESP/UWB radar technology and described in [7, 8], was proposed for:

- Finding people in conditions of poor visibility or buried in clutter conditions
- Estimation of breathing frequency for non-moving and static people buried in clutter conditions

The operating principle of a ESP/UWB radar system to search for people under rubble or ruins is as follows. The transmitter sends an ESP (e.g., UWB) signal into the rubble or ruins. The receiver takes scattering signals from different objects within the rubble or ruins, and the processing system then compares the signals over the observation time. The criterion is the following: if the shape of the signals changes, this means that there was movement or breathing, depending on the rate of fading of the recording signals.

ESP/UWB radar has a bi-static antenna system. The transmitter generates a narrow voltage step with a duration of several nanoseconds and an amplitude (in volts) of about 300 V. Under the action of such an excited waveform, the transmitting antenna radiates a sounding waveform, which looks like one period of a sine wave. The radar, with the help of the receiving antenna, receives the scattered signals in a given time window (i.e., from the given distance range). An important feature is that such a system does not have any blind zone. In the proposed system, the feature by which to detect people is their motion. Any person's motion or wiggling can be detected, including breathing. Special software allows comparison of several time realizations of received signals. Breathing can be detected even when a living object is sleeping.

An example of the use of the ESP/UWB radar to detect the movements and breathing of people behind a rock is shown in Figure 11.6a, and its

286 ■ *Electromagnetic and Acoustic Wave Tomography*

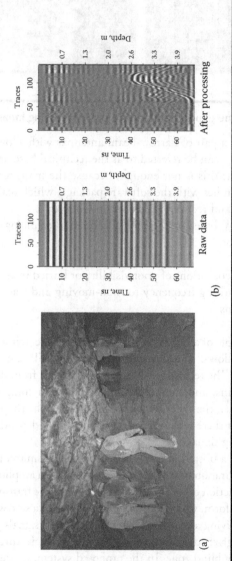

Figure 11.6 (a) Positions of moving and stationary people under the rock; (b) recording signal in the cases of raw data and after pre-processing.

pre-processing 2-D time-depth distribution of the signal amplitude (voltage) is shown in Figure 11.6b.

As is clearly seen from Figure 11.6b, where the 2-D time-depth signal distribution along the radar trace above the rock (in meters) is presented, in a time range of 45 to 60 ns (which corresponds to a depth of 2.6–3.9 m), the random quasi-sinusoidal deviations of the recording pulses reflect the existence of living people due to either their motion or their breathing. The results of such a campaign of testing showed the following [7–10]:

- Possibility of detecting moving objects behind thick wall
- Range resolution achieved using UWB signal up to 0.3 m
- Relatively big depth of penetration through thick walls provided by low frequency
- Maximum depth of detection (thickness of a wall) up to 5 m

The high sensitivity of motion detection was provided by special signal pre-processing software.

11.2 Detection and Identification of Living People behind Walls

In references [1–6], an ESP/UWB radar system was proposed for searching for living people both under rubble and behind walls. The possibilities of such an ESP/UWB radar system were the same as manifested in [7, 8], that is: (1) 2-D visualization of hidden moving objects behind constructions and walls; (2) finding of living people in conditions of poor visibility or buried in clutter conditions; and (3) estimation of breathing frequency for non-moving and static people buried in clutter conditions. The prototype of such a radar system is shown in Figure 11.7 (according to [1, 2]).

Figure 11.7 UWB radar system with control and processing block (middle).

288 ■ *Electromagnetic and Acoustic Wave Tomography*

Figure 11.8 Plan of experiment (left); radar operation (bottom right) and tomogram on the screen after post-processing (top right).

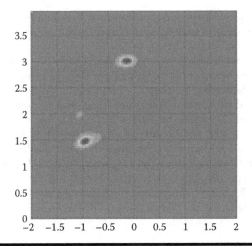

Figure 11.9 Radio tomogram of two stationary people breathing, seen from the left-hand side of Figure 11.8.

The principle of operation is shown in Figure 11.8.

Figure 11.9 presents the tomogram of definition of non-moving people behind walls, which, according to the left-hand side of Figure 11.8, are two people speaking.

In this chapter, we briefly presented only a few among numerous experimental results manifested in [1–10] where the possibility of using various kinds of subsurface radar was vividly proved, including ESP/UWB radar to search for, detect, and identify living people in various clutter conditions: in rocks, rubble, ruins, or behind walls, recognition of which can be achieved by the novel signal processing technologies and methods described in Chapters 2–4.

References

1. Yakubov, V. P., S. E. Shipilov, D. Ya. Sukhanov, and A. V. Klokov, *Radiowave Tomography: Achievements and Perspectives*, Tomsk, Russia: Scientific-Technical Literature (STL) Edition Group, 2016 (in Russian).
2. Yakubov, V. P., S. E. Shipilov, D. Ya. Sukhanov, and A. V. Klokov, *Wave Tomography*, Tomsk, Russia: Scientific Technology Publishing House, 2017.
3. Yakubov, V. P., D. Ya. Sukhanov, and A. V. Klokov, Radiotomography from ultrawideband monostatic measurements on an uneven surface, *Russian Physics Journal*, vol. 56, no. 9, 2014, pp. 1053–1061.
4. Yakubov, V. P., A. S. Omar, D. Ya. Sukhanov, et al., New fast SAR method for 3-D subsurface radiotomography, *Proceedings of 10th International Conference on Ground Penetrating Radar*, 21–24 June, 2004, Delft, the Netherlands, pp. 103–106.
5. Yakubov V. P. and D. Ya. Sukhanov, Solution of a subsurface radio-imaging inverse problem in the approximation of a strongly refractive medium, *Radiophysics and Quantum Electronics*, vol. 50, no. 4, 2007, pp. 299–307.

6. Klokov A. V., A. S. Zapasnoy, A. S. Mironchev, V. P. Yakubov, and S. S. Shipilova, A comprehensive study of underground animals habitat, *Journal of Physics: Conference Series*, vol. 671, no. 1, 2016, Art. No. 012028.
7. Yanovsky F. J., Ivashchuk V. E., and Prokhorenko V. P., Through-wall surveillance technologies, *Proceeding of Ultrawideband and Ultrashort Impulse Signals (UWBUSIS -2012)*, 17–21 September, Sevastopol, Ukraine, 2012, pp. 30–33.
8. Yanovsky, F., M. Bereznitsky, G. Lihtenshtein, and N. Blaunsteinm, Preliminary measurements of alive people behind the wall, *DPES, Report #33/2012* (Israel).
9. Ivashchuk, V. E., V. P. Prokhorenko, A. A. Pitertsev, F. J. Yanovsky, Evaluation of combined ground penetrating and through-the-wall surveillance UWB technology, *Proceedings of European Microwave Conference*, NCC Nürnberg, Germany, 6–11 October, 2013, pp. 384–387.
10. Ivashchuk, V. E., V. P. Prokhorenko, A. A. Pitertsev, and F. J. Yanovsky, Through-the-wall moving target surveillance using GPR, *Proceedings of EuMW-2013*, EuRad, 2013.

NON-CONTACTING ACOUSTIC AND COMBINED RADIO-ACOUSTIC TOMOGRAPHY

IV

Chapter 12

Applications of Radio-Acoustic Tomography

Vladimir Yakubov, Sergey Shipilov,
Dmitry Sukhanov, and Andrey Klokov

Contents

12.1 Experimental Results 294
12.2 Complexing of Radio and Ultrasound Tomography Techniques 299
12.3 Visualization of Small Defects Hidden Inside Metallic
    Constructions Using Acoustic Tomography 303
References 309

Electromagnetic radiation interacts with electro-physical inhomogeneities in the propagation medium. In contrast to electromagnetic radiation, acoustic radiation interacts mainly with density contrasts in the sounding medium [1–4]. In this context, using ultrasound in the tomography of inhomogeneous media provides additional opportunities, in particular for detecting the type of material that hidden objects are made of. Integration of radio and acoustic sounding provides opportunities, for example, for the detection of explosives.

A search on remote ultrasonic sounding techniques reveals that most industrial methods are based on contact measurements. The principal reason for this is that ultrasound attenuation in air is quite high. Ultrasound is used in parking sensors and for control of industrial robots. The ultrasound system (echolocation or biosonar) used by bats is perhaps the only ultrasonic ranging system actually effective in air (see Figure 12.1).

294 ■ Electromagnetic and Acoustic Wave Tomography

Figure 12.1 Bats are effective ultrasonic stations in air.

Some insects also use ultrasound, but rather for active jamming of bats' sounding. There is no doubt of the efficiency of ultrasound for sounding in liquids and through immersion liquids. This applies in the fields of industrial metalworking, medicine, and submarine echolocation. In nature, this also pertains to dolphins and fish. However, ultrasound can be efficiently used for tomography in air in safety systems and in non-destructive testing. If the object is radiopaque, radiation can hardly penetrate it, and only acoustic radiation will allow the recovery of its internal structure.

The overall purposes of the present chapter are: the development of physico-mathematical models of a system for reconstruction of images of inhomogeneous media based on tomographic processing of multi-angle measurements of scattered acoustic radiation; the development of the key elements of a simulation system and the development of experimental measuring techniques; and finally, an evaluation of the potential and actual characteristics of an acoustic wave tomography in different measurement and sounding schemes. The chapter summarizes the results discussed in [5–8].

12.1 Experimental Results

From a mathematical perspective, acoustic wave propagation is similar to the propagation of electromagnetic waves. From a physical point of view, electromagnetic and acoustic oscillations are fundamentally different. For a first approximation, the difference lies in the fact that an electromagnetic wave is a vector transverse wave, while an acoustic wave is a scalar longitudinal wave. It is important that the phase velocity of electromagnetic waves is lower in a more condensed material, while the phase velocity of acoustic waves is higher in air. Moreover, the attenuation of electromagnetic waves in a water-filled medium is rapid (exponential), unlike acoustic waves, which hardly attenuate in such a medium, as water is practically incompressible and transmits acoustic oscillations without significant attenuation. Most importantly, electromagnetic waves can propagate through a vacuum while acoustic waves cannot.

A common factor is that both electromagnetic and acoustic waves are described by wave equations, and both give rise to reflection, scattering, and diffraction effects

Applications of Radio-Acoustic Tomography ▪ 295

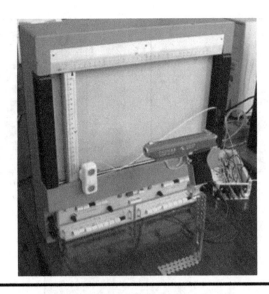

Figure 12.2 Experimental setup for ultrasonic tomography.

when the wave interacts with inhomogeneities in the medium. In this respect, all of the mathematical methods and results elaborated in previous chapters are applicable for ultrasound.

The simplest setup for our experiments in ultrasound tomography is shown in Figure 12.2. Standard piezo-ceramic transmitters and ultrasound receivers with a resonant frequency of 40 kHz are set on an x-y recorder, and a computer sound card can be employed as a signal generator and receiver. A Creative Audigy SE PCI SB0570 sound card enables signal generation and recording with a sample rate up to 96 kHz. An MA40S4S ultrasonic transmitter was used as well as an EM9767 electret microphone. The object of sounding (a toy gun) was placed at a certain distance from the transmitter. Scanning was performed through a simple computer-aided controller.

A broadband ultrasonic scanner setup was built and used to experimentally recover the ultrasonic image of the toy gun. The result is presented in Figure 12.3, which demonstrates variations of the real part of the complex amplitudes at different frequencies of the spectrum of the measured signal. Typical wave images of an ultrasonic field scattered by a test object are clearly apparent in this figure.

To recover the images of an object, the inverse focusing was used in the frequency domain at three different frequencies according to [1]. The reconstructed image of the test object is presented in Figure 12.4. The image of a gun can be clearly distinguished in the result of single-frequency sounding; however, there are minor artifacts. The number of artifacts was markedly decreased when the sounding was performed at all frequencies within the frequency band 37–43 kHz. It should be noted that in air a wavelength of 8.3 mm corresponds to a frequency of 40 kHz, which makes it possible to distinguish fine details in the image.

296 ■ *Electromagnetic and Acoustic Wave Tomography*

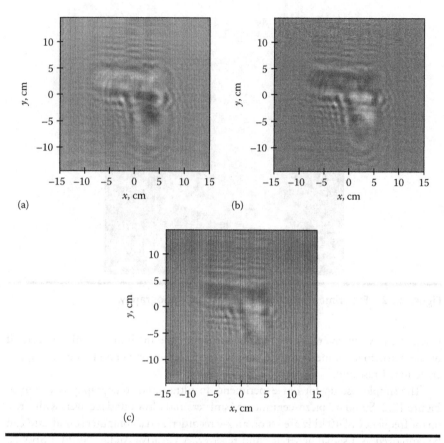

Figure 12.3 Ultrasonic echo of a toy gun located at a range of 12 cm at frequencies of: (a) 37 kHz, (b) 40 kHz, and (c) 43 kHz.

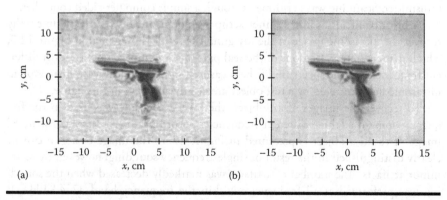

Figure 12.4 Recovered image of the object (tomogram): (a) at a frequency of 40 kHz; (b) within the frequency band 37–43 kHz.

Applications of Radio-Acoustic Tomography ■ 297

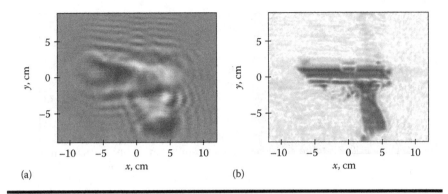

Figure 12.5 Plastic gun in air at a distance of 10 cm: (a) measured data and (b) result of focusing.

The resolution of details improves as the distance to the object is decreased. This is because the condition of Fresnel diffraction is being met more precisely (see Figure 12.5). The result of sounding of a test object in the form of a stepped triangle with a hole is displayed in Figure 12.6. The resolution in this case is close to the radiating wavelength, and even higher.

The reported results confirm the efficacy of the use of ultrasound in location (detection) tomography of small-sized objects. Furthermore, no special permit is required to use ultrasound, so it can be used for covert surveillance.

Objects that are hidden behind opaque but acoustically transparent screens (see Figure 12.7) can be visualized. The image is blurred, of course, but it remains recognizable. The resolution could be improved significantly with the use of multiple frequencies.

Mechanical scanning of an acoustic field requires up to several tens of minutes and the object must remain motionless, which is unattainable, for example, when

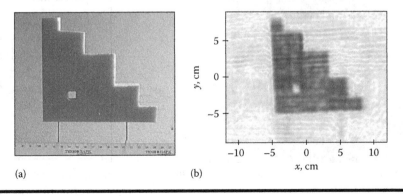

Figure 12.6 Stepped triangle in air at a distance of 10 cm: (a) photograph and (b) result of ultrasonic imaging.

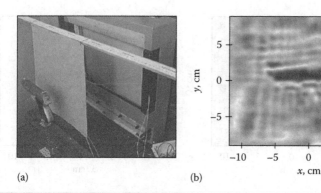

Figure 12.7 Plastic gun behind a curtain at a distance of 10 cm: (a) photograph of experimental setup and (b) result of focusing.

screening people. Thus, it would be interesting to consider how sounding could be speeded up by electronic switching of the transmit/receive arrays. In the view of the authors, cross-shaped sounding systems are of interest. The prototype of a sonar was therefore developed that included 32 ultrasonic transmitters and 32 receivers arranged in the form of a cross, situated 1 cm apart from each other (see Figure 12.8).

Measurements were carried out in the clocked mode. The time for a complete measurement, data processing, and visualization of the results on a standard computer does not exceed 2 s, which is acceptable for many applications. An image obtained using the prototype of a sonar is presented in Figure 12.9.

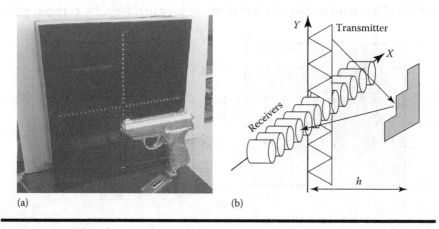

Figure 12.8 Sonar for contactless ultrasonic imaging: (a) external view; (b) configuration scheme.

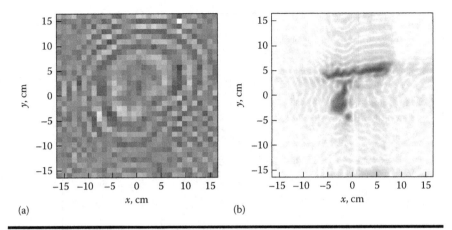

Figure 12.9 Testing results of sonar: (a) measured field; (b) recovered image.

12.2 Complexing of Radio and Ultrasound Tomography Techniques

Ultrasonic radiation is mostly scattered at the inhomogeneities of the density ρ. In remote sounding, it provides the information about the material an inhomogeneity is made of. Using electromagnetic waves in sounding makes it possible to monitor the electro-physical properties of matter. The integration of electromagnetic and acoustic waves for sounding enables us to distinguish solid and liquid media. The following empirical relation is true for solids:

$$\rho = 2(n-1), \quad \text{or} \quad n = \left[\rho/2 + 1\right]$$

where:
- ρ is the density
- $n = \sqrt{\varepsilon}$ is the refractive index of the material, defined in Section 3.1.5

A plot of the refractive index versus the density is shown in Figure 12.10, where the sloping line plots the dependence.

The values for certain solids are indicated with dots. Each value is listed in Table 12.1. Materials of higher density have a higher refractive index. Note that TNT (explosive material), which is distinctly different from other materials, takes a very low position in the diagram. The position of polytetrafluoroethylene (Teflon) is close to that of TNT. This means that Teflon can be used to simulate TNT in the laboratory; in practice it is necessary to use a reconstruction of the shape of such objects to distinguish these materials.

The refractive index of liquids is generally higher than of solids (see Table 12.2). Liquids are easily identified in the diagram. Furthermore, the electro-physical

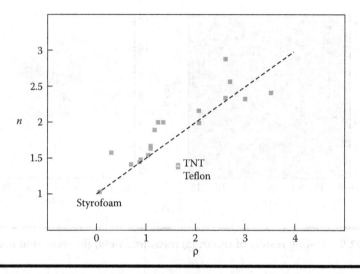

Figure 12.10 Refractive index: density diagram of several solid materials.

parameters of liquids have a definite temperature dependence that can be used for their identification.

One way or another, integration of ultrasonic sounding with radio waves provides additional information about a material and is therefore considered to be a promising direction for research. Moreover, the use of ultrasound is harmless to human health and it requires no special permit to be used. That is why the development of technology based on the synthetic aperture radar technique is considered to be important for ultrasonic tomography of inhomogeneous media and objects. Problems arising in this context and their possible solutions are discussed in the current section.

A large-sized scanner, where the radio- and ultrasonic elements are combined in a transceiver head, is shown in Figure 12.11.

A series of experiments with different objects was carried out (Figures 12.12 through 12.14). The ultrasound frequency was tuned within the band 35–45 kHz, and the duration of the ultra-wide band (UWB) pulses was 200 ps.

It was established that the images of metallic objects in radio-wave tomography have more contrast than in ultrasonic tomography. Apparently, the reason for this is that a part of the energy is spent to excite secondary waves in the metal and behind it. In contrast, radio waves have weaker reflection from dielectrics, and penetrate into dielectric objects. This was clearly demonstrated in experiments with a metallized grid (the reflection coefficient of radio waves was 0.9, whereas for ultrasonic waves it was 0.1) and with a foam–plastic object (where the radio waves did not reflect, and the reflection coefficient of ultrasound was 0.7). Ultrasound provided higher resolution (of 2–3 mm) than the radio waves (1–2 cm) at identical distances (<1.2 m) to the test object.

Table 12.1 Density and Refractive Index of Solids

Material	Density (g/cm³)	Refractive Index
Foam plastic (Styrofoam)	0.06	1.022741
Wood	0.3	1.581139
Paper	0.7	1.414214
Paraffin	0.87	1.449138
Polyethylene	0.9	1.48324
Polystyrene	1.05	1.549193
Asphalt	1.1	1.637071
Amber	1.1	1.67332
Plexiglas	1.18	1.897367
Fabric-reinforced laminate	1.25	2
Vinyl plastic (PVC)	1.35	2
PTFE (Teflon)	1.65	1.378405
TNT	1.654	1.4
Quartz	2.07	2.165641
Sulfur	2.07	1.994994
Marble	2.6	2.880972
Phlogopite	2.6	2.345208
Slate	2.7	2.569047
Glass	3	2.32379
Diamond	3.515	2.42

The reason for this is the difference between the average operating wavelength for ultrasound (8 mm) and for a radio wave (2 cm). It is important here that the Fresnel zone (the focusing region) for ultrasound has about twice the diameter as for radio waves given the same size of synthetic aperture.

Overlaying the radio-wave image and the ultrasonic image of the test object makes it possible to estimate the material of the test object without additional processing. A similar effect is observed in an element-by-element multiplication of images with a significant increase of the resolution as well.

Table 12.2 Density and Permittivity of Liquids

Liquid	Density (g/cm³)	Permittivity
Ethanol	0.7894	25.8
Methanol	0.7915	31.2
Acetone	0.792	26.6
Oil	0.86	2.13
Benzene	0.879	2.29
Flaxseed oil	0.94	3.35
Castor oil	0.96	4.67
Water	0.9982	81.1
Nitrobenzene	1.229	2.41
Glycerin	1.26	56.2
Nitroglycerin	1.601	2.196

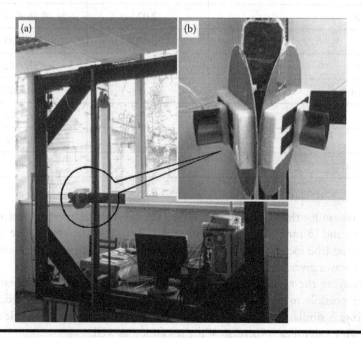

Figure 12.11 (a) Radio and ultrasonic scanner and (b) scanner detector head.

Applications of Radio-Acoustic Tomography ■ 303

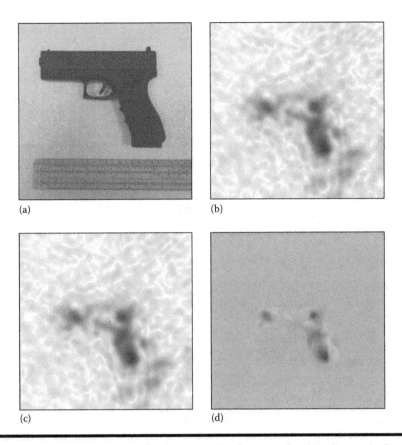

Figure 12.12 Image of a plastic gun: (a) photograph, (b) ultrasonic image, (c) UWB image, and (d) image from integrated tomography.

12.3 Visualization of Small Defects Hidden Inside Metallic Constructions Using Acoustic Tomography

Another area of ultrasonic tomography that deserves our attention is contactless tomography of flaws in metal structures. If a stimulator (source of ultrasonic vibrations) is set up on a metal surface or some other sound-conducting structure, then propagating waves are excited in it. When these waves encounter non-uniformities in their path (cuts, flaws, sags, etc.), they are scattered and reradiated at all angles and into the air. The edges (boundaries) of the object can also reradiate.

Considering these non-uniformities as sources of ultrasonic waves, a tomographic visualization of such sources can be performed using inverse focusing. As a result, the position and shape of even the small defects will be reconstructed without contact. An example is displayed in Figure 12.15.

304 ■ *Electromagnetic and Acoustic Wave Tomography*

Figure 12.13 Image of a stepped triangle made of plasterboard: (a) photograph, (b) ultrasonic image, (c) UWB image, (d) image from integrated tomography.

A similar experiment was carried out with a large metal sheet (see Figure 12.16). If a sound-conducting object is excited by an external stimulator, for example, by impact, the distribution of defects as well as the position of the stimulator, even in the other half-space, can be reconstructed by the secondary acoustic field generated by this object.

A stimulator in the form of an electric micromotor with an eccentric (vibration) motor, mounted to a fiberglass plate with dimensions 15 × 11 cm, is shown in Figure 12.17. The frequency of rotation was about 23 kHz. The size of the area for monitoring the distribution of ultrasonic oscillations was 40 × 40 cm.

The result of tomographic processing of the recorded field is shown in Figure 12.18. When using a wide spectrum of recorded frequencies, good localization of the radiation source is observed, and the shape of the plate is also reconstructed.

Applications of Radio-Acoustic Tomography ■ 305

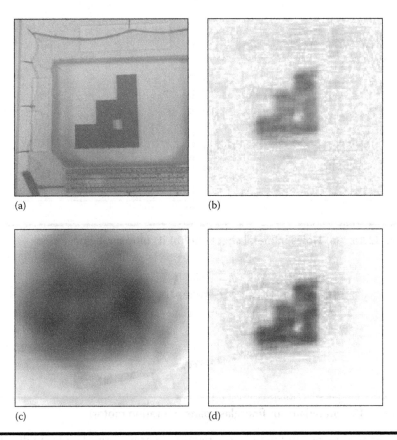

Figure 12.14 Image of a stepped triangle made of plasterboard behind a metallized grid: (a) photograph, (b) ultrasonic image, (c) UWB image, and (d) image from integrated tomography.

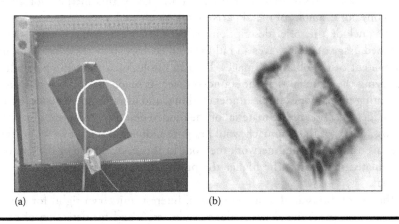

Figure 12.15 (a) Cut in thin metal sheet and (b) its ultrasonic image.

306 ■ *Electromagnetic and Acoustic Wave Tomography*

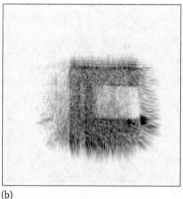

Figure 12.16 (a) Holes in metal sheet and (b) its ultrasonic image.

Figure 12.17 Stimulator of fiberglass plate (vibration motor).

Field measurements of the acoustic signals can be performed quite rapidly using a two-dimensional array of 256 EM9767 microphones arrayed over a hexagonal grid with a 1 cm step (see Figure 12.19).

The obtained result demonstrates the potential of this method for acoustic introscopy, for example, of metal parts (doors, safes, etc.).

To conclude this brief description of the matter, we should outline that the combined usage of radio waves and ultrasound enables an increase in the information content of the obtained images, both in resolution and in the possibility of identifying the material that the sounded object is made of. Radio waves provide penetrating power, for example, under clothing, and ultrasound increases the accuracy of identification of the material of the hidden object. A complete discovery of the possibilities of this approach would require individual study of a wide array of materials using the neural network method. In this case, the larger the size of the accumulated data bank, the higher the probability of valid identification of the material.

The use of ultrasound tomography is of interest in its own right, for example, in the introscopy of metal parts (doors, safes, etc.), and for detection of cavities and defects in them. For internal and near-surface introscopy of sound-conducting

Applications of Radio-Acoustic Tomography ■ 307

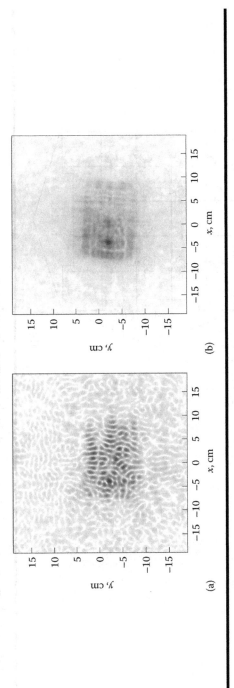

Figure 12.18 Reconstructed image of a fiberglass plate: (a) at a frequency of 23 kHz; (b) at 32 frequencies in the range 10–30 kHz.

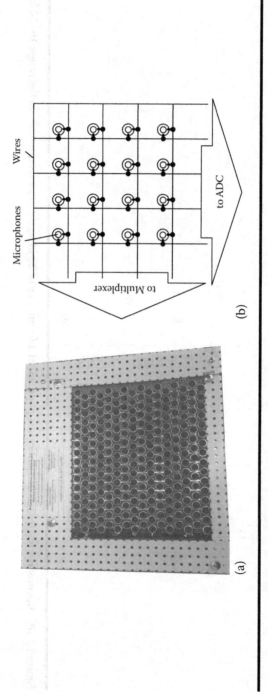

Figure 12.19 (a) Ultrasonic microphone array and (b) its interrogation scheme.

objects or fragments of them, contact-free ultrasound tomography with additional coherent stimulation can be effective. For tomographic processing of secondary-radiation data, the technology of large aperture synthesis can be used. The system is a development of known foreign and domestic technologies, but it does not require disassembly of the object for inspection or immersion in an immersion liquid. In this case, both X-rays and radio waves are simply ineffective.

References

1. Yakubov, V. P., S. E. Shipilov, D. Ya. Sukhanov, and A. V. Klokov, *Radiowave Tomography: Achievements and Perspectives*, Tomsk, Russia: Scientific-Technical Literature (STL) Edition Group, 2016 (in Russian).
2. Yakubov, V. P., S. E. Shipilov, D. Ya. Sukhanov, and A. V. Klokov, *Wave Tomography*, Tomsk, Russia: Scientific Technology Publishing House, 2017.
3. Daniels, D. J., *Surface-Penetrating Radar*, London: The Institute of Electrical Engineering, 1996.
4. Brekhovskikh, L. M., *Waves in Layered Media*, 2nd edn, New York: Academic, 1980.
5. Greguss, P., *Ultrasonic Imaging: Seeing by Sound*, New York: Focal Press, 1980.
6. Yakubov, V. P., S. E. Shipilov, and D. Ya. Sukhanov, Radio and ultrasound tomography of hidden objects, *Russian Physics Journal*, vol. 55, no. 8, 2013, pp. 878–883.
7. Sukhanov, D. Ya. and M. A. Kalashnikova, Remote ultrasonic defectoscopy of sound radiating objects through the air, *Acoustical Physics*, vol. 60, no. 3, 2014, pp. 304–308.
8. Sukhanov, D. Ya. and K. V. Barysheva, Transmission incoherent ultrasonic imaging of planar objects, *Acoustical Physics*, vol. 56, no. 4, 2010, pp. 501–505.

APPLICATIONS OF LOW-FREQUENCY MAGNETIC AND EDDY CURRENT TOMOGRAPHY

V

Chapter 13
Applications of Low-Frequency Magnetic Tomography

Vladimir Yakubov and Dmitry Sukhanov

Contents

13.1 Eddy Currents and Accompanied Magnetic Fields 313
13.2 Reconstruction of Metallic Objects by Low-Frequency Magnetic Tomography 314
13.3 Magnetic Tomography of Metallic Objects Hidden behind Metallic Screens 318
References 322

13.1 Eddy Currents and Accompanied Magnetic Fields

Magnetic and magnetic induction methods were almost entirely worked out in the twentieth century and now offer a comprehensive range of feasible solutions. These methods are not subjected to revision in the present chapter, but they are supplemented with respect to their use for magnetic tomography [1–4]. The theoretical background of the method of low-frequency magnetic tomography was introduced in Section 4.1 and that of eddy current tomography in Section 4.3.

At the same time, as was mentioned in references [5–10], the eddy current flaw detection is based on the induction of an alternating current in a metal test object by an external magnetic field. Induction currents give rise to a distortion in the external field, which is detected by an induction coil. Depending on the electrophysical properties of the metal and the origin of the flaw, the induction currents have different amplitude and phase distributions, which underlies the use of eddy currents for flaw detection (see Section 4.1). The source and receiver coils can be combined.

There is a wide selection of eddy current flaw detectors of different modifications available at the present time. These include, for example, VD-12NFM, VD-26P (Ultracon Co.), VD-89R (Association "Spektr"), D-95 Expert (AKA) (0.5–128 kHz), and so on. Eddy current flaw detectors mainly make use of a single transmitter–receiver module that functions as a sensor of the alternating magnetic field. They measure the signal phase and amplitude in the receiver coil. Flaw detection is based on an analysis of the signal amplitude and phase (see Section 4.1). A special feature of eddy current flaw detectors is the confinement of the magnetic field due to the small size of the induction coils (about 5 mm). This provides more accurate flaw detection, but decreases the operating range of the detector. The operating range of eddy current flaw detectors is comparable to the size of the sounding coil.

Metal detectors make use of the same physical phenomena as eddy current flaw detectors. They make use of large-size magnetic coils (about 30 mm) to provide an increase in the range of metal object detection. In this regard, it should be noted that the detection range of objects buried in the ground depends on the signal frequency. The higher the frequency, the smaller the penetrating depth because of losses in the conductive medium. Conversely, if the frequency is low, the penetration depth is greater, but the detection sensitivity of small inhomogeneities is decreased. Metal detectors used at depths of about 1 m operate at low frequencies (about 1 kHz).

Metal detectors can differ in their design: some of them contain a source coil and a receiver coil, other use a single transmitter–receiver coil. The use of signals of different types, harmonic, multi-frequency, frequency-modulated, compound pulse signals, and so on, is possible. Object detection is based on an analysis of the signal level in the receiver coil; or, in the case of a single transmitter–receiver coil, the coil is integrated into an oscillating LC circuit, whose frequency is known in the absence of inhomogeneities. In the presence of an inhomogeneity, the resonant circuit frequency changes, which signals the presence of a metal object; that is, it is detected.

13.2 Reconstruction of Metallic Objects by Low-Frequency Magnetic Tomography

As mentioned in Section 3.1, unlike microwaves, low-frequency magnetic fields have a high penetrating power and can penetrate even a conductive medium. However, the components of the magnetic field have a narrow spatial spectrum far from the source, so measurements with a high dynamic range and high signal-to-noise ratio are required to obtain high-resolution images from sounding with low-frequency

Applications of Low-Frequency Magnetic Tomography ■ 315

magnetic fields. The corresponding theoretical framework regarding the effects of low-frequency magnetic field tomography was briefly presented in Section 5.1. In accordance with references [1–4], we will show some practical applications of the low frequency band magnetic field tomography.

It is possible to increase the accuracy of measurement by moving the sensing coil closer to the test object. However, the most significant impact on the accuracy of measurement is achieved by deducting the background value of the magnetic field (the value without the test object), which is far larger than the value of the secondary magnetic field.

In this section, we shall discuss the possible use of self-compensated sensing coils to visualize conductive objects, according to references [1–4]. As a self-compensated source of the magnetic field, it is proposed to use two flat spiral coils of different size located in the same plane with a common center, so that they create opposite magnetic induction vectors at this common center (see Figure 13.1). The sensing coil is placed at the center. Thus, if there is no field distortion due to objects in the medium, an induction current will not be induced in the receiving coil. We call the proposed system of coils a self-compensated sensing coil (SCSC).

A self-compensated magnetic coil was printed on a circuit board (Figure 13.1b) for the purpose of these experiments. The external coil had 100 turns, the internal one 42 turns, and the sensing coil in the center had 18 turns. The helix pitch distance was 500 μm. The two outer coils (one circuit) comprised the source coil, and the innermost coil—the sensing coil—comprised the receiving coil (for a schematic depiction of this arrangement, see Figure 13.1a). The size of the circuit board was 20×20 cm, and the thickness of the conductors was 250 μm.

Calculation of the size and number of turns was based on the Biot–Savart law. First, the field of the rectilinear conductors was calculated by integration. Then, the

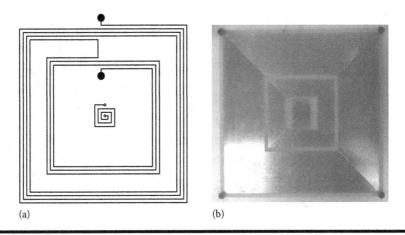

Figure 13.1 Self-compensated sensing coil: (a) connection diagram and (b) printed circuit.

fields of all the rectilinear conductors were summed. As the source coil consisted of a single wire, the current within it was the same along all segments. The number of turns was determined to meet the condition that the magnetic field be equal to zero at the center of the system.

An SB Creative Audigy SE PCI SB0570 sound card was used as the generator and the recorder. The SCSC circuit board was moved by a dual-axis scanner. During the experiments, the test objects were situated at a distance of 1 cm from the scanning plane.

The first test object was made of three aluminum foil strips of different lengths: 5, 10, and 15 cm. They were 30 μm in thickness and 5 cm in width. There was an air gap of 1 mm between the strips and no electrical contact between them. In the experiments, the cosine and sine quadratures of the output signal of the receiver coil at a frequency of 40 kHz were measured for different positions of the SCSC with a step of 5 mm over an area of 40×40 cm. The result of the amplitude measurements is displayed in Figure 13.2a, where the darker regions correspond to higher values of the amplitude. A plot of the cosine quadrature of the measured signal is shown in Figure 13.2b.

All of the strips are clearly distinguishable, even though the gap between them is just 1 mm. This means that the system is sensitive to the absence of electrical contacts. The cosine quadrature image demonstrates that the signal phase reverses sign when crossing the boundary of a metal object, which enables an accurate detection of object outlines.

To demonstrate the possibility of visualizing objects composed of several layers with different distributions of conductive regions, an experiment was carried out in which two layers of aluminum strips, each configured in the shape of the previous object, were overlaid crosswise with respect to each other. The results of the

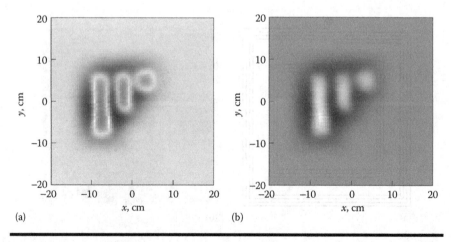

Figure 13.2 Measurements of an object consisting of three aluminum strips: (a) amplitude; (b) cosine quadrature.

Applications of Low-Frequency Magnetic Tomography ■ 317

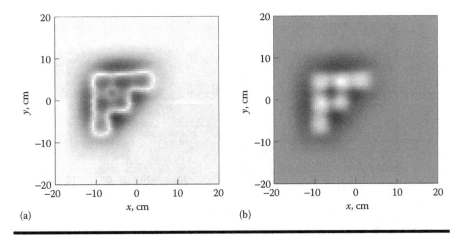

Figure 13.3 Measurement result for the object consisting of two layers of crossed aluminum strips: (a) amplitude; (b) cosine quadrature.

measurements are displayed in Figure 13.3, from which it is possible to distinguish the structure of the object.

Next, an experiment was carried out with five flat objects taped to a flat piece of styrofoam (see Figure 13.4). The first object was a square with a side length of 5 cm. The second object was a square with a side length of 5 cm and a notch 4.5 cm long down the center. The notch creates a break in the contour along the perimeter of the square-shaped object, which should decrease the level of inductive currents.

The third object was a square with a side length of 2.5 cm. This object should demonstrate a substantial decrease in the level of induction currents due to a decrease in its outer contour. The fourth object was a square with a side length of 5 cm with a square piece cut out of its middle, of such size that the width of the remaining square metal outline was 5 mm. This object should demonstrate that retaining the outer contour while removing the inner part only insubstantially decreases the level of induction currents. The fifth object consisted of four aluminum strips 50 mm long and 5 mm wide, placed 5, 10, and 15 mm apart.

The results of the experiment demonstrated that:

■ The object with a large piece cut out of its center generates a signal comparable to the signal of the corresponding intact object, that is, the greatest contribution to the signal comes from the outer contour of the conductive region.
■ A notch in the outer contour substantially decreases the signal level.
■ Objects whose outer contours circumscribe a small area are not visualized.

To evaluate the possibility of fault detection, specifically detection of defects in the form of a notch, an experiment was carried out with a brass plate having dimensions of 12×7.5 cm and 0.5 mm thickness, in which an oblique cut was

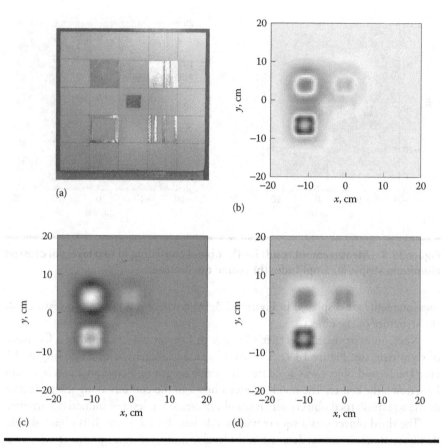

Figure 13.4 Measurement results for the object composed of five elements: (a) external view, (b) amplitude, (c) cosine quadrature, and (d) sine quadrature.

incised (Figure 13.5). Measurements were performed at a frequency 40 kHz at an offset distance of 1 cm.

The region of the notch (cut) is clearly visualized in the displayed images. The induction current skirts the cut so it can be visualized by the SCSC.

13.3 Magnetic Tomography of Metallic Objects Hidden behind Metallic Screens

A series of experiments was carried out with an SCSC to evaluate the possibility of detection of metal objects hidden behind conductive screens. A study of the possibility of detecting a metal object hidden behind a metal screen was conducted in the course of those experiments.

Applications of Low-Frequency Magnetic Tomography ■ 319

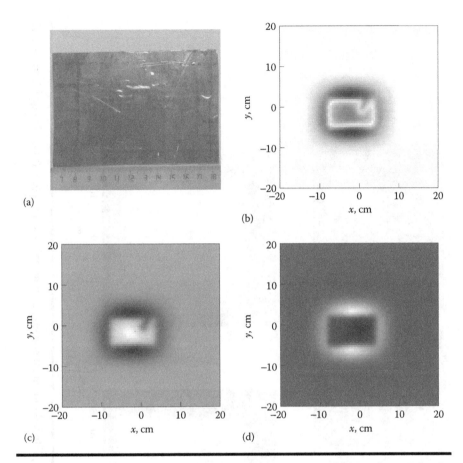

Figure 13.5 Measurements on a brass plate: (a) external view, (b) amplitude, (c) cosine quadrature, and (d) sine quadrature.

Aluminum, copper, and steel sheets were used as the metal screens. As the objects to be detected, we used brass, aluminum and steel plates.

An experiment on the detection of a brass plate hidden behind a copper screen was carried out, in which a fiberglass sheet covered with copper foil on both sides was employed as the metal screen (Figure 13.6a). The brass plate was placed 1 cm above the screen, which, in turn, was hung 2 cm above the plane of translation of the SCSC.

Figure 13.6b and c display the results of measurements at a frequency of 5 kHz. The image of the cosine quadrature of the measured signal reveals an inhomogeneity having the shape of the brass plate. Thus, an object hidden behind a barrier was detected. Since the barrier was thin, the alternating magnetic field was only insignificantly attenuated, which made it possible to detect objects behind a barrier. In the given case, the hidden object was visualized in the cosine quadrature better

320 ■ *Electromagnetic and Acoustic Wave Tomography*

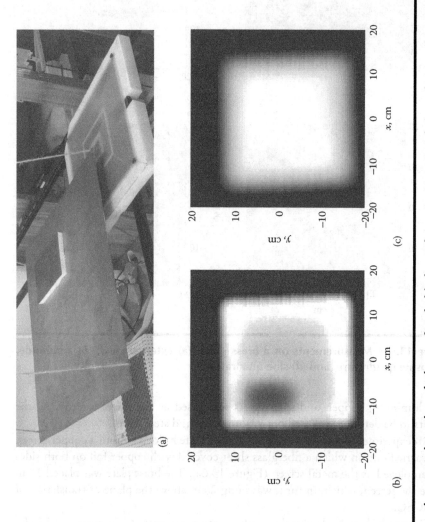

Figure 13.6 Experiment on detection of a brass plate behind metal screen: (a) external view of setup, (b) cosine quadrature, and (c) sine quadrature.

Applications of Low-Frequency Magnetic Tomography ■ 321

than in the amplitude, since the phases of the fields of the induction currents are different for different objects.

The same experiment was carried out with a steel plate 1 mm in thickness with dimensions of 65 × 180 mm. The steel plate was placed behind the copper screen (see Figure 13.7). Measurements were performed at a frequency of 3 kHz.

Similar results were also obtained for an aluminum plate behind the copper screen. When the copper screen was replaced by a thin aluminum screen, the result remained the same, but when the screen thickness was increased, the detection effect was lowered. Detection was not achieved behind an iron screen. This fact can be explained by the ferromagnetic properties of iron, which concentrate the magnetic field lines, thereby shielding the oscillating magnetic fields.

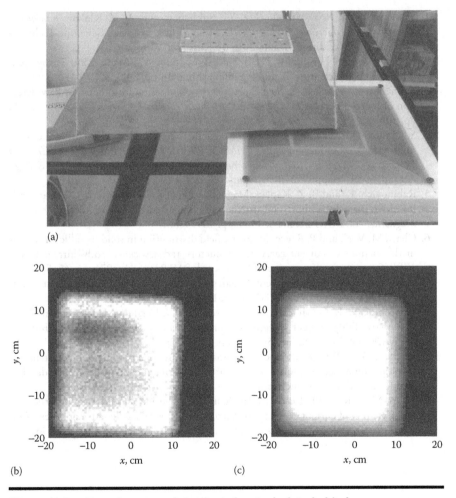

Figure 13.7 Experiment on detection of a steel plate behind a copper screen: (a) external view of setup, (b) cosine quadrature, and (c) sine quadrature.

Low-frequency magnetic fields can be used in the tomography of flat metal objects. The basis of its use consists in measurements of secondary magnetic fields generated by eddy currents induced on the object's boundaries. Scanning with a self-compensated sensitive coil is considered to be a promising tool for measurements. The SCSC should be placed very close to the test object, since the secondary magnetic field generated by this object attenuates rapidly with distance. Metal objects and their defects are clearly detected when they are placed behind dielectric screens, and also behind thin non-magnetic metal screens.

References

1. Yakubov, V. P., S. E. Shipilov, D. Ya. Sukhanov, and A. V. Klokov, *Radiowave Tomography: Achievements and Perspectives*, Tomsk, Russia: Scientific-Technical Literature (STL) Edition Group, 2016 (in Russian).
2. Yakubov, V. P., S. E. Shipilov, D. Ya. Sukhanov, and A. V. Klokov, *Wave Tomography*, Tomsk, Russia: Scientific Technology Publishing House, 2017.
3. Sukhanov, D. Y. and E. S. Sovpel', A magnetic-induction introscope for flaw detection of metal objects, *Russian Journal of Nondestructive Testing*, vol. 51, no. 5, 2015, pp. 308–314.
4. Sukhanov, D. Ya. and E. S. Berzina, Magnetic induction introscopy using the scanning matrix of induction coils, *CriMiCo 2014–2014 24th International Crimean Conference Microwave and Telecommunication Technology, Conference Proceedings*. Art. no. 6959710, pp. 952–953, 2014.
5. Rabinovich, R. and B. Z. Kaplan, Magnetization of a nonlinear ferromagnetic semi-infinitive plate by a short coil, in the presence of eddy-currents, *IEEE Trans. on Magnetics*, vol. MAG-21, no. 5, 1985, pp. 1859–1861.
6. Chari, M. V. K. and P. Reece, Magnetic field distribution in solid metallic structures in the vicinity of current carrying conductors and associated eddy-current losses, *IEEE Trans. Power Apparatus and Systems*, vol. PAS-93, no. 1, 1974, pp. 45–56.
7. Haugland, S. Mark, Fundamental analysis of the remote-field eddy-current effect, *IEEE Trans. on Magnetics*, vol. 32, no. 4, 1996, pp. 3195–3211.
8. Joynson, R. F., R. O. McCary, D. N. Oliver, K. M. Silverstein-Medengren, and L. L. Thumhart, Eddy current imaging of surface-breaking structures, *IEEE Trans. on Magnetics*, vol. MAG-22, no. 5, 1986, pp. 1260–1262.
9. Vertiy, A. A., S. P. Gavrilov, and N. Blaunstein, Eddy current tomography systems for investigation of defects (holes, cracks, wares) occurring on and inside the conductive, composite and non-conductive structures, *Report # 03/1999, DPES*, Israel.
10. Chari, M. V. K. and P. Reece, Magnetic field distribution in solid metallic structures in the vicinity of current carrying conductors and associated eddy-current losses, *IEEE Trans. Power Apparat. and Systems*, vol. PAS-93, no. 1, Jan/Feb. 1974, pp. 45–56.

Chapter 14
Eddy Current Tomography Applications

Alexey Vertiy and Nathan Blaunstein

Contents

14.1 Experimental Setup..323
14.2 Eddy Current Tomography for Reconstruction of Sub-Surface Defects....324
References..331

14.1 Experimental Setup

Based on theoretical results briefly described in Section 3.4, and Chapters 3 and 5, according to references [1–10], a non-destructive imaging system was developed based on the eddy current (EC) tomography technique, allowing the location of internal defects (cracks, holes, wears) in various conductive, non-conductive, and composite materials. The main task of such a 2-D EC image of any defects hidden in such structures was to examine the possibility of recovering and reconstructing the shapes of the defects and their exact location, and giving their depth, width, and height within the material under scanning process. In addition, the system, unlike existing X-ray systems, is not harmful in any way to the operator or environmentally damaging. The imaging system can be used for testing of various metals including aluminum, copper, titanium, and so on. It can also be used for testing and imaging of multicomponent (composite) materials.

The novelty of the system is that, for the first time, we have been able to develop a unique system that, unlike any existing system [1–7], images cracks in the 3-D domain and not in a graphic form derived from function processing by the computer [8–10]. More than 700 experimental tests have been carried out using text plates obtained from many U.S. and European firms (namely, from Nortec and General Electric).

The experimental EC tomography setup is presented in Figure 14.1. The setup consists of a personal computer (PC) laptop, step motor control block, EC device (multi-frequency mode), and slow 2-D linear scanner. The PC laptop controls all peripheral equipment. The digitized raw data of 16 frequencies are sent to the PC laptop, where EC tomography is processed. Two-way scanning is carried out. The principle of scanning is shown in Figure 14.1.

The scanning field was separated for each rivet junction investigated and was 10 mm by following the x-coordinate and 15 mm by following the y-coordinate. The paths of the EC probe are shown as black solid lines.

14.2 Eddy Current Tomography for Reconstruction of Sub-Surface Defects

The results of testing using such examples are shown in Figures 14.2a and b, 14.3, 14.4a and b, respectively.

Another example of testing was with an aluminum plate, as a sample layer closing a slot (see Figure 14.5).

It is well known that the modern EC industry has many special techniques to detect different cracks in complex metallic elements such as, for example, rivet junctions [9, 10]. The experimental results presented here were obtained using a standard Nortec (U.S.) EC absolute probe and slow 3-D linear scanning. The EC prototype works at a multi-frequency mode regime, with 16 frequencies for tomography processing. A demo panel shows the positions of the EC probe and of the investigated rivet's junction with all dimensions, shown in Figure 14.6.

Experimental measurements were carried out in several stages. Firstly, the possibility of the presented EC tomography experimental setup for testing of the kind of rivet junction has been investigated using only tomography technologies. The actual scanning procedure (step-by-step depth slicing) is presented in Figure 14.7a–c.

Further measurements of metallic sub-surface defects, such as cracks, were carried out by use of a miniature prototype. This "demo" prototype has been used for testing cracks on the sub-surface of an engine in a reconstruction factory in the U.K. The view of such a miniature prototype is shown in Figure 14.8. That prototype consists of a cylindrical miniature probe, electronic block and a notebook computer, which contains the software for imaging of cracks by use of theoretical framework on EC tomography technique.

It is able to detect and to produce 3-D images and exact data about hidden cracks within the engine block and crankshafts, as shown in Figure 14.8.

Eddy Current Tomography Applications ■ 325

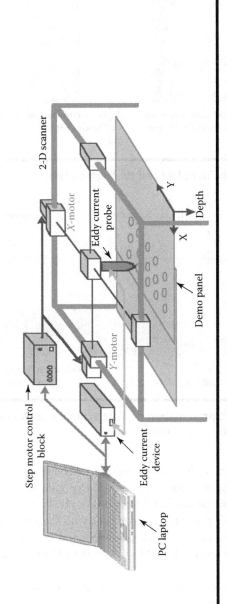

Figure 14.1 Scheme of EC tomography experimental setup.

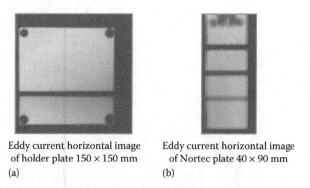

Figure 14.2 (a) EC of holder plate and (b) Nortec plate.

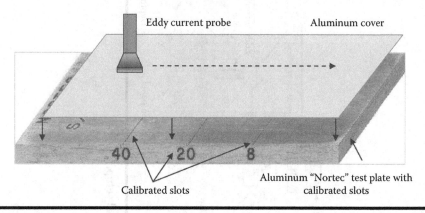

Figure 14.3 Schematic presentation of plate scanning with aluminum cover of slots denoted by numbers 8, 20, and 40.

The same application of EC tomography for detecting defects, such as cracks, in high-conductive structures (the same effects that can be found by use of the ultrasound approach mentioned in Chapter 13) is vividly seen from Figure 14.9, where the right-hand side presents a piece of tube with the probe on its top surface, the middle shows the location of the measurement and control block, and the left-hand side demonstrates one of the 2-D slices of the EC image of the crack found by the probe.

The instruments constructed for EC tomography shown in these figures were designed for wider purposes, that is, to detect the presence, location, and shape of both surface and sub-surface defects of various constructions, such as flaws, voids, cracks, holes, and so forth, in conductive, composite, and dielectric materials. Conductivity may vary from high (aluminum, copper) down to quite low (carbon–carbon composites and wood).

Eddy Current Tomography Applications ■ 327

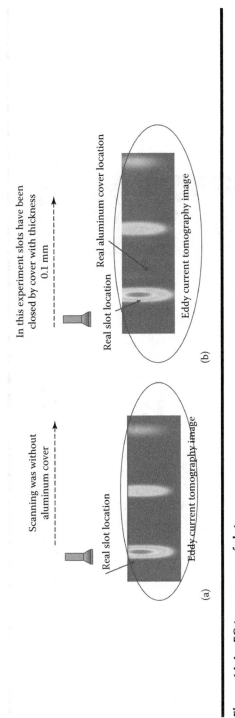

Figure 14.4 EC tomograms of slots.

328 ■ *Electromagnetic and Acoustic Wave Tomography*

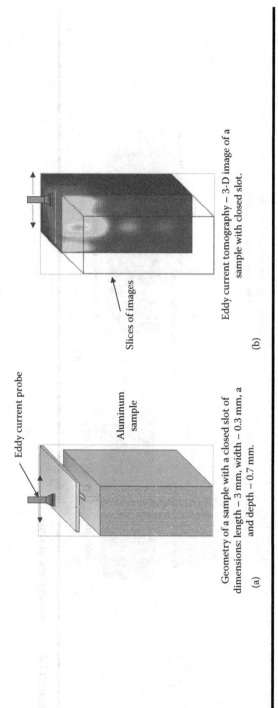

Figure 14.5 (a) Scheme of experiment; (b) EC slicing tomogram.

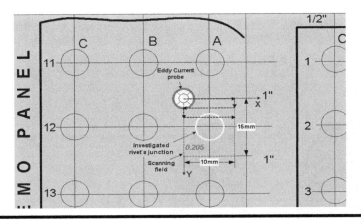

Figure 14.6 Two-way EC.

Detection is based on the usage of an induced EC, a widely and successfully used non-destructive testing (NDT) method, but a unique feature of this instrument is the ability to obtain 2-D and 3-D images of the test area, provided by the application of a special data collection technique and following tomography processing of this data.

According to the target destination of this prototype to be used in the field, detection is possible while manual scanning of the probe over the test area is performed. During scanning, the instrument quasi-simultaneously excites the multi-frequency current in the probe's coil and concurrently measures the multi-frequency response. The current in the coil induces a current in the tested material.

The exact values of the detection parameters, such as maximum depth of detection, minimum size of detectable flaw, and productivity are a strong function of the selected frequency range, material under test and probe parameters (this information is available in additional data sheets).

The presence of defects disturbs normal current distribution in the material, thus changing the response in the receiving coil. The instrument extracts this small complex useful signal from the total input signal at each frequency, applies pre-processing (amplification, filtration, and digitization) and transfers the data to the PC for the tomography processing that follows. It does not take much time to process the data, so the operator may see the results on the screen almost immediately. For this instrument, the result is a 2-D image at the depth of the area just scanned. The location and shape of all detected defects may be visually observed on this image.

The primary areas of application of this new method and instruments based on it are critical elements and materials testing in the aerospace, automobile, and energy industries.

330 ■ *Electromagnetic and Acoustic Wave Tomography*

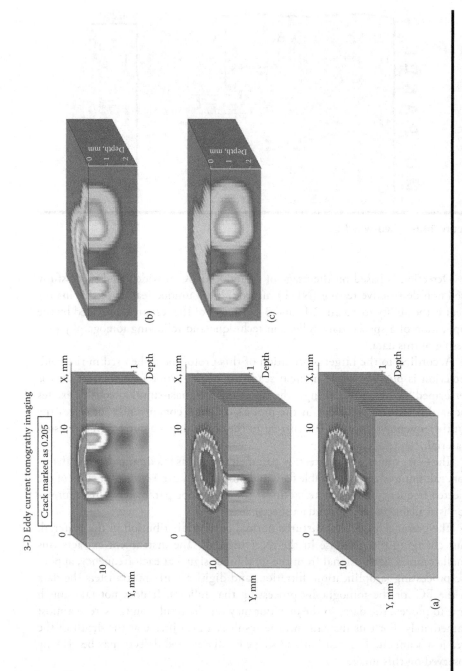

Figure 14.7 (a) EC tomography imaging of rivet junction during scanning; (b) 3-D cross-section along crack that has been detected in open rivet junction; (c) 3-D cross-section along crack that has been detected in covered rivet junction.

Eddy Current Tomography Applications ■ 331

Figure 14.8 Experimental miniature setup for detection of engine (left) and corresponding image seen on computer screen (right).

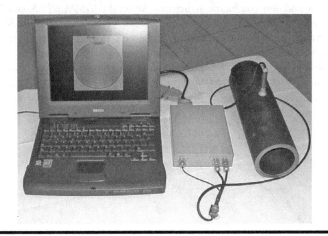

Figure 14.9 An example of detection and imaging of cracks on tube surface by use of the EC tomography technique.

References

1. Mook, G. and J. Simonin, Surface and subsurface material characterization using eddy current arrays, *Proceedings of 19th World Conference on Non-Destructive Testing*, 13–17 June 2016, München DGZfP, vol. BB 158, We.2. p. 2.
2. Peng, X., H. Songling, and Z. Wei, A new differential eddy current testing sensor used for detecting crack extension direction, *Proceedings of NDT&E International*, vol. 44, 2011, pp. 339–343.
3. Rabinovich, R. and B. Z. Kaplan, Magnetization of a nonlinear ferromagnetic semi-infinitive plate by a short coil, in the presence of eddy-currents, *IEEE Transactions on Magnetics*, vol. MAG-21, no. 5, 1985, pp. 1859–1861.

4. Chari, M. V. K. and P. Reece, Magnetic field distribution in solid metallic structures in the vicinity of current carrying conductors and associated eddy-current losses, *IEEE Transactions on Power Apparatus and Systems*, vol. PAS-93, no. 1, 1974, pp. 45–56.
5. Lebrun, B., Y. Jayet, and J.-C. Baboux, Pulsed eddy current signal analysis: application to the experimental detection and characterization of deep flaws in highly conductive materials, *Proceedings of NDT&E International*, vol. 30, no. 3, 1997, pp. 163–170.
6. Fukutomi, H., H. Huang, T. Takagi, and J. Tani, Identification of crack eddy current testing signal, *IEEE Transactions on Magnetics*, vol. 34, no. 5, 1998, pp. 2893–2896.
7. Chommeloux, L., Ch. Pichot, J.-Ch. Bolomey, Electromagnetic modeling for microwave imaging of cylindrical buried inhomogeneities, *IEEE Transactions on Microwave Theory Techniques*, vol. MTT– 34, no. 10, 1986, pp. 1064–1076.
8. Vertiy, A. A. and S. P. Gavrilov, Modelling of microwave images of buried cylindrical objects, *International Journal of Infrared and Millimeter Waves*, vol. 19, no. 9, 1998, pp. 1201–1220.
9. Vertiy, A. A., S. P. Gavrilov, and N. Blaunstein, Eddy current tomography systems for investigation of defects (holes, cracks, wares) occurring on and inside the conductive, composite and non-conductive structures, *Report # 03/1999, DPES*, Israel.
10. Vertiy, A. A, S. P. Gavrilov, S. Aksoy, I. V. Voynovskyy, A. M. Kudelya, and V. N. Stepanyuk, Reconstruction of microwave images of the subsurface objects by diffraction tomography and stepped-frequency radar methods, *Zarubejnaya Radioelektronika. Uspehi Sovremennoy Radioelektroniki [Foreign Radioelectronics. Success in Modern Radio Electronics]* (in Russian), no. 7, 2001, pp. 17–52.

VI

METHODS OF VISUALIZATION AND RECONSTRUCTION OF OBJECTS

VI

METHODS OF VISUALIZATION AND RECONSTRUCTION OF OBJECTS

Chapter 15

Visualization and Reconstruction of Objects

Vladimir Yakubov, Sergey Shipilov,
Dmitry Sukhanov, and Andrey Klokov

Contents

15.1 Visualization of 3-D Tomograms by Polyscreen Technique......................335
15.2 Visualization of 3-D Tomograms by Equipotential Surfaces 336
15.3 Visualization of 3-D Tomograms by Orthogonal Slices 337
15.4 Inverse Problems in Object Reconstruction..338
       15.4.1 Elimination of Image Spreading 338
       15.4.2 Inverse Problem for Signal Analysis 340
15.5 Inverse Problem of Source Localization.. 342
15.6 Inverse Problem of Micro-Strip Sensors Reconstruction.......................... 343
References 345

One task of radiowave, acoustic (ultrasonic), magnetic, and eddy current tomography is to generate 3-D bitmap images of test objects. However, there are no currently available displays that are capable of presenting 3-D bitmap images on the screen. As a consequence of this fact, as discussed in references [1, 2], it is necessary to develop techniques for presenting 3-D bitmap images in 2-D displays.

15.1 Visualization of 3-D Tomograms by Polyscreen Technique

One technique for displaying 3-D images is to construct a polyscreen. A polyscreen is a 2-D bitmap image composed of a set of images corresponding to different cross-sections of the sounded volume along a predefined direction. One advantage of such a visualization scheme is the simplicity inherent within it of the interpretation of the displayed data. In such a scheme, the distribution of inhomogeneities inside the test object (e.g., cavities and flaws) becomes accessible for detection and analysis. A similar visualization technique is used in medical computer-aided diagnostics.

The tomogram for a numerically simulated image of a stepped test object is presented in Figure 15.1. The polyscreen presents images of 16 layers with a 0.5 cm increment in range (depth). The first layer corresponds to a depth of 0.5 cm; the last corresponds to a depth of 8 cm. The dark areas correspond to higher amplitude values. As is readily obvious, the test object is located in the ninth layer, corresponding to a depth of 4.5 cm.

15.2 Visualization of 3-D Tomograms by Equipotential Surfaces

The contouring of equipotential surfaces is one of the techniques available for the visualization of 3-D images. The result of contouring is a three-dimensional surface, which can be visualized by the standard techniques of 3-D polygon graphics. But, this scheme requires multi-staged processing of the 3-D image: first, the polygonal model must be constructed; then, this model must be visualized, which requires additional computing resources.

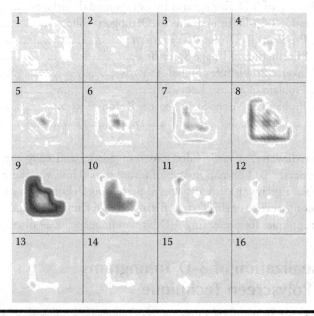

Figure 15.1   Test object tomogram.

Figure 15.2 Equipotential surface of 3-D field as seen from different angles.

The backward ray tracing method is proposed as a means for displaying equipotential surfaces of 3-D images. Each image point is obtained by scanning the 3-D image with a direct ray originating from the observer. The intersection point of the ray and the equipotential surface is indicated in the display.

Figure 15.2 displays images of the equipotential surface of a numerically simulated radio image (tomogram) of a test object of stepped shape.

An advantage of the backward ray tracing method as compared with the polygonal technique is a more accurate display of the equipotential surface and simplicity of the corresponding imaging algorithms.

15.3 Visualization of 3-D Tomograms by Orthogonal Slices

The basic form for presenting 3-D images in geolocation and radio tomography is a 3-D cube. This cube is constructed using data from the sounded volume and is composed of six planar surfaces, each of which is orthogonal to the next one. In this way, we can fully visualize the sounding data of interest to us (Figure 15.3). The cross-sections can be made at any angle and direction.

To make the 3-D image more readily understandable, we construct the cross-sections of the cube in three orthogonal planes: x, y, and z, which correspond to the use of a Cartesian coordinate system. The number of cross-sections of the cube and the spacing between them can be selected arbitrarily. The result of the cross-sections of the cube for each of the planes is recorded in a separate video file.

Figure 15.4 shows cross-sections in the x-, y-, and z-planes. These cross-sections were obtained from processing of actual experimental data. A complete 3-D picture of the test volume can be assembled by scanning the recorded video files.

The given test scene contains several localized objects (inhomogeneities), whose positions and dimensions can be estimated. The circled object most likely represents a metal pipe. The coordinates of the starting point of this object are 1.7, 0.8, −1.6 m and the object ends at the point with coordinates 3.9, 2.8, −2.0 m.

The advantage of this latter method is its ease of realization, good visual presentation, and the ability to easily scan the recorded video files by means of various MATLAB applications.

338 ■ *Electromagnetic and Acoustic Wave Tomography*

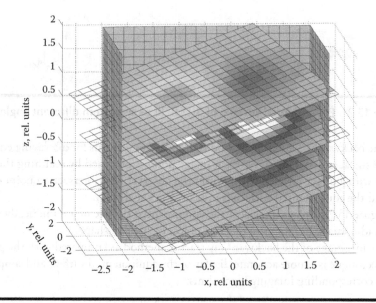

Figure 15.3   3-D cube and its cross-section sections.

The proposed techniques for visualization of 3-D tomograms entail the presentation of results as a polyscreen, equipotential surfaces, and orthogonal cross-sections in the form of bitmap images.

However, the results of 3-D tomography in the form of 2-D bitmap images can be displayed in a visually more accessible form as a video record or an animation built up from successive sections, layers, or records of camera motions around the object. Cartoon animation technology is required in this case. Applications for 3-D game developers, such as Blender 2.59, can be considered as possible options as well.

A lot of examples regarding radio tomography were discussed in previous chapters. In this chapter, we will consider typical examples of inverse problems solutions occurring in radio physics related to other applications. For these problems, the corresponding models of direct problems solutions are usually used [1–4]. In this case, the selected model of the direct problem will be simpler; the final solution of the inverse problem will be very stable.

15.4 Inverse Problems in Object Reconstruction

15.4.1 Elimination of Image Spreading

A classical inverse problem takes place when an image is reconstructed with spreading, for example, due to incorrect initial settings of the photographic apparatus

Visualization and Reconstruction of Objects ■ 339

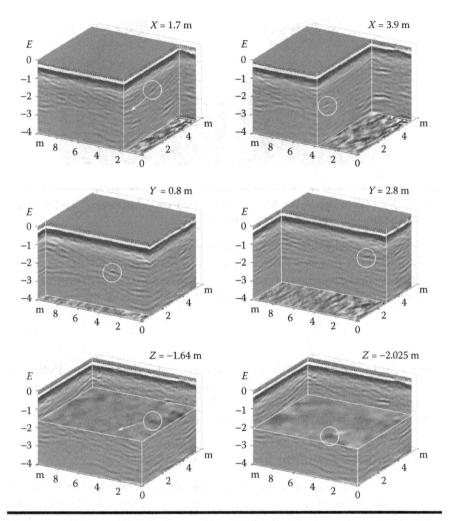

Figure 15.4 Orthogonal cross-sections of the cube in the *x*-, *y*-, *z*-plane at various distances.

(see Figure 15.5). The simplest model of such a spreading of image is the operation of 2-D integral convolution:

$$\mathbf{y}(\mathbf{\rho}) = \iint \mathbf{A}(\mathbf{\rho}')\mathbf{x}(\mathbf{\rho}-\mathbf{\rho}')\mathrm{d}^2\,\mathbf{\rho}'. \quad (15.1)$$

Here, **x** is an initial image, and **y** is a spreading image. The matrix **A**, introduced in Chapter 1, describes the so-called point spread function (PSF), which is supposed to be known. In Figure 15.5, this function overlaps six points. Such a problem can be met in optical image processing, for example, with the occurrence of so-called blurring (as shown in Figure 15.5a).

340 ■ Electromagnetic and Acoustic Wave Tomography

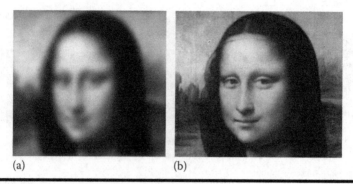

Figure 15.5 (a) Image of the Mona Lisa portrait with "spreading" and (b) after reconstruction with rejection of "blurring".

An inverse problem deals with reconstruction of the initial image **x**. As mentioned in Chapter 1, such a problem is solved by use of a 2-D Fourier transform; after that the initial equation in the spatial spectral domain can be converted to the system of linear equations (SLE) in the form of [1]:

$$\hat{\mathbf{y}} = \hat{\mathbf{A}}\hat{\mathbf{x}} \qquad (15.2)$$

where $\hat{\mathbf{y}}$, $\hat{\mathbf{x}}$, and $\hat{\mathbf{A}}$ are the spatial spectra of the corresponding matrices. Definition of the matrix $\hat{\mathbf{x}}$ is a solution of the inverse problem, which should be solved with regularization, as is done in Chapter 1. After that, by use of the inverse Fourier transform, the initial image **x** can be finally reconstructed (see Figure 15.5b).

15.4.2 Inverse Problem for Signal Analysis

In radio-physical applications, spectral analysis plays the most important role. A standard task of spectral analysis usually is solved by the use of the Fourier transform, for which the corresponding algorithms, called fast Fourier transform (*FFT*), are performed. But, often, the equidistance of frequencies or time sampling cannot be achieved. Moreover, in this case the condition of orthogonality of the basic functions is not valid, and it cannot be used. The spectral function evaluation can be decided, as the inverse problem that can be considered as the solution of SLE [1, 2].

A lot of examples regarding radio tomography were discussed in previous chapters. Based on knowledge obtained there, let us consider a time-dependent signal $S(t)$, and we will relate it to its spectral function $S(\omega_j)$, given at the discrete map of frequencies ω_j, via the well-known expression:

$$S(t) = \int_{-\infty}^{\infty} S(\omega) \exp\{i\omega t\} d\omega \approx \sum_j S(\omega_j) \exp\{i\omega_j t\} \Delta\omega_j \qquad (15.3)$$

Spectral values of $S(\omega_j)$ determine the non-known parameters for the inverse problem.

If discretization of time-samples of the initial signal $S_m = S(t_m)$ can be performed, the task of spectral evaluation and estimation can be deduced as the SLE solution:

$$\mathbf{y} = \mathbf{A}\mathbf{x} \tag{15.4}$$

where:
$\mathbf{y} = \{S_m\}$
$\mathbf{x} = \{S(\omega_j)\}$
$\mathbf{A} = \{\exp\{i\omega_j t_m\}\}$

The definition of vector $\mathbf{x}$ is definitely a solution of the inverse problem of spectral estimation.

Let us note that a correction of the inverse problem is full if it can be realized based on discrete Fourier transform, that is, by using the orthogonal expansion. The general case is not so trivial, and it is necessary to introduce regulated algorithms. Such a type of solution is usually used for problems of extra-resolution [5] and localization of the sources of electromagnetic radiation with usage of limited receiving antenna arrays, as shown in Figure 15.6. It is also related to the computation of antenna directivity diagrams. Let us consider this aspect in more detail.

If $J(x)$ is the current distribution along some antenna aperture, then the field distribution along the angle θ is defined by the following integral:

$$E(\theta) = \frac{1}{L} \int_{-\infty}^{\infty} J(x) \exp(ikx \cos\theta) dx \tag{15.5}$$

where L is the linear dimension of the antenna aperture. The written integral in Equation 15.5 defines the directivity diagram (DD) of the antenna and is really a solution of the direct problem.

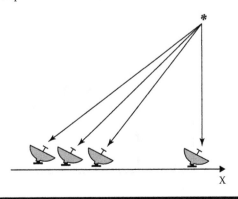

Figure 15.6 Antenna array for localization of source position.

The inverse problem relates to estimation of the current $J(x)$, which provides us with the given DD. If, in the integral (Equation 15.5), we pass the discrete values, the problem under consideration yields the SLE solution of the same form as previously, that is,

$$\mathbf{y} = \mathbf{A}\mathbf{x}$$

Here, $\mathbf{y} = \{E(\theta_m)\}$ is the vector of the measured values of the directivity diagram; $\mathbf{x} = \{J(x_n)\}$ is the vector of non-known values of current along the antenna aperture. Matrix $\mathbf{A}$ equals $\mathbf{A} \equiv \{\exp(ikx_n \cos\theta_m)\}$ and k is the wavenumber. The method of the solution of the obtained SLE can be selected from those described in Chapter 1.

The inverse problem formulated here can be completed if we set a problem of recognition of the full DD according to its recovering fragment measured over some segment of the angle θ; this is enough to put the obtained values of the current inside the integral from Equation 15.5 and to calculate it over the full range of observation angles.

15.5 Inverse Problem of Source Localization

If the source is placed at point 0 of a plane (x, y), as shown in Figure 15.7, then for the definition of its position it is enough to measure difference in time delays of signals, received at points 1–4 arbitrarily distributed on the plane (see Figure 15.7, points 1–4). Such a type of problem occurs in the definition of the position of the source of radiation located above the terrain [3, 4]. A similar method is used in non-destructive testing of electronic plates, based on acoustic emission occurring around the points of bad contact during mechanical deformation of these plates.

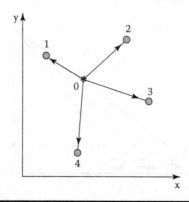

Figure 15.7 Geometry of problem of positioning of radiation source.

Let us consider the existence of the four receiving points, as shown in Figure 15.7. This is the minimum number of points that is sufficient for the definition of the real position of the source. If the speed of the waves—let us say, acoustic waves—is **v**, then whether this number of receivers is really enough should be checked. If **r** is the position of the source, and $\mathbf{r}_1$, $\mathbf{r}_2$, $\mathbf{r}_3$, $\mathbf{r}_4$ are the positions of the receiving points, then the delay of radiation $t_{i,j}$ arriving at the receiving points can be determined as:

$$t_{1,2} = \frac{|\mathbf{r}-\mathbf{r}_1|}{c} - \frac{|\mathbf{r}-\mathbf{r}_2|}{c}, t_{1,3} = \frac{|\mathbf{r}-\mathbf{r}_1|}{c} - \frac{|\mathbf{r}-\mathbf{r}_3|}{c}, t_{1,4} = \frac{|\mathbf{r}-\mathbf{r}_1|}{c} - \frac{|\mathbf{r}-\mathbf{r}_4|}{c} \quad (15.6)$$

Written in such a manner, a system of three equations allows us to precisely estimate the position of the source and additionally correct the waves' speed. This system (Equation 15.6) is purely non-linear and can be solved by one of the methods described in Chapter 1.

15.6 Inverse Problem of Micro-Strip Sensors Reconstruction

For production of multilayered micro-strip detectors, there exists a problem of mutual interactions between layered structures of charge collection [3, 4]. Such a problem occurs, for example, during the carrying out of the Atlas project at the Large Hadron Collider (LHC) at the European Organization for Nuclear Research (CERN) in Geneva, Switzerland, the outer view of which is shown in Figure 15.8a.

During the performance of such a test, it was supposed that a detector on top (see Figure 15.8b) has ideal parallel structure. The detector placed underneath is also ideal, but due to the non-ideal technology of the mechanical equipment, it can be shifted by distance h along the axis Oz and turn by the angle α around the axis Oy with respect to the detector on top. It is necessary to construct the algorithm of estimation of these parameters according to numerous measurements of transverse responses (along axis Oz) of the quasi-point vertical impact. Knowledge of these parameters allows a sufficient increase of the accuracy of tomographic measurements of trajectories of all possible particles registered by the LHC.

If the coordinates of quasi-point interactions on the top layer of the micro-strips (X_0, Z_0) are known, the local coordinates of these interactions (X, Z) located below the current layer can be found as:

$$X = X_0 \cos\alpha - (Z_0 + h)\sin\alpha, Z = (Z_0 + h)\cos\alpha + X_0 \sin\alpha \quad (15.7)$$

These coordinates' transform is a superposition of the linear shift and the rotation. With numerous measurements of only one transverse coordinate

344 ■ *Electromagnetic and Acoustic Wave Tomography*

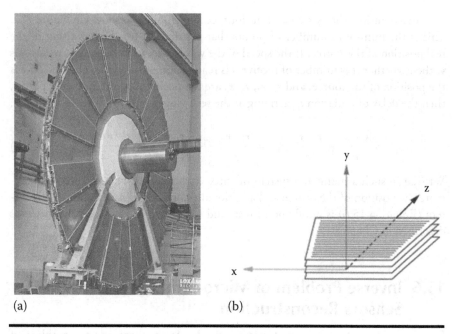

Figure 15.8 Problem of micro-strips in the Atlas Project, Geneva, (a) outer view; (b) scheme of arrangement of micro-strip detectors.

$$Z_j = (Z_{0j} + h)\cos\alpha + X_{0j}\sin\alpha \qquad (15.8)$$

for different interactions (X_{0j}, Z_{0j}), the parameters (h, α) can be found. For the following layers, the problem is solved in the same manner.

The solution of such a formulated inverse problem can be obtained with the help of the least squares method mentioned in Chapter 1. The accuracy of the solution of the inverse problem increases with the number of independent measurements. In Figure 15.9, an example of distribution of responses for 50 independent interactions is shown.

In Figure 15.9, the position of point interactions is shown (top layer) by points, and responses to these interactions (bottom layer) by squares. In this case, the accuracy achieved for the solution of the inverse problem according to estimation of the parameters defined by (h, α) is not less than 10^{-7} cm and $10^{-7°}$, respectively. At the same time, using only two independent interactions, the accuracy decreases by approximately four orders.

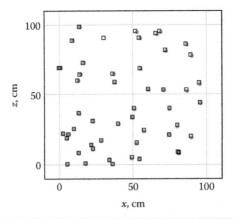

Figure 15.9 Solution of the inverse problem of particle collection.

References

1. Yakubov, V. P., S. E. Shipilov, D. Ya. Sukhanov, and A. V. Klokov, *Radiowave Tomography: Achievements and Perspectives*, Tomsk, Russia: Scientific-Technical Literature (STL) Edition Group, 2016 (in Russian).
2. Yakubov, V. P., S. E. Shipilov, D. Ya. Sukhanov, and A. V. Klokov, *Wave Tomography*, Tomsk, Russia: Scientific Technology Publishing House, 2017.
3. http://cdsweb.cern.ch/record/1552862/
4. Yakovlev, O. I., V. P. Yakubov, V. P. Uryadov, and A. G. Pavel'ev, *Radiowave Propagation*, Moscow: Lenand, 209 (in Russian).
5. Ivanov, V. K., V. V. Vasin, and V. P. Tanana, *Theory of Linear Incorrect Problems and Its Applications*, Moscow: Science, 1978 (in Russian).

Symbols and Abbreviations

2-D	two-dimensional
3-D	three-dimensional
$\mathbf{A} = [a_{ij}]$	matrix of dimension $N \times M$, where $M \geq N$
$\mathbf{A} = \{\exp\{i\omega_j t_m\}\}$	series of matrices of time-dependent signals in time domain
$\hat{\mathbf{A}}$	spatial spectrum of matrix $\mathbf{A}$
$\mathbf{A}^{-1}$	inverse matrix or operator representation
$\mathbf{A}^+$	Moore–Penrose matrix
$\mathbf{A} = \dfrac{\partial \mathbf{F}(\mathbf{x_0})}{\partial \mathbf{x_0}} = \left[\dfrac{\partial F_i(\mathbf{x_0})}{\partial \mathbf{x}_{0j}}\right]$	Jacobian matrix; $i = 1, ..., M$; $j = 1, ..., N$ – numbers of corresponding vectors' elements
$\mathbf{A}^H$	conjugate operator (matrix)
$\mathbf{B}$	vector of induction of magnetic field component of electromagnetic wave
CERN	European Organization for Nuclear Research (Switzerland)
CF	characteristic function
CR	corner reflector
$c = \dfrac{1}{\sqrt{\varepsilon_0 \mu_0}}$	velocity of light in free space
$curl \equiv rot = \nabla \times$	Rotor of arbitrary vector field or vector product of operator nabla
$\mathbf{D}$	vector of induction of electric field component of electromagnetic wave
DD	directivity diagram
DFT	direct Fourier transform
DP	directive pattern of antenna
dl	differential of vector of line l

d**S**	differential of vector of surface S
d**V**	differential of vector of volume V
$\sin\theta d\theta d\varphi = d\Omega$	element of spatial angle measured in steradians
$div = \nabla$	scalar product ("nabla dot") of field or flow of arbitrary vector
E	vector of electric field component of electromagnetic wave
e	emissivity
ESP	extremely-short-pulse (signals)
E(z,t)	2-D distribution of vector of electrical component of electromagnetic wave
FD	Fresnel diffraction
FEM	finite element method
FFF	fast Fourier transform
FS	Fourier synthesis method
$F(\mathbf{r}_s)$	characteristic function
F(**x**)	any vector function of vector **x**
$F(y,\varphi)$	intensity of radiation absorption
$f = ck/2\pi$	radiated frequency
$F(\mathbf{K},\varphi)$	spectral function of shadow projection
GLE	Gorky–Luxemburg effect
GPR	ground-penetrating radar
$g(X,Y)$	energy density of object
$G(u,v)$	spatial spectrum of body under illumination
$G_0(\mathbf{r}) = \exp\{ik|\mathbf{r}|\}/(4\pi|\mathbf{r}|)$	Green function for outgoing to infinite wave
$G_0^*(\mathbf{r}) = \exp\{-ik|\mathbf{r}|\}/(4\pi|\mathbf{r}|)$	Green function for incoming from infinite wave
$G(\mathbf{R}|\mathbf{R}_0)$	conditional Green function
$g(\mathbf{R},\tau|\mathbf{R}_0,\tau_0)$	normalized conditional Green function
$G_{ij}(\mathbf{r}|\mathbf{r}')$	dyadic conditional Green function
$\text{grad}\Phi = \nabla\Phi$	Gradient of arbitrary scalar field or effect of nabla operator on scalar field
H	vector of magnetic field component of electromagnetic wave
H(z,t)	2-D distribution of vector of magnetic component of electromagnetic wave
$H(\omega)$	transfer function of filter
$H(\mathbf{K}) = \exp\{ik_z Z\}$	transfer function
$H(\tau)$	impulse response function of linear filter
$H_0^{(1)}$	Hankel function of first kind of zero order
I	identity matrix
IFT	inverse Fourier transform
IFR	inverse focusing of radiation

IR	infrared (radiation)
I_0	constant current in circuits, transmission lines and cables
$I(z,t)$	wave of current in circuits, transmission lines and cables
$i = \sqrt{-1}$	ort of imaginary part of complex number
$\mathbf{i}_x, \mathbf{i}_y, \mathbf{i}_z$	unit vectors (orts) in Cartesian coordinate system
$\mathbf{i}_\rho, \mathbf{i}_\phi, \mathbf{i}_z$	unit vectors (orts) in cylindrical coordinate system
$\mathbf{i}_r, \mathbf{i}_\phi, \mathbf{i}_\theta$	unit vectors (orts) in spherical coordinate system
IBC	impedance boundary conditions
$\mathbf{j}$	vector of electric current density
$\mathbf{J}$	vector of full current in medium/circuit
$J(\mathbf{r}_1)$	spatial distribution of outer currents
$\mathbf{j}_c$	conductivity current density
$\mathbf{j}_d$	displacement current density
$k_1 \equiv k = 2\pi f/c$	wave number of background environment
$K(x', y')$	normalized polarization current
$k_{1z} = \sqrt{k^2 n^2 - \kappa_\perp^2} = kn = k_1$	longitudinal component of total wave vector
LFM	linear-frequency-modulated wave
LHC	Large Hadron Collider
$\mathbf{M}$	momentum of magnetic ambient source
MathCad	special computer program
MMW	millimeter-wave (radar)
$M(\mathbf{r}_\perp, f) = \exp\left\{-ik_0 \left(2\sqrt{\mathbf{r}_\perp^2 + h^2}\right)\right\}$	focusing function
$M(\mathbf{r}_{\perp F}, h_1, f) = \exp\left\{-i2kn\sqrt{\mathbf{r}_{\perp F}^2 + h_1^2}\right\}$	weighting function of focusing procedure
NDT	non-destructive testing
NFI	null field integral
NLJD	non-linear junction detector
NMR	nuclear magnetic resonance
NoteBook	personal computer
NRD	non-linear radio detection
$n(\mathbf{r}) = 1 + n_\delta(\mathbf{r}) \equiv 1 + f$	refractive index; $f = 0$ outside of refracting object
$n_\delta(\mathbf{r})$	deviations of refractive index
$\mathbf{P}$	vector of polarization
PC	personal computer

Symbols and Abbreviations

$P = \|v\|^{-1}$	space spectrum component
PSF	point spread function
$Q_\perp(\mathbf{r}_\perp)$	transverse transfer function
$Q_1(\mathbf{r}_{\perp 1}, h_1, z_1)$	transfer function of focusing system
RFEC	remote-field eddy current technique
RMM	radar-based motion detector module
RP	radiation pattern
RWTS	radio-wave tomographic synthesis
$\mathbf{r} = \{x,y,z\}$	spatial vector in Cartesian coordinate system
$\mathbf{r}_F$	point of focusing
$\mathbf{r}_m$	point of detection
$\|\mathbf{r}_\perp\| = \sqrt{r^2 - z^2} \equiv \sqrt{x^2 + y^2}$	transverse projection of vector $\mathbf{r}$ in Cartesian coordinate system; z – its longitudinal projection
$r_\perp = \|\mathbf{r}_{\perp F} - \mathbf{r}_{\perp 1}\|$	distance between focusing and observation point
$R_\alpha(\omega) = \dfrac{H(\omega) H^*(\omega)}{H(\omega) H^*(\omega) + \alpha}$	Tikhonov's regularization product
S	area of arbitrary surface
SAR	synthesized aperture radar
SCSC	self-compensated sensing coil
SF	system function
SLE	system of linear equations
SP	shadow projection
SS	spatial spectrum
$S(t)$	time-dependent signal
$S(\omega_j)$	spectral function of time-dependent signal
$S_m = S(t_m)$	series of time-dependent signals
$\bar{S}(l)$	effective cross-section area (in square meters)
$S_{12} \equiv \bar{S}(L)$	effective cross-section of current tube
$\mathbf{S} = [s_i \delta_{i,j}]$	quasi-diagonal matrix, consisting of singular values s_i
$s_i = \sqrt{\lambda_i};\ \lambda_i$	eigenvalues of matrix $\mathbf{A}$
T	transmission coefficient
TF	transfer function
$u_{zi} = \sqrt{(2kn_j)^2 - \mathbf{u}_\perp^2}$	longitudinal projection of wave number in j-layer
$\mathbf{u}_\perp = (u_x, u_y)$	normal projection of radiated field in transverse direction
$U_I(\mathbf{r}) = \exp[ik(\boldsymbol{\theta} \cdot \mathbf{r})]$	incident plane harmonic ($\sim \exp(-i\omega t)$) wave; $\boldsymbol{\theta}$ – unit vector pointing direction of wave propagation
$U_s(\mathbf{r})$	scattered wave
UWB	ultra-wide band wave
UWB/ESP	ultra-wide band/extremely-short-pulse (radar)

Symbols and Abbreviations ■ 351

V	velocity of object relative to sensor
V	volume of arbitrary surface
VIE	volume integral equation
VSWR	voltage standing wave ratio
V_0	constant voltage or potential difference
$V(z,t)$	wave of voltage in medium
v_{ph}	wave phase velocity
v_{gr}	wave group velocity
W	energy of arbitrary field
WDT	weighted differential technique
$W(\mathbf{r}_F, \mathbf{r}_1)$	arbitrary focusing function
$W_0 = \langle |U(k)|^2 \rangle = A^2 \Gamma^2$	average scattered field intensity; A – amplitude of wave; Γ – coefficient of reflection from surface of illuminated structure
$W(f) = S_y(f)/S_x(f)$	transfer function of a planar layer; $S_x(f)$ – incident signal; $S_y(f)$ – transmitted signal
$\mathbf{x}$	initial image; $\mathbf{x} = \{S(\omega_j)\}$
$\hat{\mathbf{x}}$	spatial spectrum of initial image
$\boldsymbol{x} = [x_1, x_2, \ldots, x_N]^T$	column vector of dimension N of input values
$\|\mathbf{x}\| = \sqrt{\mathbf{x}^H \cdot \mathbf{x}}$	length of norm $\|\mathbf{x}\|$
X-ray	radioactive tomography
$\mathbf{y}$	spreading image; $\mathbf{y} = \{S_m\}$
$\hat{\mathbf{y}}$	spatial spectrum of spreading image
$\boldsymbol{y} = [y_1, y_2, \ldots, y_N]^T$	column vector of dimension M of output values
$z_s(\boldsymbol{\kappa}_\perp) = \phi(\boldsymbol{\kappa}_\perp)/k$	reconstructed surface' profile
$\{x,y,z\}$	Cartesian coordinate system
$\{\rho,\phi,z\}$	cylindrical coordinate system
$\{r,\phi,\theta\}$	spherical coordinate system
α	parameter of wave attenuation in arbitrary medium
α	parameter of regularization
β	parameter of phase velocity deviation in arbitrary medium
Γ	coefficient of reflection from boundary of two media
$\Gamma(\mathbf{r}_\perp)$	transverse distribution of current over aperture of antenna system; $\mathbf{r}_\perp = (x,y)$
$\gamma = \alpha + i\beta$	parameter of propagation in arbitrary material medium
$\Delta = \nabla^2$	Laplacian of vector or scalar field
∇	operator nabla of arbitrary scalar field
$\Delta\varepsilon_r$	distribution of sounded inhomogeneity's permittivity
$\Delta\varepsilon_r(\mathbf{r}_1)$	relative permittivity values deviation
$\Delta\varepsilon(\mathbf{r}_{\perp 1}, z_1)$	spatial distribution of inhomogeneities

Symbols and Abbreviations

$\delta = 1/\alpha$	skin layer in arbitrary material medium
$\delta_{i,j}$	Kronecker symbols
$\delta(|\mathbf{r}_1 - \mathbf{r}_2|)$	spatial Dirac δ-function
$\delta(\mathbf{r})$	point spread function or sinc-function
$\varepsilon = \varepsilon' + i\varepsilon''$	complex permittivity of arbitrary medium
$\varepsilon_r = \varepsilon_r' + i\varepsilon_r''' = \varepsilon_{Re}' + i\varepsilon_{Im}'''$	relative permittivity of arbitrary medium; $\varepsilon_{Re}' = \varepsilon'/\varepsilon_0$ – its real part, $\varepsilon_{Im}''' = \varepsilon''/\varepsilon_0$ – its imaginary part
$\varepsilon_0 = \dfrac{1}{36\pi} 10^{-9}$ (F/m)	dielectric constant of free space
$\varepsilon(\mathbf{R}) = 1 + \tilde{\varepsilon}(\mathbf{R})$	random permittivity distribution
η	wave impedance in arbitrary medium
$\eta_0 = 120\pi\Omega = 377\ \Omega$	wave impedance of free space
$(\kappa_x, \kappa_y, \kappa_z)$	wave vector coordinate system
$\|\kappa_\perp\| = \sqrt{k^2 - k_z^2} \equiv \sqrt{k_x^2 + k_y^2}$	transverse projection of total wave vector; k_z – its longitudinal projection
λ	wave length in arbitrary medium
λg	wave length in arbitrary waveguide structure
$\mu = \mu' + \mu''$	complex permeability of arbitrary medium
$\mu_r = \mu_r' + i\mu_r'' = \mu_{Re}' + i\mu_{Im}'''$	relative permittivity of arbitrary medium; $\mu_{Re}' = \mu'/\mu_0$ – its real part, $\mu_{Im}''' = \mu''/\mu_0$ – its imaginary part
$\mu_0 = 4\pi \cdot 10^{-9} (H/m)$	magnetic constant of free space
ρ	charge density in medium
ρ	normalized resistivity
σ	standard deviation
$\sigma = 1/\rho$	normalized conductivity
$\phi(\kappa_\perp)$	phase distribution of recovered image
$\chi(s_i - \tau)$	Heaviside step function: $\chi(s_i - \tau) = 1$, if $s_i \geq \tau$; $\chi(s_i - \tau) = 0$, if $s_i < \tau$
χ_e	electrical sensitivity
χ_m	magnetic sensitivity
ψ	flux of electric or magnetic field
$\psi(x, y_1)$	scattered scalar field
$\breve{\psi}(\nu, y_1)$	Fourier image of $\psi(x, y_1)$

Index

3-D cube, 337–338
3-D tomograms visualization, 213, 216
   by equipotential surfaces, 336–337
   by orthogonal slices, 337–339
   by polyscreen technique, 335–336

Acoustic tomography, defects visualization using, 303–309
Active zone, 150
Air–sand interface reflection, 208
Algebraic approach, 4–5
Aluminum strips, 316–317
Amplitude-phase distribution, 149–150
Anodization function, 65
Antenna arrays, 10
   for localization, 341
   UWB, 128–130, 139
   UWB/ESP, 145
Antennas, 122–129
   combined reflector, 158–163
   dipole, 122
   electric, 122
   equivalent aperture, 66, 275
   parabolic, 160
   RP, 127
   uses, 125
   UWB, 126
   VSWR, 124, 129
Atlas project, 343–344
Attenuation of radiation, 239

Babinet's principle, 37
Backward ray tracing method, 337
BEM (boundary element method), 90–91
Bipolar generator, 247
Black-body law of radiation, 258–259
Born approximation, 48, 63, 70–71, 106
Boundary conditions, 90
Boundary element method (BEM), 90–91
Boundary integral formulation, 90
Brass plate, 317, 319–320
Brick wall structure, 132–133
Brightness temperature, 259

Cartesian coordinate systems, 20, 36
Characteristic function (CF), 21–22, 37, 157
Clutter noise, 15
Convolution integral equation, solution, 14–15
Coordinate systems, 38
   cartesian, 20, 36
   polar, 20–21
Corner reflector (CR), 267, 270
   on azimuthal orientation, 273
   scattering diagram, 274
Cracks detection and imaging, 324, 326, 331
Crossing focusing procedure, 27
Cross-modulation, 180–181
Curl vector, 96

Data acquisition/processing program, 241
Data processing, 195–196
   algorithm, 39
Deconvolution, 24, 28, 268
Degree of localization, 59
Density and permittivity of liquids, 302
Density and refractive index solids, 301
DFT (direct Fourier transforms), 19
Diffraction hyperbolas, 49–50, 52, 215, 238
   summation by, 55
   trace, 239
Diffraction imaging, 70
Diffraction tomography
   approach, 102–115
   elements, 226–227
   on Feynman's path integral, 73–75
   mine detection/imaging using, 218–221
   post-processing on, 262
   on Rytov approximation, 72–73

353

techniques, 70–75
theoretical consideration, 102–115
Direct Fourier transforms (DFT), 19
Directional antenna, 168
Directivity pattern (DP), 66–67, 162, 266–267, 275–276
   anisotropy, 179
   and aperture of antennas, 66, 275
   reflector antenna, 162
Direct problem, 3, 23
   and inverse problem, 4–5
   solution, 191, 229
   stability, 6
Dispersion prism, 173
Doppler motion sensor, 196
   CON-RSM1700, 185–186
Doppler scanning device, 191–195
Doppler sub-surface tomography, 195–197
Doppler tomography experiment, 185
   disadvantage, 196
   Doppler sub-surface tomography, 195–197
   microwave Doppler sensor/location sounding, 185–188
   objects, reconstruction, 189–191
   positioning system/Doppler scanning device, 191–195
   wooden slat, 196
Double focusing method, 23–28, 147
Double reflector focusing system, 247
DP, *see* Directivity pattern (DP)
Dry-and-wet weight method, 236
Dry density, 236–237

EC, *see* Eddy currents (EC)
ECT, *see* Eddy current testing (ECT)
EC tomography, 323, 330–331
   experimental setup, 323–325
   holder and Nortec plate, 326
   slicing tomogram, 328
   sub-surface defects, 324, 326–331
   tomograms of slots, 327
   two-way, 329
Eddy currents (EC), 80–81, 89–90
   density distribution, 101
   distortion, 94
   distribution, 83–84
   effect, 89
   flaw detectors, 313–314
   flaw reconstruction, 97
   interaction, 101
   and magnetic fields, 313–314
   NDE, 100

   overview, 89–94
   time-dependent electromagnetic field in, 92–93
   time-harmonic solution, 90
   tomography, 89–116; *see also* EC tomography
Eddy current testing (ECT), 100
   crack and initial shape, 100
   inverse problem using, 94–102
Electric antennas, 122
Electric charge, 84
Electric field, 96
Electromagnetic radiation, 293
Electromagnetic waves, 294, 299
Electrostatic tomography, impedance methods, 84–86
Emissivity, 258–259
Equalization function, 86
Equipotential surfaces, 336–337
ESP/UWB radar system, 283, 285, 287
Euclid's definition, 5
Extremely-short-pulse (ESP), 129, 257, 283

Fast Fourier transform (FFT), 60, 174, 340
Feynman's path integrals, 73–75
Fiberglass plate, 304, 306–307
Finite element method (FEM), 90–92
Finite element solution, 91
Foam concrete wall, 132, 136–137
Focusing effect, 28, 54, 57
Focusing function, 25, 31, 36, 38, 54
Focusing method in rural environments, 273–278
Focusing system, 149–150
Forest areas, 266
Fourier interpolation, 212
Fourier synthesis (FS) method, 19
   advantage, 22
   problem, 20
   on SP, 19–22
Frequency clocking technique, 173–174
Frequency domain, 6, 58, 129
   time-tact of objects, 172–177
Fresnel zone of diffraction, 28, 68, 134
FS method, *see* Fourier synthesis (FS) method
Full-angle image, 33, 35, 151

Geolocation
   rough/uneven terrain surfaces, 226–232
   sub-surface man-made objects, 247–257
Geological survey, 236–240
Geometrical optics approximation, 35–36

Index ■ 355

Geomorphological slope/land view, 233–234
Geo-radar, 204, 236, 238, 281
Gorky–Luxembourg effect (GLE), 181
GPR, *see* Ground-penetrating radar (GPR)
Green's function, 65
    dyuadic, 95
Ground-penetrating radar (GPR), 204, 226
    data, 209
    detecting signal/spectrum shape, 208
    and measurement scheme, 207
    OKO-2, 227, 233
    random/incomplete, 206–211
    reconstructed system function, 216
    for subsoil mine detection, 219
Group focusing technique, 63–65

Hadamard's definition, 5–6
Helmholtz's equation, 23, 73, 75, 91
Hermitian transpose, 5
Huygens–Fresnel principle, 31
    interpolation, 231–232
Hybrid method, 90
Hyperbola method, summation over, 210, 212

IFR (inverse focusing of radiation), 24
IFT (inverse Fourier transform), 6, 19, 43, 102
Image spreading, elimination, 338–340
Impedance boundary conditions (IBC), 90
Impedance tomography, 84
Incoherent tomography, 42–45, 168
Induction currents, 313–314
Inhomogeneities, 190
    cases, 229
    detection, 56
    point-like, 213–214, 230, 232
    tomogram, 277
In-phase summation, 56
Integrated tomography
    plasterboard, 304–305
    plastic gun, image, 303
Interpolation method, 212
Inverse focusing method, 37–38, 226, 229–230
Inverse focusing of radiation (IFR), 24
Inverse Fourier transform (IFT), 6, 19, 43, 102
Inverse problems, 3–15, 52, 81
    direct and, 4–5
    iteration algorithms for, 12–113
    micro-strip sensors reconstruction, 343–345
    in object reconstruction, 338–342
    particle collection, 345
    for signal analysis, 340–342
    source localization, 342–343

Inverse projections method, 18–19
Inverse scattering, 70
Iterative algorithm, 169

Kirchhoff approximation, 33–36, 70
Kirchhoff method, 147–148
Kronecker symbols, 9

Lagrange's product, 7
Least squares method, 8
Levenberg–Marquardt algorithm, 11–12
LGE (Luxembourg–Gorky effect), 180
Linear frequency modulation (LFM), 146
    radiation tomography technique, 39–42
    radio waves, 39
    tomography on, 146–147
    transceiver and object, 146
Living people in clutter conditions
    behind walls, detection/identification, 287–289
    in sub-soil media, 281–287
Logarithmic amplifier, 270
Low-frequency magnetic tomography, 79–84
    EC and magnetic fields, 313–314
    metallic objects, magnetic tomography, 318–322
    reconstruction of metallic objects, 314–318
Luxembourg–Gorky effect (LGE), 180

Magnetic dipole, 115
Magnetic fields, 79–82, 84
Magnetic tomography of metallic objects, 318–322
Man-made structures, 247–257
Manual microwave scanner, 194
Matched filtering method for radiation sources distribution, 50–51
Matrix
    diagonal, 9
    diagonal unit, 8
    inverse, 8
    Jacobian, 12
    Moore–Penrose, 9–10
    pseudo-inverse, 8–9, 11–12
    quasi-diagonal, 9
    square, 8
Maxwell's equation, 93, 95
Mean square error, 98
Media
    inhomogeneous, 65, 293–294, 300
    non-regular, 65
Metal detectors, 314

Index

Metallic objects, 175–177
    by low-frequency magnetic tomography, 314–318
    magnetic tomography, 318–322
Metallized cylinders experiment, 135
Methods
    diffraction tomography, 72–73
    double focusing, 23–28
    FS, 19–22
    inverse projections, 18–19
    least squares, 8
    matched filtering, 50–51
    migration, 51–54
    refocusing, 65–68
    single focusing, 56–60
    synthesized focusing, 54–55
    two-step focusing, 60–63
Micro-strip sensors reconstruction, 343–345
Microwave scanner, 194
Microwave sensors, linear array, 192–193
Migration method in spatiotemporal region, 51–54
Millimeter-wave (MMW) radar, 257
    active LFM radiation, 146–147
    passive, 258
    radiometers, application, 260–262
    sub-surface tomography applications, 257–260
Mine detection
    and imaging using diffraction tomography, 218–219, 221
    physical/numerical aspects, 210, 212–214
    in subsoil media, 220
Minerals detection/identification, 244–247
MMW radar, *see* Millimeter-wave (MMW) radar
Monostatic point reflector spreading function (MPSF), 102
Monostatic radar, 48
Monostatic radio-location scheme, 48
Moore–Penrose matrix, 9–12
MPSF (monostatic point reflector spreading function), 102
Multiple-angle tomography of opaque objects, 156–158

Nanosecond radar, 269–270
NDT (non-destructive testing) method, 329
Near field zone, 68, 122, 275
New mobile scanning systems, 278
NFI (null field integral) equations, 93
NLJD (non-linear junction detector), 178–180

Non-contacting sounding of mines, 205–218
Non-destructive testing (NDT) method, 329
Non-filled aperture, spatial testing with, 169–172
Non-linear inclusions, UWB tomography, 177–182
    signal processing and algorithm, 182–185
Non-linear junction detector (NLJD), 178–180
Non-linear markers, 180
Non-linear radiolocation (detection) (NRD), 177
    to small objects, 180
    solution to, 177–180
Non-transparent objects; *see also* Opaque objects
    radio tomography, 152–163
    single-side monostatic tomography, 69–70
NRD, *see* Non-linear radiolocation (detection) (NRD)
Null field integral (NFI) equations, 93
Nyquist–Shannon–Kotelnikov theorem, 169

Object reconstruction, inverse problems in
    image spreading, elimination, 338–340
    for signal analysis, 340–342
Opaque objects; *see also* Non-transparent objects
    computed imaging, 38
    multiple-angle tomography, 152–163
    transmission tomography, 36–39
    triangular cylinder and diffraction field, 39
    wave scattering on, 37
Orthogonal cross-sections, 217, 339
Orthogonal slices, 337–338

Parabolic reflector, 70
    with deformation/aperture of antenna, 72
    quadrature components, 71
Parameter of regularization, 7–8
Parseval's theorem, 14
Petroleum reservoir
    permittivity values, 252–253
    shape of reflected pulse, 255
    structure in, 250
Phase screen approximation, 230–231
Planck's law, 258–259
Plaster object image, 175–176
Point-like inhomogeneities, 213–214
Point spread function (PSF), 25–26, 53, 102, 339
Polar coordinate systems, 20–21
Polarization current distribution, 105–106

Index

Poly-metallic elements, 246–247
Polyscreen technique, 335–336
Post-focus processing, 102
Pre-processing/post-processing, 214–216
 data of soil layers, 24
 on diffraction tomography, 262
 radar operation/tomogram on, 288
Principle of reciprocity, 27
Pseudo-fiber, 235
PSF, *see* Point spread function (PSF)
Pulsed EC techniques, 93
Pulsed radio-wave scanning, 278
Pulse generator, 123
Pulse radiation, 27

Radar-based motion detector module (RMM), 185, 187–188
Radar cross section, 240
Radar systems
 ESP radar, 226, 283
 LFM, 39–42
 SAR, 28, 56
 sub-surface radar, 218, 244
 UWB (*see* UWB radar)
Radiation pattern (RP) of antenna, 123, 125, 127
Radiation sources/receivers, planes, 30
Radio-acoustic tomography, 293–294
 acoustic tomography, defects visualization using, 303–309
 experimental results, 294–299
 radio/ultrasound tomography techniques, 299–303
Radio-holographic method, 169–170
Radio imaging, 134, 144
Radio-location tomography solution, 48–49
Radiometers, 257
 MMW application, 259–262
 operation, 258–260
Radiometry, monostatic methods, 125
Radiopaque objects
 transmission tomography, 153–156
 unilateral location tomography, 158
Radio tomogram, 41–42, 206
 non-moving people, 289
 test object and, 174
Radio-tomographic image, 272
Radio tomography
 hidden objects, 121–163
 knife/plastic pistol, 148
 non-transparent objects, 152–163
 plastic pistol in jacket, 147
 on UWB tomographic synthesis, 130–145

Radio/ultrasonic scanner, 302
Radio/ultrasound tomography techniques, 299–303
RadioView program interface, 184
Radio waves, 185, 306
Radio wave tomographic synthesis (RWTS), 28–33
Ratio, signal-to-noise (S/N), 51, 212, 268, 314–315
Reconstruction of objects
 imitational modeling, 189–191
 incoherent tomography for, 42–45
 visualization and, 338–345
Reference-frequency generator, 187
Reference wave, 168–169
Reflection coefficient, 300
Reflectivity, 153, 273
Reflector antennas, recovery of focusing property, 158–163
Refocusing method, 65–68
Refractive index
 versus density, 299–300
 solids, density and, 301
Regularization product, 15
Remote-field eddy current (RFEC) technique, 93–94
Remote sensing system, 245
Rural environments, focusing method in, 273–278
RWTS (radio wave tomographic synthesis), 28–33
Rytov approximation, 71–73, 75

SAR, *see* Synthesis aperture radar (SAR)
Scalar product, 14
Scattered radiation, 204
Self-compensated sensing coil (SCSC), 315–316, 322
Semi-transparent object, 22
 transmission tomography of, 147–152
SF, *see* System function (SF)
Shadow function, 19
Shadow projection (SP), 18–20
 FS method on, 19–22
 performance, 20
Signal analysis, inverse problems for, 340–342
Signal processing, 214–216
 post-processing, 214–218
 pre-processing, 214–218
Signal-to-noise (S/N) ratio, 51, 212, 268, 314–315
Siliceous sand, 247

Simulation modeling, 190
    location sounding problem, 32
    results, 191
    SF, 190
Singular regularization, 10–11
Singular-value decomposition, 9–11
Skin depth, 91–92
Snell's law of reflection, 213
S/N (signal-to-noise) ratio, 51, 212, 268, 314–315
Sod-podzolic soil, 233, 235, 237
Soil characteristics, measurements, 233, 235–236
Sonar, prototype, 298–299
Sounding path, radar, 240
Source localization, inverse problems, 342–343
SP, *see* Shadow projection (SP)
Space spectrum modulus, dependences on, 110–112
Spatial frequency spectrum, 161–162, 216
Spatial (transverse) resolution, 67
Spatial spectrum, 159
Spatiotemporal region, migration method in, 51–54
Stability, 5
Stolt's migration method, 28–33, 210, 212
Stratton–Chu formula, 80
Sub-soil media, living people in, 281–287
Subsoil medium, object buried in, 240–244
    metropolitan tunnels, 242
    structure, 243
Subsoil structures detection/visualization
    and reconstruction of images, 233–240
    rough/uneven terrain surfaces, geolocation, 226–232
    tunnels/tubes (pipelines), 226
Sub-surface defects, EC tomography for, 324, 326–331
Sub-surface radar, 218, 244
Sub-surface tomography applications
    minerals detection/identification, 244–247
    MMW, 257–262
    subsoil medium, object buried in, 240–244
    subsoil structures detection/visualization, 226–240
    sub-surface man-made structures/objects, geolocation, 247–257
Surface integra, 94
Synthesis aperture radar (SAR), 56, 172, 190–191
    data processing algorithm in, 39
    processing, 195–196
    tomography of metal plates, 197
Synthesized focusing method, 54–55
System function (SF), 53, 189
    2-D form, 193
    experimental data and, 215
    reconstruction, 193, 195
    simulation model, 190

Tapering function, 65
TEM (transverse electromagnetic), 122
TF, *see* Transfer function (TF)
Tikhonov regularization, 5–8, 14
Tikhonov's methodology, 14
Time-harmonic fields, 90
TMR8140 stroboscopic oscilloscope, 123
Tomograms
    box, 143
    EC, 327–328
    forest, 268–269
    inhomogeneities, 277
    metallic test object, 177
    plastic tube, 228
    radio, 174, 206, 289
    test, 215
    with tested luggage, 138
    test object, 159, 336
    triangular cylinder, 154
    two orthogonal, 217
    ultrasonic, 156
Tomographic processing, 183, 267, 278, 294, 309
Tomography, 17, 151
    acoustic, 303–309
    diffraction (*see* Diffraction tomography)
    Doppler, 185–197
    EC, 89–116, 323–331
    electrostatic, 84–86
    forested areas, 265–269
    impedance, 84
    incoherent, 42–45, 168
    on LFM radiation, 39–42, 146–147
    low-frequency magnetic, 79–84
    magnetic, 318–322
    over incomplete data, 218
    radio (*see* Radio tomography)
    radio-acoustic (*see* Radio-acoustic tomography)
    semitransparent objects, 151–152
    sub-surface (*see* Sub-surface tomography applications)
    thin metallic object, 131

Index

transmission, 33, 147–152
ultrasound, 299–303
wave location (*see* Wave location tomography)
wooden structures/constructions, 269–273
Transceiver, 48
  module, 123
Transfer function (TF), 51, 57
  3-D, 59–60
  approximate representation, 62
  in frequency domain, 58
  in two-step focusing, 62
Transmission tomography, 33
  opaque objects, 36–39, 153
  optical circuit and, 149
  radiopaque objects, 153–156
  scheme, 33
  semitransparent media on Kirchhoff approximation, 33–36
  semitransparent objects, 147–151
Transmitter, 170
  and receiving antennas, 173
  test object for, 172
Transverse direction, 134
Transverse electromagnetic (TEM), 122
Two-step focusing, 60–63

Ultrasonic imaging
  echo of toy gun, 296
  metal sheet and, 305–306
  plasterboard, 304–305
  plastic gun, 303
  sonar for contactless, 298
  stepped triangle in, 297
Ultrasonic microphone array, 308
Ultrasonic scanner, 295, 302
Ultrasonic tomogram
  square and round cylinder, 156
  triangular cylinder, 156
Ultrasonic tomography, 295, 300, 303
Ultrasound, 293–294
  and radio tomography techniques, 299–303
  used by bats, 294
Ultra-wide band (UWB), 122, 205
  antenna, 126, 129, 139
  collimated shape, 251
  ESP antenna array, 145
  forested/rural environments, 265–278
  incoherent tomography, 168–177
  with luggage, 140
  plasterboard, 304–305

plastic gun, image, 303
pulse envelope and attenuation, 254
radiated pulse pattern by, 129
radiation, attenuation, 267
radio tomography, 206, 249
scanner and box, 142
for sounding/spectra, 127, 137
tomogram, 138
tomography, 125
two-dimensional scanner for, 122
Unit vectors (orts), 5
UWB, *see* Ultra-wide band (UWB)
UWB radar, 226, 266, 283
  with control/processing block, 287
  CW/ESP radar systems, 282
  to detect living people, 283
  ESP, 240–241, 285
  for forest tomography, 266
  parameters of ESP, 283
  pattern, 267
UWB signal, 172
  antennas, 175
  incoherent tomography, 168–177
  from object, 128
  signal shapes in, 182
  tomography of non-linear inclusions, 177–185
UWB tomographic synthesis
  bags and luggage, 138–139, 141
  building constructions, 130–137
  object with metallic inclusions, 138, 140–145

Vector equations, 12
VIM (volume integral equation method), 91
Visualization of 3-D tomograms
  by equipotential surfaces, 336–337
  by orthogonal slices, 337–339
  by polyscreen technique, 335–336
Visualization of objects, 335–338
Visualization scheme, 336
Voltage standing wave ratio (VSWR), 123–124, 129
Volume integral equation method (VIM), 91

Wave location tomography
  group focusing technique, 63–65
  matched filtering method for, 50–51
  migration method in spatiotemporal region, 51–54
  radio-location tomography solution, 48–49

synthesized focusing method, 54–55
two media, single focusing at boundary, 56–60
two-step focusing, 60–63
Waves plane, 2
Wave tomography, theoretical view, 17–45

Weighted differential technique (WDT), 252
Well-posed problem, Tikhonov definition, 5–8
Weyl's formula, 29, 31, 50, 82
Wiener filtering, solution, 14–15, 102
Window function, 65
Wooden structures and constructions, 269–273